Local Zeta Regularization
and the
Scalar Casimir Effect

A General Approach based on Integral Kernels

Local Zeta Regularization
and the
Scalar Casimir Effect

A General Approach based on Integral Kernels

$$\zeta(s) := \sum_{\ell=1}^{+\infty} \frac{1}{\ell^s}$$

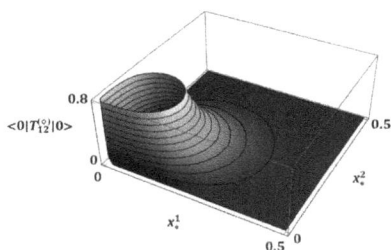

Davide Fermi
Livio Pizzocchero

Università degli Studi di Milano, Italy
and Istituto Nazionale di Fisica Nucleare, Italy

World Scientific

NEW JERSEY · LONDON · SINGAPORE · BEIJING · SHANGHAI · HONG KONG · TAIPEI · CHENNAI · TOKYO

Published by

World Scientific Publishing Co. Pte. Ltd.

5 Toh Tuck Link, Singapore 596224

USA office: 27 Warren Street, Suite 401-402, Hackensack, NJ 07601

UK office: 57 Shelton Street, Covent Garden, London WC2H 9HE

Library of Congress Cataloging-in-Publication Data
Names: Fermi, Davide, 1988– author. | Pizzocchero, Livio, 1962– author.
Title: Local zeta regularization and the scalar Casimir effect : a general approach based on
    integral kernels / Davide Fermi (Università degli Studi di Milano, Italy & Istituto Nazionale di
    Fisica Nucleare, Italy), Livio Pizzocchero (Università degli Studi di Milano, Italy & Istituto
    Nazionale di Fisica Nucleare, Italy).
Description: Singapore ; Hackensack, NJ : World Scientific, [2018] |
    Includes bibliographical references.
Identifiers: LCCN 2017034088| ISBN 9789813224995 (hardcover ; alk. paper) |
    ISBN 9813224991 (hardcover ; alk. paper)
Subjects: LCSH: Functions, Zeta. | Casimir effect.
Classification: LCC QC20.7.Z47 F37 2018 | DDC 515/.56--dc23
LC record available at https://lccn.loc.gov/2017034088

British Library Cataloguing-in-Publication Data
A catalogue record for this book is available from the British Library.

Preface

Zeta regularization is a method to treat the divergent quantities appearing in several areas of mathematics and physics; these are managed by introducing a complex parameter, with the role of a regulator, and defining their "renormalized versions" in terms of the analytic continuation with respect to the regulator. The standard textbook example of this procedure deals with the divergent series

$$\sum_{\ell=1}^{+\infty} \ell \tag{1}$$

which, in the zeta approach, is interpreted as the analytic continuation at $s = -1$ of the regularized series $\zeta(s) := \sum_{\ell=1}^{+\infty} \ell^{-s}$. The latter series only converges for $\Re s > 1$ but the function $s \mapsto \zeta(s)$ defined in this way, the well-known Riemann zeta function, possesses a unique analytic extension to $\mathbf{C} \setminus \{1\}$ and $\zeta(-1) = -1/12$; in this sense, the sum (1) "equals" $-1/12$. In a more pictorial language, one could say that $-1/12$ is the "renormalized" value of the series (1).

The application of zeta regularization to the divergences of quantum field theory was first proposed by Dowker and Critchley [46], Hawking [84] and Wald [146] to renormalize local observables, especially the vacuum expectation value (VEV) of the stress-energy tensor. The ultimate purpose was the semiclassical treatment of quantum effects in general relativity, e.g. using the stress-energy VEV as a source term in Einstein's equations. Due to the attention to vacuum states, the zeta approach was connected from its very beginning to Casimir physics.

However, the above mentioned pioneers focused their investigations on the conceptual validity of the method, rather than on its implementation for actual computations in specific configurations. For example, the elegant formula of Dowker–Critchley–Hawking, which gives the VEV of the stress-energy tensor as the functional derivative of the (renormalized) effective field action with respect to the space-time metric is not useful for actual computations in a given geometry, apart from very special cases. This is due to the fact that its use would require computing the effective action for all spacetime metrics close the one under consideration. For this reason, Birrel and Davies have described the application of this formula as "impossibly difficult" (see [18], page 190).

Moreover, papers [46, 84, 146] deal with a quantized field on the whole spacetime manifold and do not consider the possibility of confining the field to a given space domain by means of suitable boundary conditions, a situation which has nontrivial aspects, even in the case of a space domain in flat Minkowski space-time.

After the previously cited pioneers, the zeta approach to the treatment of local observables did not receive so much attention in the literature, apart from some notable exceptions. To be more specific: Cognola, Zerbini, Elizalde [39, 40] and Moretti [30, 81, 86, 108–111] analyzed the general aspects of local zeta regularization in some particular configurations, working mainly on curved spacetimes with no spatial boundary conditions; Actor, Svaiter *et al.* [2, 3, 5, 126, 127] worked on spatial domains with boundaries in flat spacetime, performing computations in some special cases.

Contrary to what happened in the analysis of local aspects, the application of the zeta strategy has become much more popular in the treatment of global observables, such as the VEV of the total energy. The literature on global zeta regularization is enormous; here we only cite the classical papers [129, 130] by Zimerman *et al.*, the work [19] by Blau, Visser and Wipf, along with the monographs of Elizalde *et al.* [30, 53, 54], Bordag *et al.* [21, 22], Kirsten [88] (see also [35, 65, 66, 83, 89, 93]) and Milton [105].

The beautiful book [105] by Milton well exemplifies a typical attitude towards the zeta strategy. Therein, global zeta regularization is presented as one of the most important approaches to the renormalization of the total energy VEV; on the contrary, when considering the stress-energy tensor, the possibility of using zeta regularization is not even mentioned and instead, employed systematically is a point-splitting technique (which amounts to represent, e.g. the square of a scalar quantum field as the limit of a product where the field is evaluated at two different points).

In view of the above considerations, we think that it is not useless to propose a comprehensive illustration of zeta regularization, using methodically this technique to renormalize both local and global observables in the vacuum state and presenting explicit computations in a number of configurations with given spatial boundary conditions for the field. We hope that this book will contribute to show the effectiveness of the zeta strategy; in our opinion, this method is competitive (both conceptually and computationally) in comparison with other approaches to quantum vacuum effects. Let us mention, in particular, the original derivation of these effects by Casimir [34], relying on the Euler–Maclaurin summation formula, the point-splitting strategy [18, 29, 105] along with the related algebraic, microlocal approach [14, 43, 67, 118, 121] (see also the citations therein), the exponential cutoff method systematically employed by Fulling *et al.* [69–71, 73, 103] and the Born series expansion technique of Graham, Jaffe *et al.* [76–78] (see also the citations therein), where the divergences of one loop Feynman diagrams are renormalized by addition of suitable counterterms.

In this book we consider a quantized neutral scalar field, typically on flat Minkowski spacetime of arbitrary dimension $d + 1$. The field is confined within a d-dimensional spatial domain Ω and suitable boundary conditions are prescribed for it on the boundary $\partial\Omega$; besides, we admit the presence of a classical external potential $V = V(\mathbf{x})$, which could be understood as an effective representation for the interaction with other fields or as the model for a "soft boundary". We will only occasionally mention the possibility of replacing the flat domain Ω with a Riemannian manifold, or an open subset of it with prescribed boundary conditions. This suggests the possibility to extend the proposed general formalism to the case of a non-flat ultrastatic spacetime [30] where the line element reads $ds^2 = -dt^2 + d\ell^2$, with $d\ell^2$ the Riemannian line element of Ω.

Our attention is mainly focused on the VEV of the stress-energy tensor, of which we consider both the conformal and the non-conformal parts; nevertheless, we also determine the total energy and the boundary forces. Our implementation of zeta regularization follows a scheme that we already proposed in [58] for the special, very "classical" case of a scalar field confined between parallel Dirichlet planes.

Both in [58] and in the present book, the field is treated using canonical quantization in a genuine Lorentzian framework; the zeta approach is formulated so as to fit to this setting. This aspect makes a difference with respect to the previously cited works of Cognola, Zerbini, Elizalde, Moretti, Actor, Svaiter and coworkers, where zeta regularization is implemented in the framework of Euclidean quantum field theory.

Another aspect of this book that we wish to stress is that we aim to describe the zeta strategy as a collection of algorithms, to be employed nearly automatically in applications. To this purpose we refer to the fundamental fact indicated hereafter: the analytic continuations involved in the zeta approach are closely related to certain integral kernels determined by the basic elliptic operator $\mathcal{A} := -\Delta + V(\mathbf{x})$ (in $L^2(\Omega)$) which governs the spatial part of the field equation. Among these objects we mention, in particular, the Dirichlet and heat kernels corresponding, respectively, to complex powers of $\mathcal{A}$ and to the exponential $e^{-t\mathcal{A}}$. The formulation presented here emphasizes as far as possible the basic role of these and other kernels in view of zeta regularization; in the end, we write down few very general and "mechanical" rules to construct the required analytic continuations using these tools.

Apart from technical differences (e.g. the Lorentzian setting) and from our insistence on algorithmic aspects, the idea that zeta regularization is related to certain integral kernels is present in the previous literature. When we implement it in this book, we give appropriate credit for this idea to some of the previously cited references. In addition, we also mention the earlier works of mathematicians who, independently from any application to quantum field theory, discussed analytic continuation problems in connection with some basic integral kernels associated to elliptic operators: let us recall, in particular, the seminal papers on this subject by Minakshisundaram and Pleijel [106, 107], Seeley [138], Ray and Singer [122].

The book is divided into two parts. Part 1 is mainly devoted to the general formulation of zeta regularization, and to the set of algorithms mentioned before.

In Chapter 1 we introduce the general (Lorentzian) framework for the canonical quantization of a scalar field, and the zeta approach to renormalize the VEVs of its local or global observables. Chapter 2 is devoted to the general theory of the previously mentioned integral kernels, and to their connections with the local observables of the field theory. Chapter 3 discusses the total energy and the boundary forces in the vacuum state. In Chapter 4 we present some variations of the previous schemes; these variations concern the technically delicate case where 0 is in the spectrum of $\mathcal{A} := -\Delta + V(\mathbf{x})$, or extend the framework of the previous chapters to flat spatial domains with non Cartesian coordinates, or even sketch the treatment of curved space domains. To conclude the description of Part 1 we mention the Appendices A–F, related to some technical issues.

In the subsequent Part 2 we show how the general techniques of Part 1 work in a number of concrete cases.

For "pedagogical" reasons, we start Part 2 with the simple, one-dimensional case of a massless field on a segment; this is discussed in Chapter 5. Next, in Chapters 6 and 7 we analyze, respectively, cases involving a scalar field confined by parallel and perpendicular planes in any space dimension. Chapter 8 treats a scalar field living inside a three-dimensional wedge, with arbitrary boundary conditions and an arbitrary choice of the conformal parameter.

The above configurations have been previously considered in the literature using alternative regularization schemes (say, point-splitting methods for the stress-energy VEV), or the zeta approach in some version different from the one presented in this book (in particular, Euclidean zeta regularization); for each one of the above cases, we make a detailed comparison with the existing literature. One of the reasons for reconsidering these classical cases in the present book is that we aim at a unified viewpoint on these configurations, based solely on the zeta strategy as formulated in Part 1; we hope this will make a difference with respect to the material found in the previous literature, presenting different approaches to different cases.

In the final Chapters 9 and 10 of Part 2, following our recent works [60, 61], we present two more engaging applications of the methods of Part 1; these require a mixture of analytical and numerical calculations, that were performed using Mathematica. These applications concern a quantized field in the presence of an external harmonic potential (a form of "soft" confinement) or living within a rectangular box in any space dimension. One would expect that everything had been computed a long time ago for these configurations; nonetheless, to the best of our knowledge, this is not the case and some results of our papers [60, 61] have never been derived previously, neither by zeta approach nor by other methods. In particular, the renormalized stress-energy VEV for the harmonic case, determined in [60], had been left as an open problem in the seminal papers [1, 4] of Actor and Bender on the total energy VEV in the presence of harmonic potentials. On the other hand, our computation of the renormalized stress-energy VEV for a field confined within a rectangular box [61] has some contact with a paper of Svaiter *et al.* [127]; however, the latter refers to the case of an infinite rectangular wave guide and presents for this configuration a series expansion of the stress-energy VEV with a polynomial convergence rate, while the expansion of [61] for the same quantity converges exponentially.

In our opinion the "zeta idea" to cure by analytic continuation the divergences of quantum fields promises a number of intriguing developments, possibly including the rigorous definition of path integrals for some field theories with interaction. Investigation of such developments is left to the future; for the moment, we hope this book will provide a toolbox for people interested in applying this beautiful technique to the physics of quantum vacuum.

Milano,   *Davide Fermi*
June 2017   *Livio Pizzocchero*

Acknowledgements

This work was partly supported by INdAM, INFN and by MIUR, PRIN 2010 Research Project "Geometric and analytic theory of Hamiltonian systems in finite and infinite dimensions". We are grateful to Claudio Dappiaggi and Valter Moretti for encouraging us in our studies on zeta regularization.

We also thank the staff of World Scientific Publishing, especially Ms. Lakshmi Narayanan, for kind and skillful work in editing and producing the book.

Contents

PART 1
General theory

Chapter 1

Zeta regularization for a scalar field

In the present chapter and in the related Appendix A, we introduce the general framework for a neutral scalar field on a d-dimensional spatial domain Ω, including a brief discussion of the stress-energy tensor with its conformal and non-conformal parts. This is an occasion to fix the attention on the basic elliptic operator $\mathcal{A} := -\Delta + V$ on Ω, where Δ is the Laplacian and $V : \Omega \to \mathbf{R}$ is an external potential. Next, we proceed to introduce zeta regularization in the formulation first given in [58]: the basic idea is to replace the quantized field $\widehat{\phi}$ with its regularized version $\widehat{\phi}^u := (\mathcal{A}/\kappa^2)^{-u/4}\widehat{\phi}$, where $u \in \mathbf{C}$ is the regulating parameter (or regulator) and $\kappa > 0$ is a mass parameter introduced for dimensional reasons. Formally, $\widehat{\phi}^u$ becomes $\widehat{\phi}$ for $u = 0$; the zeta approach implements this idea in terms of analytic continuation. This means that for any local or global observable (say, the VEV of either the stress-energy tensor or of the total energy), after introducing a regularized version of it based on $\widehat{\phi}^u$, the corresponding renormalized value is defined via the analytic continuation at $u = 0$. We distinguish between two versions of this prescription: the restricted zeta approach, in which the regularized observable is analytic at $u = 0$, and the extended approach, where a singularity appears at $u = 0$ and is eliminated by a minimal subtraction prescription, removing the negative powers of u in the corresponding Laurent expansion.

1.1   Conventions and notations

Throughout the whole book we use natural units, so that

$$c = 1 , \qquad \hbar = 1 , \tag{1.1}$$

and work in $(d+1)$-dimensional Minkowski spacetime; the latter is identified with $\mathbf{R}^{d+1}$ using a set of inertial coordinates

$$x = (x^\mu)_{\mu=0,1,\dots,d} \equiv (x^0, \mathbf{x}) \equiv (t, \mathbf{x}) , \tag{1.2}$$

in which the Minkowski metric has coefficients

$$(\eta_{\mu\nu}) = \mathrm{diag}(-1, 1, \dots, 1) . \tag{1.3}$$

We denote with Ω an open spatial domain in $\mathbf{R}^d$, and we often consider the Hilbert space $L^2(\Omega)$ of complex-valued functions on Ω which are square integrable with respect to the standard Lebesgue measure $d\mathbf{x}$. The boundary of the spatial domain and the area element on it are indicated with $\partial\Omega$ and da, respectively. We also write $\mathbf{n}(\mathbf{x}) \equiv (n^\ell(\mathbf{x}))_{\ell=1,\dots,d}$ for the outer unit normal at a point $\mathbf{x} \in \partial\Omega$; this is assumed to exist everywhere (which happens if $\partial\Omega$ is globally smooth) or almost everywhere (which happens if $\partial\Omega$ has edges or corners).

Moreover, the following conventions are employed:

(i) $\ln : (0, +\infty) \to \mathbf{R}$ is the elementary logarithm;

(ii) for any $\alpha \in \mathbf{C}$, we systematically refer to the standard definition

$$x^\alpha := e^{\alpha \ln x} \qquad \text{for all } x \in (0, +\infty) ; \tag{1.4}$$

(iii) for any $\alpha \in \mathbf{C}$ and any z in a convenient subset $\mathbf{C}^\times$ of the complex plane, we define

$$z^\alpha := e^{\alpha \ln |z| + i\alpha \arg z} , \tag{1.5}$$

where $\arg : \mathbf{C}^\times \to \mathbf{R}$ is some determination of the argument; this determination depends on the domain $\mathbf{C}^\times$ and must be specified in each case of interest. In most applications considered hereafter we set

$$\mathbf{C}^\times := \mathbf{C} \setminus [0, +\infty) ;$$

$\arg :=$ the unique determination of the argument with values in $(0, 2\pi)$.

$$\tag{1.6}$$

1.2 The general setting

In this section we summarize the zeta regularization method for the propagator and the vacuum expectation value (VEV) of the stress-energy tensor of a quantized, neutral scalar field, in the formulation first presented in [58] (see also [59]). Here the scheme of [58] is slightly generalized, admitting the presence of a classical background potential V and arbitrary spacetime dimensions.

Let us fix a spatial domain $\Omega \subset \mathbf{R}^d$ where we consider a quantized neutral, scalar field $\widehat{\phi}$ in presence of a classical background static potential V; so, we have $V : \Omega \to \mathbf{R}$, $\mathbf{x} \mapsto V(\mathbf{x})$ and

$$\widehat{\phi} : \mathbf{R} \times \Omega \to \mathcal{L}_{sa}(\mathfrak{F}) \; ; \qquad 0 = (-\partial_{tt} + \Delta - V(\mathbf{x}))\widehat{\phi}(t, \mathbf{x}) \; . \qquad (1.7)$$

Here we are referring to the space $\mathcal{L}(\mathfrak{F})$ of linear operators on the Fock space $\mathfrak{F}$ and to the subset $\mathcal{L}_{sa}(\mathfrak{F})$ of self-adjoint operators.[a] We indicate with $\Delta := \sum_{i=1}^d \partial_{ii}$ the d-dimensional Laplacian; besides, we assume the potential V to be smooth and prescribe appropriate boundary conditions (e.g. the Dirichlet conditions $\widehat{\phi}(t, \mathbf{x}) = 0$ for $\mathbf{x} \in \partial\Omega$).

The *fundamental operator* is

$$\mathcal{A} := -\Delta + V \; , \qquad (1.8)$$

intending that the previously mentioned boundary conditions are accounted for in the above definition; we assume the framework under analysis to grant that $\mathcal{A}$ is a *self-adjoint* operator in the Hilbert space $L^2(\Omega)$, with the inner product

$$\langle f | g \rangle := \int_\Omega d\mathbf{x} \, \overline{f}(\mathbf{x}) \, g(\mathbf{x}) \qquad (1.9)$$

($d\mathbf{x}$ denoting the standard Lebesgue measure on Ω).[b] Moreover, we assume $\mathcal{A}$ to be *strictly positive*, by which we mean that the *spectrum* $\sigma(\mathcal{A})$ is contained in $[\varepsilon^2, +\infty)$ for some $\varepsilon > 0$. Let us mention that

---

[a]Of course the notation $\widehat{\phi} : \mathbf{R} \times \Omega \to \mathcal{L}_{sa}(\mathfrak{F})$, $(t, \mathbf{x}) \mapsto \widehat{\phi}(t, \mathbf{x})$ is used here in connection with a *generalized* operator-valued function; in fact, as well known, $\widehat{\phi}$ is an operator-valued *distribution*. Throughout the book we use the classical functional notations even for generalized functions.

[b]In passing, we recall that the Fock space $\mathfrak{F}$ is a Hilbert space that can be realized as the direct sum of all symmetrized tensor powers of $L^2(\Omega)$.

$\mathcal{A}$ may have continuous spectrum, a fact typically occurring when Ω is unbounded. Therefore, when we speak of the eigenvectors of $\mathcal{A}$ we always intend them in a generalized sense, possibly including improper eigenfunctions.

Let us consider a complete orthonormal set $(F_k)_{k \in \mathcal{K}}$ of (generalized) eigenfunctions of $\mathcal{A}$,[c] indexed by an unspecified set of labels $\mathcal{K}$, and let us write the corresponding eigenvalues in the form $(\omega_k^2)_{k \in \mathcal{K}}$ ($\omega_k \geqslant \varepsilon$ for all $k \in \mathcal{K}$). Thus, we have

$$F_k : \Omega \to \mathbf{C}; \qquad \mathcal{A} F_k = \omega_k^2 F_k \; ;$$
$$\langle F_k | F_h \rangle = \delta(k, h) \quad \text{for all } k, h \in \mathcal{K} \; . \tag{1.10}$$

Any eigenfunction label $k \in \mathcal{K}$ can include different parameters, both discrete and continuous. We generically write $\int_{\mathcal{K}} dk$ to indicate summation over all labels, meaning literal summation over discrete parameters and integration over continuous parameters (with respect to a suitable measure). Besides, we indicate the Dirac delta function for the label space $\mathcal{K}$ with $\delta(h, k) = \delta(k, h)$; this reduces to the ordinary Kronecker symbol in the case of discrete parameters. Note that the condition $\omega_k \geqslant \varepsilon > 0$ excludes the presence of infrared divergences from all the sums over k appearing in the sequel.

The functions

$$f_k : \mathbf{R} \times \Omega \to \mathbf{C} \; , \qquad x = (t, \mathbf{x}) \mapsto f_k(x) := e^{-i\omega_k t} F_k(\mathbf{x}) \tag{1.11}$$

fulfill the equation $(-\partial_{tt} - \mathcal{A}) f_k = 0$; therefore, they allow to infer for the quantized field a normal modes expansion of the form

$$\widehat{\phi}(x) = \int_{\mathcal{K}} \frac{dk}{\sqrt{2\omega_k}} \left[\widehat{a}_k \, f_k(x) + \widehat{a}_k^{\dagger} \, \overline{f_k}(x) \right] \tag{1.12}$$

(with † indicating the adjoint operator, and $^{-}$ the complex conjugate). In the above equation we are considering the annihilation and

---

[c]For a fully rigorous discussion of generalized eigenfunctions, see, e.g. Chapter IV of [74]. In the sequel, following the usual terminology, when speaking of functions (or distributions) on Ω we use the adjectives "proper" or "improper", to distinguish between objects which actually belong to $L^2(\Omega)$ or not. In this spirit we speak of proper or improper eigenfunctions, and use the same terminology for the corresponding eigenvalues. In the sequel, the adjective "generalized" in relation to eigenfunctions is sometimes omitted.

creation operators $\widehat{a}_k, \widehat{a}_k^\dagger \in \mathcal{L}(\mathfrak{F})$, which fulfill the canonical commutation relations

$$[\widehat{a}_k, \widehat{a}_h] = 0 , \quad [\widehat{a}_k, \widehat{a}_h^\dagger] = \delta(h, k) , \qquad \widehat{a}_k|0\rangle = 0 , \qquad (1.13)$$

where $|0\rangle \in \mathfrak{F}$ is the vacuum state (of unit norm). A relevant character for the sequel of our analysis will be the *propagator*, i.e. the vacuum expectation value (VEV)

$$\langle 0|\widehat{\phi}(x)\widehat{\phi}(y)|0\rangle \qquad (x, y \in \mathbf{R} \times \Omega) . \qquad (1.14)$$

Let us pass to the stress-energy tensor operator; this depends on a parameter $\xi \in \mathbf{R}$, and its components $\widehat{T}_{\mu\nu} : \mathbf{R} \times \Omega \to \mathcal{L}_{sa}(\mathfrak{F})$, for $\mu, \nu \in \{0, 1, ..., d\}$, are given by

$$\widehat{T}_{\mu\nu} := (1 - 2\xi)\, \partial_\mu\widehat{\phi}\circ\partial_\nu\widehat{\phi} - \left(\frac{1}{2} - 2\xi\right) \eta_{\mu\nu}(\partial^\lambda\widehat{\phi}\,\partial_\lambda\widehat{\phi} + V\widehat{\phi}^2) - 2\xi\,\widehat{\phi}\circ\partial_{\mu\nu}\widehat{\phi} ;$$
$$(1.15)$$

here we are using the Jordan's symmetrized operator product $\widehat{A} \circ \widehat{B} := (1/2)(\widehat{A}\widehat{B} + \widehat{B}\widehat{A})$ and all the bilinear terms in the field are evaluated along the diagonal (e.g. $\partial_\mu\widehat{\phi} \circ \partial_\nu\widehat{\phi}$ indicates the map $x \mapsto \partial_\mu\widehat{\phi}(x) \circ \partial_\nu\widehat{\phi}(x)$).

Equation (1.15) provides a natural quantization of what is often called the "improved" stress-energy tensor; this is a well-known modification of the canonical stress-energy tensor with an additive term proportional to the parameter ξ, that does not alter its divergence. For certain boundary conditions, e.g. of Dirichlet type, this addition does not even alter the corresponding momentum vector. We refer to Appendix A for further details on this topic. Here we only recall that the improved stress-energy tensor was first proposed by Callan, Coleman and Jackiw [33] in order to deal with some pathologies appearing in perturbation theory; later on, this tensor was reinterpreted in terms of the Minkowskian limit for a scalar field coupled to gravity via the curvature scalar [18, 30, 46, 115, 118].

Needless to say, the vacuum expectation value of the stress-energy tensor (1.15) is the most relevant local quantity in the theory of the Casimir effect. It is evident from Eq. (1.15) that $\langle 0|\widehat{T}_{\mu\nu}(x)|0\rangle$ can be expressed formally in terms of the propagator (1.14) (and of its derivatives) evaluated on the diagonal $y = x$. On the other hand, the propagator is known to be plagued with ultraviolet divergences along this diagonal, so that $\langle 0|\widehat{T}_{\mu\nu}(x)|0\rangle$ is a merely formal expression for a divergent quantity; our ultimate purpose is to redefine the propagator and the stress-energy VEV via a suitable regularization procedure, ultimately yielding finite values for these quantities.

1.3   Zeta regularization and the stress-energy tensor

Let $\kappa > 0$ denote a parameter, to which we attribute the dimension of a mass (or of an inverse length, since $c = 1$ and $\hbar = 1$). This parameter will be called the mass scale; it is introduced for dimensional reasons and plays the role of a normalization scale. We refer to [19, 30, 53, 86, 108] for further comments regarding this parameter and its presence or absence in the renormalized observables of the field. We will check that the final, renormalized results depend on κ only when singularities appear in the analytic continuations involved in the following construction.

The zeta strategy, in the version proposed in [58] to give meaning to the VEV of $\widehat{T}_{\mu\nu}$, relies on the powers

$$(\kappa^{-2}\mathcal{A})^{-u/4} = \kappa^{u/2}\mathcal{A}^{-u/4} \qquad (1.16)$$

where $\mathcal{A}$ is the operator (1.8) and $u \in \mathbf{C}$; these are employed to define the *smeared*, or *zeta-regularized*, *field operator*

$$\widehat{\phi}^u := (\kappa^{-2}\mathcal{A})^{-u/4}\,\widehat{\phi}\,, \qquad (1.17)$$

depending on the complex parameter u and coinciding with the usual field operator $\widehat{\phi}$ for $u = 0$. If $(F_k)_{k\in\mathcal{K}}$ is a complete orthonormal set of eigenfunctions of $\mathcal{A}$ with corresponding eigenvalues $(\omega_k^2)_{k\in\mathcal{K}}$, we have $(\kappa^{-2}\mathcal{A})^{-u/4}F_k = \kappa^{u/2}\omega_k^{-u/2}F_k$ for any $k \in \mathcal{K}$; so, the functions $f_k(t, \mathbf{x}) = e^{-i\omega_k t}F_k(\mathbf{x})$ fulfill $(\kappa^{-2}\mathcal{A})^{-u/4}f_k = \kappa^{u/2}\omega_k^{-u/2}f_k$. In view of these considerations, after application of $(\kappa^{-2}\mathcal{A})^{-u/4}$, the expansion (1.12) for $\widehat{\phi}$ becomes[d]

$$\widehat{\phi}^u(x) = \kappa^{u/2}\int_{\mathcal{K}}\frac{dk}{\sqrt{2}\,\omega_k^{1/2+u/2}}\left[\widehat{a}_k\,f_k(x) + \widehat{a}_k^\dagger\,\overline{f_k}(x)\right]. \qquad (1.18)$$

---

[d]One could even regard Eq. (1.18) as a *definition* for $\widehat{\phi}^u(x)$, motivated by the considerations which precede it.

Note that, in the limit $\omega_k \to +\infty$, the term $1/\omega_k^{1/2+u/2}$ in the above integral vanishes rapidly if $\Re u$ is large; this is a manifestation of the regularizing effect of the operator $(\kappa^{-2}\mathcal{A})^{-u/4}$ for large $\Re u$, a fact we will describe with greater detail in the sequel.

Using $\widehat{\phi}^u$, we can define a *regularized propagator*

$$\langle 0|\widehat{\phi}^u(x)\widehat{\phi}^u(y)|0\rangle \qquad (x,y \in \mathbf{R} \times \Omega) \tag{1.19}$$

and a *zeta-regularized stress-energy tensor operator*

$$\widehat{T}_{\mu\nu}^u := (1 - 2\xi)\,\partial_\mu\widehat{\phi}^u \circ \partial_\nu\widehat{\phi}^u - \left(\frac{1}{2} - 2\xi\right)\eta_{\mu\nu}\left(\partial^\lambda\widehat{\phi}^u\partial_\lambda\widehat{\phi}^u + V(\widehat{\phi}^u)^2\right)$$

$$- 2\xi\,\widehat{\phi}^u \circ \partial_{\mu\nu}\widehat{\phi}^u\,, \tag{1.20}$$

where, as in Eq. (1.15), all the bilinear terms in the field are evaluated on the diagonal.

We are interested in the VEV of this regularized stress-energy tensor, which formally gives $\langle 0|\widehat{T}_{\mu\nu}(x)|0\rangle$ for $u = 0$. Of course, we can relate the VEV of $\widehat{T}_{\mu\nu}^u(x)$ to the regularized propagator (1.19) in the following way:

$$\langle 0|\widehat{T}_{\mu\nu}^u(x)|0\rangle$$

$$= \left(\frac{1}{2}-\xi\right)(\partial_{x^\mu y^\nu} + \partial_{x^\nu y^\mu}) - \left(\frac{1}{2}-2\xi\right)\eta_{\mu\nu}\left(\partial^{x^\lambda}\partial_{y^\lambda}+V(\mathbf{x})\right)$$

$$- \xi(\partial_{x^\mu x^\nu} + \partial_{y^\mu y^\nu})\bigg|_{y=x}\langle 0|\widehat{\phi}^u(x)\widehat{\phi}^u(y)|0\rangle\,. \tag{1.21}$$

We will return later on this equation and on its use for the actual computation of the VEV $\langle 0|\widehat{T}_{\mu\nu}^u(x)|0\rangle$.

Typically, the regularized propagator and the VEV of $\widehat{T}_{\mu\nu}^u(x)$ are analytic functions of u, for $\Re u$ sufficiently large; the same can be said of many related observables, including global objects such as the total energy, which is related to the space integral of the $(0,0)$ component of the stress-energy tensor. Let us consider any one of these (local or global) objects, and denote with $\mathcal{F}(u)$ its zeta-regularized version, based on Eq. (1.17) (see, e.g. Eq. (1.19) or Eq. (1.20)); we assume that the function $u \mapsto \mathcal{F}(u)$ is well defined and analytic for u in a suitable domain $\mathcal{U}_0$ of the complex plane. The zeta approach to renormalization can be formulated in either a "restricted" or an "extended" version, both described hereafter.

(i) *Zeta approach, restricted version.* Assume that the function $\mathcal{U}_0 \to \mathbf{C}, u \mapsto \mathcal{F}(u)$ can be analytically continued to a larger open subset $\mathcal{U}$ of $\mathbf{C}$ such that $0 \in \mathcal{U}$; let us use the notation $u \mapsto \mathcal{F}(u)$ also for this extension.[e] In this case, making reference to the analytic continuation, we define the renormalized value of the observable under consideration as

$$\mathcal{F}_{ren} := \mathcal{F}(0) \; . \tag{1.22}$$

(ii) *Zeta approach, extended version.* Assume that there is an open subset $\mathcal{U}$ of $\mathbf{C}$, larger than $\mathcal{U}_0$, such that $0 \in \mathcal{U}$ and that the function $u \in \mathcal{U}_0 \mapsto \mathcal{F}(u)$ possesses an analytic continuation to $\mathcal{U} \setminus \{0\}$, still indicated with $\mathcal{F}$. In this case, since there is an isolated singularity at $u = 0$, in a neighborhood of this point we have the Laurent expansion $\mathcal{F}(u) = \sum_{k=-\infty}^{+\infty} \mathcal{F}_k u^k$. Let us consider the *regular part*

$$(RP\,\mathcal{F})(u) := \sum_{k=0}^{+\infty} \mathcal{F}_k u^k \; ; \tag{1.23}$$

we define the renormalized value of the given observable as

$$\mathcal{F}_{\text{ren}} := (RP\,\mathcal{F})(0) \tag{1.24}$$

(i.e. $\mathcal{F}_{\text{ren}} = \mathcal{F}_0$). In most applications $\mathcal{F}$ is *meromorphic* close to $u = 0$, which means that it has a *pole* at this point; in this case the previous Laurent expansion has the form $\mathcal{F}(u) = \sum_{k=-N}^{+\infty} \mathcal{F}_k u^k$, where $N \in \{1, 2, 3, ...\}$ is the order of the pole. Let us stress that the prescription (1.24) is a quite straightforward generalization of the minimal subtraction prescription considered in [19, 53], where $\mathcal{F}$ was assumed to posses a simple pole in $u = 0$ (i.e. it was assumed that $N = 1$).

---

[e]In the style of our previous work [58] we should write $\mathcal{F} : \mathcal{U}_0 \to \mathbf{C}$ for the initially given function and $AC\,\mathcal{F} : \mathcal{U} \to \mathbf{C}$ for its analytic continuation; here, we choose to simplify the notation, writing $\mathcal{F}$ for both functions. Note that the analogue of Eq. (1.22) in the style of [58] would be $\mathcal{F}_{\text{ren}} := (AC\,\mathcal{F})(0)$.

Of course, the restricted zeta approach of item (i) is equivalent to a special case of the extended approach, in which $\mathcal{F}(u)$ has a removable singularity at $u = 0$ and the Laurent expansion at this point is the usual power series expansion. Needless to say, the restricted approach, when it is possible, is a more elegant way to eliminate the divergences of the system.

A large part of this book will be devoted to the application of the previous scheme to the VEV of the stress-energy tensor. Provided that no singularity appears at $u = 0$, one can use the restricted approach (i) and put

$$\langle 0|\widehat{T}_{\mu\nu}(x)|0\rangle_{\mathrm{ren}} := \langle 0|\widehat{T}^u_{\mu\nu}(x)|0\rangle\Big|_{u=0}. \tag{1.25}$$

In presence of a singularity, one passes to the extended approach and puts

$$\langle 0|\widehat{T}_{\mu\nu}(x)|0\rangle_{\mathrm{ren}} := RP\Big|_{u=0} \langle 0|\widehat{T}^u_{\mu\nu}(x)|0\rangle. \tag{1.26}$$

In [58] we only considered the prescription (1.25) in the special case of a Dirichlet field (with $V = 0$) between two parallel planes, i.e. in the configuration corresponding to the standard theory of the Casimir effect. In that case, the approach (1.25) was implemented via a direct computation of the analytic continuation appearing therein; as already stressed, here we aim to much more generality.

A remark

In the sequel, when performing zeta regularization and the consequent renormalization, it is sometimes natural to consider, in place of u, some complex parameter s related to u by a simple transformation. In view of such situations, it is convenient to generalize some notations of the present section in the following way:

(i) Consider an analytic function $\mathcal{S}_0 \to \mathbf{C}$, $s \mapsto \mathcal{F}(s)$, where $\mathcal{S}_0$ is an open subset of $\mathbf{C}$; if this admits an analytic continuation to a larger open subset $\mathcal{S}$, the latter will be still denoted with $s \mapsto \mathcal{F}(s)$.

(ii) Suppose the analytic function $\mathcal{S}_0 \to \mathbf{C}$, $s \mapsto \mathcal{F}(s)$ has an analytic extension to $\mathcal{S} \setminus \{s_0\}$, where $\mathcal{S}$ is an open subset of $\mathbf{C}$ and $s_0 \in \mathcal{S}$. Then, the Laurent expansion $\mathcal{F}(s) = \sum_{k=-\infty}^{+\infty} \mathcal{F}_k (s - s_0)^k$

will be used to define the regular part (near s_0) of this analytic continuation as $(RP\,\mathcal{F})(s) := \sum_{k=0}^{+\infty} \mathcal{F}_k(s - s_0)^k$; of course, this implies $(RP\,\mathcal{F})(s_0) = \mathcal{F}_0$.

1.4 More about the stress-energy tensor and its VEV

1.4.1 *Staticity features of the stress-energy VEV*

Let us return to the regularized stress-energy tensor; for $x = (t, \mathbf{x}) \in \mathbf{R} \times \Omega$, we claim that

$$\langle 0|\widehat{T}^u_{\mu\nu}(x)|0\rangle \ \text{is independent of } t \ ,$$
$$\langle 0|\widehat{T}^u_{0i}(x)|0\rangle = \langle 0|\widehat{T}^u_{i0}(x)|0\rangle = 0 \qquad \text{for } i \in \{1, ..., d\} \ . \tag{1.27}$$

These statements are not surprising, due to the staticity of the general framework considered in this book; a formal proof will be given in Section 2.5 (see Eqs. (2.28)–(2.30) and the considerations which follow them). Of course the features of Eq. (1.27) are preserved by analytic continuation, so that an analogue of this equation holds for the renormalized VEV $\langle 0|\widehat{T}_{\mu\nu}(x)|0\rangle_{\mathrm{ren}}$ as well.

1.4.2 *Conformal and non-conformal parts of the stress-energy VEV*

In the literature (see, e.g. [18, 30, 147]) it is customary to write the stress-energy tensor (here to be intended as one of the operators $\widehat{T}_{\mu\nu}$, $\widehat{T}^u_{\mu\nu}$, or either one of the VEVs $\langle 0|\widehat{T}^u_{\mu\nu}|0\rangle$, $\langle 0|\widehat{T}_{\mu\nu}|0\rangle_{\mathrm{ren}}$) as the sum of a *conformal* and a *non-conformal* part. In order to define these quantities, let us consider for ξ the critical value

$$\xi_d := \frac{d-1}{4d} \ . \tag{1.28}$$

It is known that, when taking into account the coupling of the scalar field to gravity, the theory happens to be invariant (for $V = 0$) under conformal transformations of the spacetime line element if ξ has the above critical value (see, e.g. [147], page 447). In the sequel we adopt the notations

$$\Diamond \equiv \text{conformal} \ , \qquad \blacksquare \equiv \text{non-conformal} \tag{1.29}$$

and define, for example, the conformal and non-conformal parts of the renormalized stress-energy VEV $\langle 0|\widehat{T}_{\mu\nu}|0\rangle_{\rm ren}$ in the following way:

$$\langle 0|\widehat{T}_{\mu\nu}^{(\Diamond)}|0\rangle_{\rm ren} := \langle 0|\widehat{T}_{\mu\nu}|0\rangle_{\rm ren}\Big|_{\xi=\xi_d} , \qquad (1.30)$$

$$\langle 0|\widehat{T}_{\mu\nu}^{(\blacksquare)}|0\rangle_{\rm ren} := \frac{1}{\xi-\xi_d}\left(\langle 0|\widehat{T}_{\mu\nu}|0\rangle_{\rm ren} - \langle 0|\widehat{T}_{\mu\nu}^{(\Diamond)}|0\rangle_{\rm ren}\right) . \qquad (1.31)$$

Of course, the above definitions yield

$$\langle 0|\widehat{T}_{\mu\nu}|0\rangle_{\rm ren} = \langle 0|\widehat{T}_{\mu\nu}^{(\Diamond)}|0\rangle_{\rm ren} + (\xi-\xi_d)\,\langle 0|\widehat{T}_{\mu\nu}^{(\blacksquare)}|0\rangle_{\rm ren} . \qquad (1.32)$$

In the applications to be considered in Part 2, when presenting the final results for the renormalized stress-energy VEV, we will either write them in the form (1.32) or give separately the conformal and non-conformal parts (1.30), (1.31).

1.4.3 *A few remarks on the total energy and pressure on the boundary*

The *total energy* is, by definition, the integral of $\langle 0|\widehat{T}_{00}(x)|0\rangle$ over the spatial domain Ω. We defer the detailed discussion of this topic to Chapter 3; therein we will describe the representation of the total energy as the sum of a bulk term and a boundary term, in the framework of zeta regularization.

In the same chapter we will use zeta regularization to treat the *pressure* on the boundary $\partial\Omega$ of the spatial domain; this quantity can be defined in terms of the VEV of the spatial components $\widehat{T}_{ij}$. There is an alternative characterization of the pressure in terms of the variation of the bulk energy with respect to deformations of the spatial domain Ω (see Eq. (3.6)). The equivalence of this definition with the previous one has often been assumed uncritically in the literature, so we think it can be useful to produce a formal proof, for which we refer to Chapter 3.

Chapter 2

The zeta-regularized stress-energy VEV in terms of integral kernels

In this chapter and in the related Appendices C and D, we introduce a number of integral kernels associated to the operator $\mathcal{A} := -\Delta + V$. The Dirichlet kernel $D_s(\mathbf{x}, \mathbf{y}) := \langle \delta_{\mathbf{x}} | \mathcal{A}^{-s} \delta_{\mathbf{y}} \rangle$ (s in a suitable complex domain, $\mathbf{x}, \mathbf{y} \in \Omega$, $\delta_{\mathbf{x}}, \delta_{\mathbf{y}}$ the Dirac delta functions at these points) is closely related to the stress-energy VEV. More precisely, the regularized stress-energy VEV built using $\widehat{\phi}^u$ can be expressed in terms of the Dirichlet kernel $D_s(\mathbf{x}, \mathbf{y})$ (and of its spatial derivatives) at $\mathbf{y} = \mathbf{x}$, $s = (u \pm 1)/2$; so, the renormalized version of this VEV is determined by the analytic continuation of D_s near $s = \pm 1/2$. We show that the continuation of the Dirichlet kernel can be computed algorithmically relating it to the heat kernel $K(\mathfrak{t}; \mathbf{x}, \mathbf{y}) := \langle \delta_{\mathbf{x}} | e^{-\mathfrak{t}\mathcal{A}} \delta_{\mathbf{y}} \rangle$ or to other kernels (among which is the so-called cylinder kernel $T(\mathfrak{t}; \mathbf{x}, \mathbf{y}) := \langle \delta_{\mathbf{x}} | e^{-\mathfrak{t}\sqrt{\mathcal{A}}} \delta_{\mathbf{y}} \rangle$, often considered by Fulling [56, 68]). This procedure ultimately gives a set of mechanical rules for zeta regularization.

2.1 Basics on integral kernels

Let us consider a linear operator $\mathcal{B}$ acting on $L^2(\Omega)$. The *integral kernel* of $\mathcal{B}$ is the (generalized) function

$$\mathcal{B}(\ ,\ ) : \Omega \times \Omega \to \mathbf{C}, \qquad (\mathbf{x}, \mathbf{y}) \mapsto \mathcal{B}(\mathbf{x}, \mathbf{y}) := \langle \delta_{\mathbf{x}} | \mathcal{B} \, \delta_{\mathbf{y}} \rangle \qquad (2.1)$$

where $\delta_{\mathbf{x}}$ and $\delta_{\mathbf{y}}$ are the Dirac delta functions centered at $\mathbf{x}$ and $\mathbf{y}$, respectively, here viewed as improper vectors of the Hilbert space

$L^2(\Omega)$.[a] Equivalently, the integral kernel of the operator $\mathcal{B}$ can be defined as the unique (generalized) function $\mathcal{B}(\ ,\ ) : \Omega \times \Omega \to \mathbf{C}$ such that

$$(\mathcal{B}\psi)(\mathbf{x}) = \int_\Omega d\mathbf{y}\, \mathcal{B}(\mathbf{x}, \mathbf{y})\, \psi(\mathbf{y})\ , \qquad (2.2)$$

for all sufficiently regular $\psi : \Omega \to \mathbf{C}$. If $\mathcal{B}$ possesses a complete orthonormal set of eigenfunctions $(F_k)_{k \in \mathcal{K}}$ with corresponding eigenvalues $\beta_k \in \mathbf{C}$ ($\mathcal{B}F_k = \beta_k F_k$), then

$$\mathcal{B}(\mathbf{x}, \mathbf{y}) = \int_\mathcal{K} dk\, \beta_k\, F_k(\mathbf{x})\overline{F_k}(\mathbf{y}) \qquad (2.3)$$

(since the function in the right-hand side fulfills Eq. (2.2) for all ψ). The precise sense in which the eigenfunction expansion (2.3) converges depends on the specific features of the operator $\mathcal{B}$; we generally assume distributional convergence,[b] leaving to future sections

---

[a]Let us repeat what we already declared in the footnote a in Chapter 1: throughout this book we use functional notations even for objects which are not ordinary functions. Making reference to the theory of Schwartz distributions, one can understand the statement "$\mathcal{B}(\ ,\ ) : \Omega \times \Omega \to \mathbf{C}$" as a way to indicate that we are considering a (complex) distribution on $\Omega \times \Omega$; let us describe how to intend Eq. (2.1) from this viewpoint. We refer to the space of test functions $D(\Omega)$, formed by the C^∞, compactly supported functions $\varphi : \Omega \to \mathbf{C}$ and equipped with its standard inductive limit topology; $D(\Omega \times \Omega)$ has a similar meaning. We wish to define rigorously $(\mathbf{x}, \mathbf{y}) \mapsto \langle \delta_\mathbf{x} | \mathcal{B}\, \delta_\mathbf{y} \rangle$ as a distribution on $\Omega \times \Omega$, i.e. as a continuous linear form on $D(\Omega \times \Omega)$. To this purpose we consider $\varphi, \psi \in D(\Omega)$ and note that the identities $\varphi = \int_\Omega d\mathbf{x}\, \varphi(\mathbf{x})\delta_\mathbf{x}$ and $\psi = \int_\Omega d\mathbf{x}\, \psi(\mathbf{x})\delta_\mathbf{x}$ give formally

$$\int_{\Omega \times \Omega} d\mathbf{x}\, d\mathbf{y}\, \overline{\varphi}(\mathbf{x})\langle \delta_\mathbf{x} | \mathcal{B}\delta_\mathbf{y} \rangle \psi(\mathbf{y}) = \langle \varphi | \mathcal{B}\psi \rangle\ .$$

On the other hand, if the domain of the operator $\mathcal{B}$ contains $D(\Omega)$ and the sesquilinear map $D(\Omega) \times D(\Omega) \to \mathbf{C}$, $(\psi, \varphi) \mapsto \langle \varphi | \mathcal{B}\psi \rangle$ is continuous, using the nuclear theorem of Schwartz (see [135], page 223, Théorème II) one proves rigorously the existence of a unique distribution on $\Omega \times \Omega$, denoted with $(\mathbf{x}, \mathbf{y}) \mapsto \langle \delta_\mathbf{x} | \mathcal{B}\, \delta_\mathbf{y} \rangle$, such that the above relation holds for all $\varphi, \psi \in D(\Omega)$, intending the integral in the left hand side as the action of this distribution on the test function $(\mathbf{x}, \mathbf{y}) \mapsto \overline{\varphi}(\mathbf{x})\psi(\mathbf{y})$. One can use similar considerations to give a distributional meaning to Eq. (2.2) (see again [135]). In this book, whenever we speak of the integral kernel of an operator $\mathcal{B}$ we implicitly assume $\mathcal{B}$ to possess the regularity features mentioned before, so that $\langle \delta_\mathbf{x} | \mathcal{B}\delta_\mathbf{y} \rangle$ makes sense at least as a distribution. In some cases described explicitly in the sequel, stronger assumptions on $\mathcal{B}$ ensure that $\langle \delta_\mathbf{x} | \mathcal{B}\delta_\mathbf{y} \rangle$ can be understood as an ordinary function, continuous with its derivatives up to a certain order; this is the situation outlined in Appendix B, often mentioned in the sequel in relation to several kernels of interest for us.

[b]Meaning that $\int_{\Omega \times \Omega} d\mathbf{x}\, d\mathbf{y}\, \mathcal{B}(\mathbf{x}, \mathbf{y})\varphi(\mathbf{x}, \mathbf{y}) = \int_\mathcal{K} dk\, \beta_k \int_{\Omega \times \Omega} d\mathbf{x}\, d\mathbf{y}\, F_k(\mathbf{x})\overline{F_k}(\mathbf{y})\varphi(\mathbf{x}, \mathbf{y})$ for all test functions φ on $\Omega \times \Omega$.

the analysis of special cases where convergence can be intended in a stronger sense.

Incidentally, let us mention the relation existing between the kernel $\mathcal{B}(\ ,\ )$ and the *trace of* $\mathcal{B}$; the latter, if it exists, is the number $\operatorname{Tr}\mathcal{B} := \int_{\mathcal{K}} dk\, \langle F_k | \mathcal{B}F_k \rangle \in \mathbf{C}$, where $(F_k)_{k\in\mathcal{K}}$ is any complete orthonormal set of $L^2(\Omega)$. The right-hand side does not depend on the choice of $(F_k)_{k\in\mathcal{K}}$; in particular, if $\mathcal{B}$ has purely point spectrum, $(F_k)_{k\in\mathcal{K}}$ is a complete orthonormal set of proper eigenfunctions labeled by a countable set $\mathcal{K}$ and $(\beta_k)_{k\in\mathcal{K}}$ are the corresponding eigenvalues, we have $\operatorname{Tr}\mathcal{B} = \sum_{k\in\mathcal{K}} \beta_k$ (assuming the series on the right-hand side to converge). Returning to the definition (2.1) of the kernel $\mathcal{B}(\ ,\ )$, we see that

$$\operatorname{Tr}\mathcal{B} = \int_\Omega d\mathbf{x}\, \mathcal{B}(\mathbf{x}, \mathbf{x}) \qquad (2.4)$$

(since $(\delta_\mathbf{x})_{\mathbf{x}\in\Omega}$ is a generalized complete orthonormal set).

Let us move on and note that the boundary conditions possibly involved in the definition of $\mathcal{B}$ have implications for the kernel $\mathcal{B}(\ ,\ )$; for example, if boundary conditions of the Dirichlet type are involved, the eigenfunctions $(F_k)_{k\in\mathcal{K}}$ in Eq. (2.3) vanish on $\partial\Omega$, thus yielding $\mathcal{B}(\mathbf{x}, \mathbf{y}) = 0$ for $\mathbf{x} \in \partial\Omega$ or $\mathbf{y} \in \partial\Omega$.

Let us also mention that from Eq. (2.1) one infers

$$\mathcal{B}^\dagger(\mathbf{x}, \mathbf{y}) = \overline{\mathcal{B}(\mathbf{y}, \mathbf{x})}\,, \qquad \overline{\mathcal{B}}(\mathbf{x}, \mathbf{y}) = \overline{\mathcal{B}(\mathbf{x}, \mathbf{y})} \qquad (2.5)$$

where $\mathcal{B}^\dagger$ is the adjoint operator of $\mathcal{B}$ with respect to the inner product of $L^2(\Omega)$ while $\overline{\mathcal{B}}$ is the complex conjugate operator, such that $\overline{\mathcal{B}\psi} = \overline{\mathcal{B}}\,\overline{\psi}$ for all ψ. These facts imply

$$\mathcal{B}(\mathbf{y}, \mathbf{x}) = \mathcal{B}(\mathbf{x}, \mathbf{y}) \qquad \text{if } \mathcal{B}^\dagger = \overline{\mathcal{B}}\,. \qquad (2.6)$$

2.2   The operator $\mathcal{A}$

Most of the integral kernels considered in the sequel will be related to a self-adjoint operator $\mathcal{A}$ acting in $L^2(\Omega)$; namely, they will be the kernels associated to some function of the operator $\mathcal{A}$ (say, a power $\mathcal{A}^s$). To treat these kernels, precise assumptions on $\mathcal{A}$ will be specified whenever necessary; in any case, we will typically consider three situations of decreasing generality, described by the forthcoming Eqs. (2.7)–(2.9).

In the first situation, we will simply assume that

$$\mathcal{A} \text{ is a strictly positive, self-adjoint operator in } L^2(\Omega) \qquad (2.7)$$

(let us recall that strict positivity means that $\sigma(A) \subset [\varepsilon^2, +\infty)$ for some $\varepsilon > 0$).

In the second situation, the assumptions are the following:

$$\mathcal{A} = -\Delta + V \text{ in } \Omega, \text{ with } V \text{ a real } C^\infty \text{ potential on } \Omega \text{ and boundary}$$
conditions on $\partial\Omega$ such that $\mathcal{A}$ is self-adjoint and strictly positive.
$$(2.8)$$

In the third situation, the assumptions are:

$$\mathcal{A} = -\Delta + V \text{ on a bounded domain } \Omega$$
with $\partial\Omega$ of class C^∞ and Dirichlet boundary conditions, $\qquad$ (2.9)
V a real C^∞ potential on $\overline{\Omega}$ and $V(\mathbf{x}) \geqslant 0$ for all $\mathbf{x} \in \overline{\Omega}$.

Here and in the sequel $\overline{\Omega} = \Omega \cup \partial\Omega$ is the closure of Ω. In particular, let us stress that the assumptions (2.9) suffice to grant self-adjointness and strict positivity of $\mathcal{A}$; moreover, they imply that $\mathcal{A}$ has a purely point spectrum and one can build a complete orthonormal system of proper eigenfunctions F_k with eigenvalues ω_k^2 labeled by $\mathcal{K} = \{1, 2, 3, ...\}$ in such a way that $0 < \omega_1 \leqslant \omega_2 \leqslant \omega_3 \leqslant ...$ (with the possibility that some of these inequalities are equalities, to deal with the case of degenerate eigenvalues). It is well-known that the eigenvalues, when ordered in this manner, fulfill the Weyl asymptotic relation

$$\omega_k \sim C \, k^{1/d} \qquad \text{for } k \to +\infty , \qquad (2.10)$$

where $C := 2\sqrt{\pi}\,\Gamma(d/2 + 1)^{1/d} \operatorname{Vol}(\Omega)^{-1/d}$ (see [104], page 189, Theorem 5, and [52], pages 99–101, § 8.2 for elementary derivations). Appendix B contains a number of technical results concerning cases (2.7)–(2.9), that will be mentioned whenever necessary in the sequel of this chapter. These results often refer to the spaces $C^j(\Omega)$ and $C^j(\Omega \times \Omega)$ or (in the case (2.9)) $C^j(\overline{\Omega})$ and $C^j(\overline{\Omega} \times \overline{\Omega})$, for $j \in \mathbf{N}$ or $j = \infty$; these spaces are formed by the complex functions on $\Omega, \Omega \times \Omega$ and so on, which are continuous along with their partial derivatives of all orders $\leqslant j$. In particular, in the rest of this chapter we will present situations connected to cases (2.7)–(2.9), in which certain integral kernels $(\mathbf{x}, \mathbf{y}) \mapsto \mathcal{B}(\mathbf{x}, \mathbf{y})$ related to $\mathcal{A}$ are of class $C^j(\Omega \times \Omega)$ or $C^j(\overline{\Omega} \times \overline{\Omega})$, for suitable j.

Finally, let us anticipate that the condition of strict positivity for $\mathcal{A}$ in Eqs. (2.7) and (2.8) will be occasionally relaxed, assuming only that the eigenvalues of $\mathcal{A}$ are non-negative (and declaring this explicitly, to avoid misunderstandings); see, e.g. Subsection 2.6.3. We will declare explicitly when reference is made to these relaxed assumptions.

2.3   The Green function

Let $\mathcal{A}$ be a strictly positive self-adjoint operator in $L^2(\Omega)$. Then, we can introduce the inverse operator $\mathcal{A}^{-1}$ and the corresponding kernel

$$G(\mathbf{x}, \mathbf{y}) := \mathcal{A}^{-1}(\mathbf{x}, \mathbf{y}) \,, \tag{2.11}$$

which is called the *Green function* of $\mathcal{A}$. In terms of this kernel, the identity $\mathcal{A}\,\mathcal{A}^{-1} = \mathbf{1}$ can be re-expressed as

$$\mathcal{A}_{\mathbf{x}}\, G(\mathbf{x}, \mathbf{y}) = \delta(\mathbf{x} - \mathbf{y}) \,, \tag{2.12}$$

where $\mathcal{A}_{\mathbf{x}}$ indicates the operator $\mathcal{A}$ acting on $G(\mathbf{x}, \mathbf{y})$ as a function of $\mathbf{x}$, for fixed $\mathbf{y} \in \Omega$. Using a complete orthonormal system $(F_k)_{k \in \mathcal{K}}$ of eigenfunctions of $\mathcal{A}$ with corresponding eigenvalues $(\omega_k^2)_{k \in \mathcal{K}}$, we can express the Green function as

$$G(\mathbf{x}, \mathbf{y}) = \int_{\mathcal{K}} \frac{dk}{\omega_k^2}\, F_k(\mathbf{x})\, \overline{F_k}(\mathbf{y}) \,. \tag{2.13}$$

The Green function is among the most familiar integral kernels, especially when $\mathcal{A} = -\Delta + V$ with suitable boundary conditions (of Dirichlet, Neumann or Robin type); in this case, uniqueness results are available for the Poisson equation (perturbed with an external potential), allowing to characterize the Green function $G(\mathbf{x}, \mathbf{y})$ as the unique solution of Eq. (2.12) fulfilling the prescribed boundary conditions for $\mathbf{x} \in \partial\Omega$ or $\mathbf{y} \in \partial\Omega$. The literature on this topic is enormous and here we only mention some well-known monographs: Berezanskii [15], Krylov [90], Sauvigny [133] and Shimakura [139] give abstract and rigorous analyses, while Duffy [49], Kythe [91], Sommerfeld [143] and Stakgold and Holst [145] present more practical and explicit discussions.

2.4 The Dirichlet kernel

2.4.1 *Basic facts*

Let again $\mathcal{A}$ be a strictly positive self-adjoint operator in $L^2(\Omega)$. The power $\mathcal{A}^{-s}$ can be defined through the standard functional calculus for each $s \in \mathbf{C}$; the corresponding integral kernel

$$D_s(\mathbf{x}, \mathbf{y}) := \mathcal{A}^{-s}(\mathbf{x}, \mathbf{y}) \qquad (2.14)$$

is called the s-th *Dirichlet kernel*. In passing, let us remark that $D_{-1}(\mathbf{x}, \mathbf{y})$ coincides with the Green function $G(\mathbf{x}, \mathbf{y})$ considered in Section 2.3.

If $(F_k)_{k \in \mathcal{K}}$ is a complete orthonormal set of eigenfunctions of $\mathcal{A}$ with corresponding eigenvalues $(\omega_k^2)_{k \in \mathcal{K}}$ we have $\mathcal{A}^{-s} F_k = \omega_k^{-2s} F_k$, so that (by Eq. (2.3))

$$D_s(\mathbf{x}, \mathbf{y}) = \int_{\mathcal{K}} \frac{dk}{\omega_k^{2s}} \, F_k(\mathbf{x}) \overline{F_k}(\mathbf{y}) \ . \qquad (2.15)$$

The denomination "Dirichlet kernel" employed for D_s is suggested by the similarity between the above expansion and the Dirichlet series, considered in [82, 106, 107, 138].

For our purposes, it is important to give sufficient conditions under which $D_s(\mathbf{x}, \mathbf{y})$ is a regular function of $(\mathbf{x}, \mathbf{y})$ (even on the diagonal $\mathbf{y} = \mathbf{x}$, of special interest in the sequel). We are also interested in cases where the expansion (2.15) converges in a stronger sense, in comparison with the distributional sense that we are generically ascribing to kernel expansions. Let us present some results of this kind, which are discussed in Appendix B.

Let $\mathcal{A} = -\Delta + V$ be as in Eq. (2.8); then

$$D_s \in C^j(\Omega \times \Omega) \quad \text{for } s \in \mathbf{C}, \, j \in \mathbf{N} \text{ such that } \Re s > \frac{d}{2} + \frac{j}{2} \ . \quad (2.16)$$

Making the stronger assumptions (2.9), we have

$$D_s \in C^j(\overline{\Omega} \times \overline{\Omega}) \quad \text{for } s \in \mathbf{C}, \, j \in \mathbf{N} \text{ such that } \Re s > \frac{d}{2} + \frac{j}{2} \ . \quad (2.17)$$

Again with the assumptions (2.9), $\mathcal{A}$ possesses a complete orthonormal set of proper eigenfunctions $(F_k)_{k=1,2,3,\dots}$ (see the comments after the cited equation), and the expansion (2.15) reads

$$D_s(\mathbf{x}, \mathbf{y}) = \sum_{k=1}^{+\infty} \frac{1}{\omega_k^{2s}} \, F_k(\mathbf{x})\overline{F_k}(\mathbf{y}) \; . \tag{2.18}$$

In this case one has $F_k \in C^\infty(\overline{\Omega})$ for each k; moreover, for $\Re s > d + j/2$ the expansion (2.18) is absolutely and uniformly convergent on $\overline{\Omega} \times \overline{\Omega}$, with all its derivatives up to order j. See Appendix B (especially Eq. (B.26)) for more details.

To conclude the present section let us mention that the general statement of Eq. (2.4) on traces, here applied with $\mathcal{B} = \mathcal{A}^{-s}$, yields

$$\int_\Omega d\mathbf{x} \, D_s(\mathbf{x}, \mathbf{x}) = \operatorname{Tr} \mathcal{A}^{-s} \; , \tag{2.19}$$

provided that the above trace exists. Making the assumptions (2.9) for $\mathcal{A} = -\Delta + V$ and using the Weyl estimates (2.10) for $\omega_1 \leqslant \omega_2 \leqslant ...$, we find the following:

$$\operatorname{Tr} \mathcal{A}^{-s} = \sum_{k=1}^{+\infty} \frac{1}{\omega_k^{2s}} \quad \text{is finite for } \Re s > \frac{d}{2} \; . \tag{2.20}$$

2.4.2 *Some additional remarks*

Let us consider again a strictly positive self-adjoint operator $\mathcal{A}$ in $L^2(\Omega)$; in addition, assume this operator to be *real*, in the sense that $\overline{\mathcal{A}} = \mathcal{A}$ (i.e. $\overline{\mathcal{A}\psi} = \mathcal{A}\overline{\psi}$ for all ψ).

If $(F_k)_{k\in\mathcal{K}}$ is a complete orthonormal set of eigenfunctions of $\mathcal{A}$ with related eigenvalues $(\omega_k^2)_{k\in\mathcal{K}}$, then the conjugate system $(\overline{F_k})_{k\in\mathcal{K}}$ is as well a complete orthonormal set of eigenfunctions of $\mathcal{A}$ with the same eigenvalues; therefore, besides Eq. (2.15) we have an alternative representation of the Dirichlet kernel, based on this conjugate system. In view of these considerations, one can easily infer that, for $s \in \mathbf{C}$ with complex conjugate $\bar{s}$,

$$D_s(\mathbf{x}, \mathbf{y}) = D_s(\mathbf{y}, \mathbf{x}) \quad \text{and} \quad \overline{D_s(\mathbf{x}, \mathbf{y})} = D_{\bar{s}}(\mathbf{x}, \mathbf{y}) \; . \tag{2.21}$$

To go on we claim that, for any pair of multi-indexes α, β,

$$\partial_{\mathbf{x}}^\alpha \partial_{\mathbf{y}}^\beta D_s(\mathbf{x}, \mathbf{y})\Big|_{\mathbf{y}=\mathbf{x}} = \partial_{\mathbf{x}}^\beta \partial_{\mathbf{y}}^\alpha D_s(\mathbf{x}, \mathbf{y})\Big|_{\mathbf{y}=\mathbf{x}} \; . \tag{2.22}$$

Indeed, due to the first identity in Eq. (2.21), we have $\partial_{\mathbf{x}}^{\alpha}\partial_{\mathbf{y}}^{\beta}D_s(\mathbf{x},\mathbf{y}) = \partial_{\mathbf{x}}^{\alpha}\partial_{\mathbf{y}}^{\beta}D_s(\mathbf{y},\mathbf{x})$; when evaluating the right-hand side of this equality on the diagonal $\mathbf{y} = \mathbf{x}$, the variables can be relabeled to yield $\partial_{\mathbf{x}}^{\alpha}\partial_{\mathbf{y}}^{\beta}D_s(\mathbf{x},\mathbf{y})\big|_{\mathbf{y}=\mathbf{x}} = \partial_{\mathbf{y}}^{\alpha}\partial_{\mathbf{x}}^{\beta}D_s(\mathbf{x},\mathbf{y})\big|_{\mathbf{y}=\mathbf{x}}$, thus proving Eq. (2.22).

All the above results can be applied to the (real) operator $\mathcal{A} := -\Delta + V(\mathbf{x})$; the symmetry properties outlined in the present subsection for the corresponding Dirichlet kernel will be relevant in connection with the results of the next chapter.

2.5   The regularized propagator and the stress-energy VEV: connections with the Dirichlet kernel

Let us refer to the framework described in the previous chapter, where the operator $A = -\Delta + V$ in $L^2(\Omega)$ has been considered in connection with a quantized scalar field. In the sequel $x = (x^0, \mathbf{x})$, $y = (y^0, \mathbf{y}) \in \mathbf{R} \times \Omega$; if we use the expansion (1.18) for the regularized field $\widehat{\phi}^u$ in terms of creation and annihilation operators we obtain for the regularized propagator the expression

$$\langle 0|\widehat{\phi}^u(x)\widehat{\phi}^u(y)0\rangle = \kappa^u \int_{\mathcal{K}\times\mathcal{K}} \frac{dk\,dh}{2(\omega_k\omega_h)^{\frac{u+1}{2}}} \, \langle 0|\left[\widehat{a}_k f_k(x) + \widehat{a}_k^\dagger \overline{f_k}(x)\right]$$

$$\times \left[\widehat{a}_h f_h(y) + \widehat{a}_h^\dagger \overline{f_h}(y)\right]|0\rangle. \tag{2.23}$$

This relation, along with the identities $\langle 0|\widehat{a}_k\widehat{a}_h|0\rangle = \langle 0|\widehat{a}_k^\dagger\widehat{a}_h^\dagger|0\rangle = \langle 0|\widehat{a}_k^\dagger\widehat{a}_h|0\rangle = 0$ and $\langle 0|\widehat{a}_k\widehat{a}_h^\dagger|0\rangle = \delta(k,h)$, gives

$$\langle 0|\widehat{\phi}^u(x)\widehat{\phi}^u(y)|0\rangle = \kappa^u \int_{\mathcal{K}} \frac{dk}{2\,\omega_k^{1+u}} \, f_k(x)\overline{f_k}(y)$$

$$= \kappa^u \int_{\mathcal{K}} \frac{dk}{2\,\omega_k^{1+u}} \, F_k(\mathbf{x})\overline{F_k}(\mathbf{y}) \, e^{-i\omega_k(x^0-y^0)}. \tag{2.24}$$

From here we can easily obtain the derivatives of the propagator; for example, for $j \in \{1, ...d\}$, we have

$$\partial_{x^0 y^j}\langle 0|\widehat{\phi}^u(x)\widehat{\phi}^u(y)|0\rangle = -i\,\kappa^u \int_{\mathcal{K}} \frac{dk}{2\,\omega_k^u} \, F_k(\mathbf{x})(\partial_{y^j}\overline{F_k})(\mathbf{y}) \, e^{-i\omega_k(x^0-y^0)}. \tag{2.25}$$

In particular, if we apply Eqs. (2.24), (2.25) with $y = x$ and compare with the eigenfunction expansion (2.15) of the Dirichlet kernel, we get

$$\langle 0|\widehat{\phi}^u(x)\widehat{\phi}^u(y)|0\rangle\Big|_{y=x} = \frac{\kappa^u}{2} \, D_{\frac{u+1}{2}}(\mathbf{x},\mathbf{y})\Big|_{\mathbf{y}=\mathbf{x}} \; ; \qquad (2.26)$$

$$\partial_{x^0 y^j}\langle 0|\widehat{\phi}^u(x)\widehat{\phi}^u(y)|0\rangle\Big|_{y=x} = -\frac{i\kappa^u}{2} \, \partial_{y^j} D_{\frac{u}{2}}(\mathbf{x},\mathbf{y})\Big|_{\mathbf{y}=\mathbf{x}} \; . \qquad (2.27)$$

One can express similarly all the derivatives with $y = x$ appearing in Eq. (1.21) for the regularized stress-energy VEV. In this way, recalling the identity (2.22) we obtain the following results, where i, j, ℓ are spatial indeces ranging in $\{1, ..., d\}$:

$$\langle 0|\widehat{T}^u_{00}(t,\mathbf{x})|0\rangle = \kappa^u\Bigg[\left(\frac{1}{4}+\xi\right)D_{\frac{u-1}{2}}(\mathbf{x},\mathbf{y})$$

$$+ \left(\frac{1}{4}-\xi\right)\left(\partial^{x^\ell}\partial_{y^\ell}+V(\mathbf{x})\right)D_{\frac{u+1}{2}}(\mathbf{x},\mathbf{y})\Bigg]_{\mathbf{y}=\mathbf{x}} \, , \qquad (2.28)$$

$$\langle 0|\widehat{T}^u_{0j}(t,\mathbf{x})|0\rangle = \langle 0|\widehat{T}^u_{j0}(t,\mathbf{x})|0\rangle = 0 \, , \qquad (2.29)$$

$$\langle 0|\widehat{T}^u_{ij}(t,\mathbf{x})|0\rangle = \langle 0|\widehat{T}^u_{ji}(t,\mathbf{x})|0\rangle$$

$$= \kappa^u\Bigg[\left(\frac{1}{4} - \xi\right)\delta_{ij}\left(D_{\frac{u-1}{2}}(\mathbf{x},\mathbf{y})\right.$$

$$\left. - \left(\partial^{x^\ell}\partial_{y^\ell}+V(\mathbf{x})\right)D_{\frac{u+1}{2}}(\mathbf{x},\mathbf{y})\right)$$

$$+ \left(\left(\frac{1}{2} - \xi\right)\partial_{x^i y^j} - \xi\,\partial_{x^i x^j}\right)D_{\frac{u+1}{2}}(\mathbf{x},\mathbf{y})\Bigg]_{\mathbf{y}=\mathbf{x}} \, . \qquad (2.30)$$

The above equations[c] indicate, among else, that $\langle 0|\widehat{T}^u_{\mu\nu}(t,\mathbf{x})|0\rangle$ does not depend on the time variable t; this comes as no surprise at all, since the general framework we are considering is itself static (indeed, both the spatial domain Ω and the potential V are time independent). These features of the regularized stress-energy VEV had

---

[c]To prove Eq. (2.29), note that

$$\langle 0|\widehat{T}^u_{0j}(t,\mathbf{x})|0\rangle = -\frac{i\kappa^u}{4}\left(\partial_{y^j}D_{\frac{u}{2}}(\mathbf{x},\mathbf{y}) - \partial_{x^j}D_{\frac{u}{2}}(\mathbf{x},\mathbf{y})\right)\Big|_{\mathbf{y}=\mathbf{x}}$$

and that the last expression vanishes identically due to identity (2.22). Besides, let us been anticipated in Subsection 1.4.1; because of them, in the rest of this text we will use the notation[d]

$$\langle 0|\widehat{T}^u_{\mu\nu}(\mathbf{x})|0\rangle \equiv \langle 0|\widehat{T}^u_{\mu\nu}(t,\mathbf{x})|0\rangle\ . \qquad (2.31)$$

Up to now we have been working rather loosely, but it is not difficult to indicate the precise conditions for the validity of our manipulations. Indeed, Eqs. (2.28)–(2.30) for $\langle 0|\widehat{T}^u_{\mu\nu}(\mathbf{x})|0\rangle$ involve the Dirichlet kernels $D_{\frac{u-1}{2}}$ and $D_{\frac{u+1}{2}}$, with its second order derivatives, along the diagonal $\mathbf{y} = \mathbf{x}$. So, for these equations to be meaningful it is sufficient that $D_{\frac{u-1}{2}} \in C^0(\Omega \times \Omega)$ and $D_{\frac{u+1}{2}} \in C^2(\Omega \times \Omega)$; with the assumptions (2.8) on $\mathcal{A} = -\Delta + V$, recalling Eq. (2.16) we see that both conditions are fulfilled if

$$\Re u > d + 1\ . \qquad (2.32)$$

For $\mathcal{A}$, u as in Eqs. (2.8), (2.32), the map $\mathbf{x} \mapsto \langle 0|\widehat{T}^u_{\mu\nu}(\mathbf{x})|0\rangle$ is in $C^0(\Omega)$ for all μ,ν. With the stronger assumptions (2.9) on $\mathcal{A}$, and again with u as in (2.32), we infer from (2.17) that $\mathbf{x} \mapsto \langle 0|\widehat{T}^u_{\mu\nu}(\mathbf{x})|0\rangle$ belongs to $C^0(\overline{\Omega})$ for all $\mu,\nu \in \{0,...,d\}$ (i.e. the regularized stress-energy VEV is continuous up to the boundary).[e]

Now, let us fix a point $\mathbf{x} \in \Omega$ and, making the assumptions (2.8), consider the functions $u \mapsto \langle 0|\widehat{T}^u_{\mu\nu}(\mathbf{x})|0\rangle$ ($\mu,\nu \in \{0,...,d\}$); these are not only well defined, but even analytic on the half-plane $\{u \in \mathbf{C} \mid \Re u > d + 1\}$.[f]

---

point out that Eq. (2.30) is equivalent to the more explicitly symmetric expression

$$\langle 0|\widehat{T}^u_{ij}(t,\mathbf{x})|0\rangle = \langle 0|\widehat{T}^u_{ji}(t,\mathbf{x})|0\rangle$$

$$= \kappa^u \left[\left(\frac{1}{4} - \xi\right)\delta_{ij}\left(D_{\frac{u-1}{2}}(\mathbf{x},\mathbf{y}) - (\partial^{x^\ell}\partial_{y^\ell} + V(\mathbf{x}))D_{\frac{u+1}{2}}(\mathbf{x},\mathbf{y})\right) \right.$$

$$\left. + \left(\left(\frac{1}{4} - \frac{\xi}{2}\right)(\partial_{x^iy^j} + \partial_{x^jy^i}) - \frac{\xi}{2}(\partial_{x^ix^j} + \partial_{y^iy^j})\right)D_{\frac{u+1}{2}}(\mathbf{x},\mathbf{y})\right]_{\mathbf{y}=\mathbf{x}}\ .$$

[d]This is slightly abusive, since staticity occurs only *after* taking the VEV; the alternative notation $\langle 0|\widehat{T}^u_{\mu\nu}|0\rangle(\mathbf{x})$ is more precise, but graphically disturbing and will not be employed in the sequel.

[e]Generalizing the previous considerations, one proves the following for any $\ell \in \mathbf{N}$: with the assumptions (2.8) (resp. (2.9)), if $\Re u > d + \ell + 1$ the function $\mathbf{x} \mapsto \langle 0|\widehat{T}^u_{\mu\nu}(\mathbf{x})|0\rangle$ is in $C^\ell(\Omega)$ (resp., in $C^\ell(\overline{\Omega})$).

[f]This fact can be established proving that $D_{\frac{u-1}{2}}(\mathbf{x},\mathbf{x})$, $D_{\frac{u+1}{2}}(\mathbf{x},\mathbf{x})$ and the second order derivatives of $D_{\frac{u+1}{2}}$ at $(\mathbf{x},\mathbf{x})$ are analytic functions of the parameter u. We do not go into the details of the proofs, that are given elsewhere [62] using appropriate analyticity results for the operator functions $u \mapsto \mathcal{A}^{-(u\mp1)/2}$.

Let us recall that, according to Eqs. (1.22), (1.24), the analytic continuation of $\langle 0|\widehat{T}^u_{\mu\nu}(\mathbf{x})|0\rangle$ at $u=0$ determines the zeta renormalized VEV of the stress-energy tensor; of course, the latter does not depend on t as well and we will write

$$\langle 0|\widehat{T}_{\mu\nu}(\mathbf{x})|0\rangle_{\mathrm{ren}} \equiv \langle 0|\widehat{T}_{\mu\nu}(t,\mathbf{x})|0\rangle_{\mathrm{ren}} \ . \qquad (2.33)$$

Of course, due to Eqs. (2.28)–(2.30), the renormalized stress-energy VEV is determined by the "renormalized" functions

$$D^{(\kappa)}_{\pm\frac{1}{2}}(\mathbf{x},\mathbf{y}) := RP\Big|_{u=0}\left(\kappa^u D_{\frac{u\pm 1}{2}}(\mathbf{x},\mathbf{y})\right) \ , \qquad (2.34)$$

$$\partial_{zw}D^{(\kappa)}_{\frac{1}{2}}(\mathbf{x},\mathbf{y}) := RP\Big|_{u=0}\left(\kappa^u \partial_{zw} D_{\frac{u+1}{2}}(\mathbf{x},\mathbf{y})\right) \qquad (2.35)$$

(with z,w any two spatial variables), to be evaluated along the diagonal $\mathbf{y}=\mathbf{x}$. More precisely, we have

$$\langle 0|\widehat{T}_{00}(\mathbf{x})|0\rangle_{\mathrm{ren}} = \Bigg[\left(\frac{1}{4}+\xi\right)D^{(\kappa)}_{-\frac{1}{2}}(\mathbf{x},\mathbf{y})$$
$$+\left(\frac{1}{4}-\xi\right)\left(\partial^{x^\ell}\partial_{y^\ell}+V(\mathbf{x})\right)D^{(\kappa)}_{+\frac{1}{2}}(\mathbf{x},\mathbf{y})\Bigg]_{\mathbf{y}=\mathbf{x}} \ , \qquad (2.36)$$

$$\langle 0|\widehat{T}_{0j}(\mathbf{x})|0\rangle_{\mathrm{ren}} = \langle 0|\widehat{T}_{j0}(\mathbf{x})|0\rangle_{\mathrm{ren}} = 0 \ , \qquad (2.37)$$

$$\langle 0|\widehat{T}_{ij}(\mathbf{x})|0\rangle_{\mathrm{ren}} = \langle 0|\widehat{T}_{ji}(\mathbf{x})|0\rangle_{\mathrm{ren}}$$
$$= \Bigg[\left(\frac{1}{4}-\xi\right)\delta_{ij}\left(D^{(\kappa)}_{-\frac{1}{2}}(\mathbf{x},\mathbf{y})-\left(\partial^{x^\ell}\partial_{y^\ell}+V(\mathbf{x})\right)D^{(\kappa)}_{+\frac{1}{2}}(\mathbf{x},\mathbf{y})\right)$$
$$+\left(\left(\frac{1}{2}-\xi\right)\partial_{x^i y^j}-\xi\,\partial_{x^i x^j}\right)D^{(\kappa)}_{+\frac{1}{2}}(\mathbf{x},\mathbf{y})\Bigg]_{\mathbf{y}=\mathbf{x}} \ . \qquad (2.38)$$

Let us remark that, if $D_{\frac{u\pm 1}{2}}(\mathbf{x},\mathbf{y})$ and $\partial_{zw}D_{\frac{u+1}{2}}(\mathbf{x},\mathbf{y})$ have analytic continuations regular at $u=0$, indicated hereafter with $D_{\pm\frac{1}{2}}(\mathbf{x},\mathbf{y})$ and $\partial_{zw}D_{\frac{1}{2}}(\mathbf{x},\mathbf{y})$ and also called "renormalized", one has

$$D^{(\kappa)}_{\pm\frac{1}{2}}(\mathbf{x},\mathbf{y}) = D_{\pm\frac{1}{2}}(\mathbf{x},\mathbf{y}) \ ,$$
$$\partial_{zw}D^{(\kappa)}_{\frac{1}{2}}(\mathbf{x},\mathbf{y}) = \partial_{zw}D_{\frac{1}{2}}(\mathbf{x},\mathbf{y}) \qquad (2.39)$$

for any choice of the mass scale $\kappa \in (0, +\infty)$; clearly, in this case the renormalized stress-energy VEV is independent of κ. On the contrary, an explicit dependence on κ appears if the analytic continuations of $D_{\frac{u\pm1}{2}}(\mathbf{x}, \mathbf{x})$ or $\partial_{zw}D_{\frac{u\pm1}{2}}(\mathbf{x}, \mathbf{x})$ (or both) are singular at $u = 0$; this will occur in some specific examples, to be considered in the subsequent Part 2 (see, in particular, Chapters 7, 9 and 10).

A connection between a regularized stress-energy VEV and a Dirichlet-like kernel is mentioned by Cognola, Vanzo and Zerbini [38] in a slightly different framework, where the field theory becomes Euclidean after Wick rotation of the time coordinate $t \to ix^0$ ($x^0 \in \mathbf{R}$), and the operator $\mathcal{A}$ (in the spatial variables $\mathbf{x}$) considered in this book is replaced by the (spacetime) differential operator $-\partial_{00} - \Delta + V$.

A variant of the previous results about the Dirichlet kernel and the regularized VEV $\langle 0|\widehat{T}^u_{\mu\nu}(\mathbf{x})|0\rangle$ can be formulated in the case of a *slab*. In this case the spatial domain is $\Omega = \Omega_1 \times \mathbf{R}^{d_2}$, with Ω_1 a domain in $\mathbf{R}^{d_1}$ and $d_1 + d_2 = d$; moreover, the potential V depends only on the coordinates $\mathbf{x}_1 \in \Omega_1$. In this situation we can express $\langle 0|\widehat{T}^u_{\mu\nu}(\mathbf{x})|0\rangle$ in terms of the Dirichlet kernel associated to the operator $\mathcal{A}_1 := -\Delta_1 + V(\mathbf{x}_1)$. We defer to Section 2.10 a comprehensive discussion of slabs configurations.

In the following sections we will return to the case where Ω is an arbitrary domain in $\mathbf{R}^d$ and we connect the Dirichlet kernel to other integral kernels, in order to shed light on the analytic continuation of D_s; needless to say, these connections will also be useful, in their d_1-dimensional formulation, in the case of a slab.

2.6   The heat kernel, the cylinder kernel and some variations

2.6.1   *Basic facts*

Let us consider again a strictly positive self-adjoint operator $\mathcal{A}$ in $L^2(\Omega)$ (that will be $-\Delta + V$ in the subsequent applications). For all $t \in [0, +\infty)$, using the standard functional calculus we can define the operators

$$e^{-t\mathcal{A}}, \qquad e^{-t\sqrt{\mathcal{A}}} . \tag{2.40}$$

These fulfill the following identities:

$$\left(\frac{d}{dt} + \mathcal{A}\right) e^{-t\mathcal{A}} = 0 , \qquad e^{-t\mathcal{A}}\big|_{t=0} = 1 ; \tag{2.41}$$

$$\left(\frac{d^2}{dt^2} - \mathcal{A}\right) e^{-t\sqrt{\mathcal{A}}} = 0 , \qquad e^{-t\sqrt{\mathcal{A}}}\big|_{t=0} = 1 . \tag{2.42}$$

Moreover, due to the strict positivity of $\mathcal{A}$, both $e^{-t\mathcal{A}}$ and $e^{-t\sqrt{\mathcal{A}}}$ are expected to vanish for $t \to +\infty$, in some sense which can be made more precise in terms of integral kernels. Let us now pass to the kernels

$$K(t;\mathbf{x},\mathbf{y}) := e^{-t\mathcal{A}}(\mathbf{x},\mathbf{y}) , \qquad T(t;\mathbf{x},\mathbf{y}) = e^{-t\sqrt{\mathcal{A}}}(\mathbf{x},\mathbf{y}) , \tag{2.43}$$

which can be expressed as follows in terms of a complete orthonormal set $(F_k)_{k\in\mathcal{K}}$ of eigenfunctions of $\mathcal{A}$ and of the corresponding eigenvalues $(\omega_k^2)_{k\in\mathcal{K}}$:

$$K(t;\mathbf{x},\mathbf{y}) = \int_{\mathcal{K}} dk \, e^{-t\omega_k^2} \, F_k(\mathbf{x})\overline{F_k}(\mathbf{y}) ; \tag{2.44}$$

$$T(t;\mathbf{x},\mathbf{y}) = \int_{\mathcal{K}} dk \, e^{-t\omega_k} \, F_k(\mathbf{x})\overline{F_k}(\mathbf{y}) . \tag{2.45}$$

We notice that

$$(\partial_t + \mathcal{A}_\mathbf{x}) \, K(t;\mathbf{x},\mathbf{y}) = 0 , \qquad K(0;\mathbf{x},\mathbf{y}) = \delta(\mathbf{x} - \mathbf{y}) ; \tag{2.46}$$

$$(\partial_{tt} - \mathcal{A}_\mathbf{x}) \, T(t;\mathbf{x},\mathbf{y}) = 0 , \qquad T(0;\mathbf{x},\mathbf{y}) = \delta(\mathbf{x} - \mathbf{y}) . \tag{2.47}$$

In the above $\mathcal{A}_\mathbf{x}$ indicates the operator $\mathcal{A}$ acting on $K(t;\mathbf{x},\mathbf{y})$ and $T(t;\mathbf{x},\mathbf{y})$ as functions of the $\mathbf{x}$ variable, for any fixed $\mathbf{y} \in \Omega$. Eqs. (2.46) and (2.47) follow, respectively, from Eqs. (2.41) and (2.42). Besides, under minimal supplementary conditions one can prove that $K(t;\mathbf{x},\mathbf{y})$ and $T(t;\mathbf{x},\mathbf{y})$ vanish exponentially for any fixed $\mathbf{x},\mathbf{y} \in \Omega$ and $t \to +\infty$; we shall return on this topic in Subsection 2.6.5.

When the assumptions (2.8) are fulfilled (i.e. when $\mathcal{A} = -\Delta + V$ with V smooth) we have $\partial_t + \mathcal{A}_\mathbf{x} = \partial_t - \Delta_\mathbf{x} + V(\mathbf{x})$ and $\partial_{tt} - \mathcal{A}_\mathbf{x} = \partial_{tt} + \Delta_\mathbf{x} - V(\mathbf{x})$; therefore, Eq. (2.46) contains a heat equation and Eq. (2.47) a $(d + 1)$-dimensional Laplace equation (with an external potential). Note as well that both $K(t;\mathbf{x},\mathbf{y})$ and $T(t;\mathbf{x},\mathbf{y})$ fulfill the boundary conditions in the definition of $\mathcal{A}$ for $\mathbf{x}$ or $\mathbf{y}$ in $\partial\Omega$. For obvious reasons, K is called the *heat kernel* of $\mathcal{A}$ (even in cases where $\mathcal{A}$ is not of the form $-\Delta+V$); T is called by Fulling [69] the *cylinder kernel* of $\mathcal{A}$. It should be mentioned that, under the assumptions (2.8), there are rigorous proofs that

$$\begin{aligned}(t,\mathbf{x},\mathbf{y}) &\mapsto K(t;\mathbf{x},\mathbf{y})\,,\\(t,\mathbf{x},\mathbf{y}) &\mapsto T(t;\mathbf{x},\mathbf{y}) \ \text{ are in } C^{\infty}((0,+\infty)\times\Omega\times\Omega)\,.\end{aligned} \tag{2.48}$$

With the stronger assumptions (2.9) there is a result similar to (2.48), with Ω replaced by $\overline{\Omega}$. These and the previously stated results can be proved by a slight generalization of Theorem 5.2.1 on page 149 of [44]; for some related statements, see also Appendix B and [62].

Again with the assumptions (2.9), $\mathcal{A}$ has pure point spectrum and possesses a complete orthonormal set of proper eigenfunctions $(F_k)_{k=1,2,3,\ldots}$, for which the expansions (2.44) and (2.45) hold with $\int_{\mathcal{K}} dk = \sum_{k=1}^{+\infty}$. One can prove that, for each $t > 0$, these expansions converge absolutely and uniformly on $\overline{\Omega}\times\overline{\Omega}$ with their derivatives of any order (see again Appendix B, especially Eq. (B.35)).

Making the more general assumptions (2.8) we have $\mathcal{A}^{\dagger} = \overline{\mathcal{A}} = \mathcal{A}$, which in turn implies similar relations for $e^{-t\mathcal{A}}$, $e^{-t\sqrt{\mathcal{A}}}$; due to Eq. (2.6), this gives

$$K(t;\mathbf{y},\mathbf{x}) = K(t;\mathbf{x},\mathbf{y})\,, \qquad T(t;\mathbf{y},\mathbf{x}) = T(t;\mathbf{x},\mathbf{y})\,. \tag{2.49}$$

Needless to say, the heat kernel has been the object of intensive and detailed studies, even in much more general frameworks than the one considered in the present book; exhaustive analyses have been given, for example, by Berline *et al.* [17], Calin *et al.* [32], Chavel [36], Davies [44], Gilkey [75] and Grigor'yan [80]. On the contrary, the cylinder kernel is a less popular object; it has mainly been investigated by Fulling and co-authors [68, 69, 71].

Before moving on, let us consider the *heat* and *cylinder traces*; these are respectively defined, for $t \in (0,+\infty)$, as

$$K(t) := \operatorname{Tr} e^{-t\mathcal{A}}\,, \qquad T(t) := \operatorname{Tr} e^{-t\sqrt{\mathcal{A}}}\,. \tag{2.50}$$

Assuming the above traces to exist, the general identity (2.4) for the trace of an operator $\mathcal{B}$ (here applied with either $\mathcal{B} = e^{-t\mathcal{A}}$ or

$\mathcal{B} = e^{-t\sqrt{\mathcal{A}}}$) yields respectively

$$K(t) = \int_\Omega d\mathbf{x}\, K(t; \mathbf{x}, \mathbf{x})\,, \qquad T(t) = \int_\Omega d\mathbf{x}\, T(t; \mathbf{x}, \mathbf{x})\,. \qquad (2.51)$$

In particular, in the case (2.9) where the eigenvalues of $\mathcal{A}$ are labelled by $\mathcal{K} = \{1, 2, 3, ...\}$ and the Weyl estimate (2.10) holds, we have:

$$K(t) = \sum_{k=1}^{+\infty} e^{-t\omega_k^2} < +\infty\,, \quad T(t) = \sum_{k=1}^{+\infty} e^{-t\omega_k} < +\infty \qquad \text{for all } t > 0\,.$$
$$(2.52)$$

2.6.2 *The modified cylinder kernel*

Some considerations of Fulling (see, e.g. [73]) also involve the operator $\sqrt{\mathcal{A}}^{-1} e^{-t\sqrt{\mathcal{A}}}$ and the associated kernel

$$\tilde{T}(t; \mathbf{x}, \mathbf{y}) := (\sqrt{\mathcal{A}}^{-1} e^{-t\sqrt{\mathcal{A}}})(\mathbf{x}, \mathbf{y}) = \int_\mathcal{K} \frac{dk}{\omega_k}\, e^{-t\omega_k}\, F_k(\mathbf{x})\overline{F_k}(\mathbf{y})\,,$$
$$(2.53)$$

which we will refer to as the *modified cylinder kernel*, for reasons which become apparent hereafter (see Eq. (2.54)). Let us observe that the trivial identity $e^{-t\sqrt{\mathcal{A}}} = -\frac{d}{dt}(\sqrt{\mathcal{A}}^{-1} e^{-t\sqrt{\mathcal{A}}})$ can be reformulated in terms of integral kernels as

$$T(t; \mathbf{x}, \mathbf{y}) = -\partial_t \tilde{T}(t; \mathbf{x}, \mathbf{y})\,. \qquad (2.54)$$

Conversely, $\tilde{T}$ can be determined as the primitive of $-T$ which vanishes for $t \to +\infty$, i.e.

$$\tilde{T}(t; \mathbf{x}, \mathbf{y}) = \int_t^{+\infty} dt'\, T(t'; \mathbf{x}, \mathbf{y})\,. \qquad (2.55)$$

In some cases $\tilde{T}$ is easier to compute than T, and some identities relating the cylinder kernel T to the Dirichlet kernel D_s can be applied more efficiently if they are rephrased in terms of $\tilde{T}$; this situation will be exemplified in Chapters 5 and 8, where the space domain is a segment or a wedge.

With the assumptions (2.8) or (2.9), one could give for $\tilde{T}$ some regularity results very similar to the ones illustrated previously for T (e.g. statement (2.48) holds as well for $\tilde{T}$ under the conditions (2.8)).

2.6.3   *The case of a non-negative $\mathcal{A}$*

Let us remark that the heat and cylinder kernels can both be defined even when $\mathcal{A}$ is *non-negative*, without requiring strict positivity; by this we mean that $\sigma(\mathcal{A}) \subset [0, +\infty)$, and that we are not assuming $\sigma(\mathcal{A}) \subset [\varepsilon^2, +\infty)$ for any $\varepsilon > 0$. The non-negativity of $\mathcal{A}$ is equivalent to the existence of a complete orthonormal system $(F_k)_{k \in \mathcal{K}}$ of (either proper or improper) eigenfunctions with corresponding non-negative eigenvalues $(\omega_k^2)_{k \in \mathcal{K}}$ ($\omega_k \geqslant 0$). Most of the previous considerations still hold, in particular Eqs. (2.44), (2.45).

For example, if $\mathcal{A} = -\Delta$ and $\Omega = \mathbf{R}^d$, then $\mathcal{A}$ is non-negative with eigenfunctions $F_{\mathbf{k}}(\mathbf{x}) = (2\pi)^{-d/2} e^{i\mathbf{k} \cdot \mathbf{x}}$ and eigenvalues $\omega_{\mathbf{k}}^2 = |\mathbf{k}|^2$, labeled by $\mathbf{k} \in \mathbf{R}^d$; the measure $d\mathbf{k}$ on the set of labels is the usual Lebesgue measure of $\mathbf{R}^d$. The eigenfunction expansion (2.44) of the heat kernel yields in this case the familiar result

$$K(\mathsf{t}; \mathbf{x}, \mathbf{y}) = \frac{1}{(4\pi\mathsf{t})^{d/2}} \, e^{-\frac{|\mathbf{x}-\mathbf{y}|^2}{4\mathsf{t}}} \; ; \qquad (2.56)$$

moreover, the expansion (2.45) of the cylinder kernel gives the result

$$T(\mathsf{t}; \mathbf{x}, \mathbf{y}) = \frac{\Gamma(\frac{d+1}{2}) \, \mathsf{t}}{\pi^{\frac{d+1}{2}} (\mathsf{t}^2 + |\mathbf{x} - \mathbf{y}|^2)^{\frac{d+1}{2}}} \; , \qquad (2.57)$$

which is a bit less popular and appears, e.g. in [70].

In some subcases with non-negative spectrum we can speak as well of the modified cylinder kernel $\tilde{T}(\mathsf{t}; \mathbf{x}, \mathbf{y})$. In fact, if 0 has zero spectral measure (a fact holding when 0 belongs to the continuous spectrum, but not holding when 0 is a proper eigenvalue), $\sqrt{\mathcal{A}}^{-1} e^{-\mathsf{t}\sqrt{\mathcal{A}}}$ can be defined through the standard functional calculus for self-adjoint operators. With minimal supplementary assumptions of regularity, the kernel $\tilde{T}(\mathsf{t}; \mathbf{x}, \mathbf{y}) := \langle \delta_{\mathbf{x}} | \sqrt{\mathcal{A}}^{-1} e^{-\mathsf{t}\sqrt{\mathcal{A}}} \delta_{\mathbf{y}} \rangle$ makes sense, as well as its expansion (2.53) in terms of a complete orthonormal set of (generalized) eigenfunctions F_k with eigenvalues ω_k^2 ($k \in \mathcal{K}$); note that the assumption of zero spectral measure for 0 is equivalent to the requirement that $\omega_k = 0$ only on a zero-measure subset of $\mathcal{K}$. In these situations we have again Eq. (2.55), describing the cylinder kernel T as the primitive of $-\tilde{T}$.

For example, let us return to the case where $\mathcal{A} = -\Delta$ and $\Omega = \mathbf{R}^d$, in which the spectrum $\sigma(\mathcal{A}) = [0, +\infty)$ is purely continuous. We have mentioned previously the eigenfunctions $F_{\mathbf{k}}$ and the eigenvalues $\omega_{\mathbf{k}}^2$ (labeled by $\mathbf{k} \in \mathbf{R}^d$), where $F_{\mathbf{k}}(\mathbf{x}) = (2\pi)^{-d/2} e^{i\mathbf{k}\cdot\mathbf{x}}$ and $\omega_{\mathbf{k}} = |\mathbf{k}|$; of course $\omega_{\mathbf{k}} = 0$ only on a set of zero Lebesgue measure (consisting of the unique point $\mathbf{k} = 0$). In this case, the expansion (2.53) (involving an integral in the Lebesgue measure $d\mathbf{k}$) gives the following for $d \geqslant 2$:[g]

$$\tilde{T}(\mathfrak{t}; \mathbf{x}, \mathbf{y}) = \frac{\Gamma(\frac{d-1}{2})}{2\pi^{\frac{d+1}{2}} (\mathfrak{t}^2 + |\mathbf{x} - \mathbf{y}|^2)^{\frac{d-1}{2}}} \ . \tag{2.58}$$

2.6.4 *Connections between the cylinder kernel and a $(d+1)$-dimensional Green function*

Due to the limited popularity of the cylinder kernel T it can be useful to connect this kernel to a more familiar object, namely a Green function, even though this requires to pass to $d+1$ dimensions.

To this purpose, let $\Omega \subset \mathbf{R}^d$ be an open set, and let $\mathcal{A}$ be a strictly positive self-adjoint operator in $L^2(\Omega)$ (keeping into account suitable boundary conditions on $\partial\Omega$). Next, let us consider in $\mathbf{R}^{d+1}$ the domain $\mathcal{O} := (0, +\infty) \times \Omega \ni (\mathfrak{t}, \mathbf{x})$ and the Hilbert space $L^2(\mathcal{O})$; we introduce therein the operator

$$\mathcal{P} := -\partial_{\mathfrak{t}\mathfrak{t}} + \mathcal{A} \ , \tag{2.59}$$

with suitable boundary conditions on $\partial\mathcal{O} := (\{0\}\times\Omega)\cup((0, +\infty)\times\partial\Omega)$. More precisely, we assume Dirichlet boundary conditions on $\{0\} \times \Omega$ and the previously given boundary conditions for $\mathcal{A}$ on $(0, +\infty)\times\partial\Omega$. The operator $\mathcal{P}$ is self-adjoint; in the sequel we will prove that it is strictly positive.

To proceed, let us introduce the Green function

$$G(\mathfrak{t}, \mathbf{x}; \mathfrak{t}', \mathbf{y}) := \mathcal{P}^{-1}((\mathfrak{t}, \mathbf{x}), (\mathfrak{t}', \mathbf{y})) \equiv \langle \delta_{\mathfrak{t}} \, \delta_{\mathbf{x}} | \mathcal{P}^{-1} \delta_{\mathfrak{t}'} \, \delta_{\mathbf{y}} \rangle \ ; \tag{2.60}$$

---

[g]For $d = 1$ the right hand side of Eq. (2.53) for $\tilde{T}$ does not converge (not even distributionally); we take this as an indication that the modified cylinder kernel is ill-defined.

this is characterized by the equation

$$(-\partial_{tt} + \mathcal{A}_{\mathbf{x}})G(t, \mathbf{x}; t', \mathbf{y}) = \delta(t - t')\delta(\mathbf{x} - \mathbf{y}) , \qquad (2.61)$$

and by the boundary conditions prescribed on $\partial\mathcal{O}$. We claim that the cylinder kernel T is related to G by

$$\partial_{t'}G(t, \mathbf{x}; t', \mathbf{y})\big|_{t'=0} = T(t; \mathbf{x}, \mathbf{y}) . \qquad (2.62)$$

To prove this (and the previous statement on the strict positivity of $\mathcal{P}$), let $(F_k)_{k\in\mathcal{K}}$ be a complete orthonormal system of eigenfunctions of $\mathcal{A}$ with corresponding eigenvalues $(\omega_k^2)_{k\in\mathcal{K}}$; clearly, the functions $t \in (0, +\infty) \mapsto \sqrt{\frac{2}{\pi}} \sin(\lambda t)$ $(\lambda \in (0, +\infty))$ are a complete orthonormal system in $L^2((0, +\infty))$ and are eigenfunctions of $-\partial_{tt}$ vanishing for $t = 0$. These facts ensure that the family of functions

$$Y_{(\lambda,k)}(t, \mathbf{x}) := \sqrt{\frac{2}{\pi}} \sin(\lambda t) F_k(\mathbf{x}) \qquad \text{for } (\lambda, k) \in (0, +\infty) \times \mathcal{K} \tag{2.63}$$

is a complete orthonormal system of eigenfunctions of $\mathcal{P}$, with

$$\mathcal{P} Y_{(\lambda,k)} = (\lambda^2 + \omega_k^2) Y_{(\lambda,k)} . \qquad (2.64)$$

We recall that we are assuming $\omega_k^2 \geqslant \varepsilon^2$ for some $\varepsilon > 0$; the eigenvalues $(\lambda^2 + \omega_k^2)$ also have ε^2 as a lower bound, so $\mathcal{P}$ is strictly positive.

The Green function of Eq. (2.60) can be expressed via the equation

$$G(t, \mathbf{x}; t', \mathbf{y}) = \int_{(0,+\infty)\times\mathcal{K}} \frac{d\lambda\, dk}{\lambda^2 + \omega_k^2} \left(\sqrt{\frac{2}{\pi}} \sin(\lambda t) F_k(\mathbf{x}) \right)$$

$$\times \left(\sqrt{\frac{2}{\pi}} \sin(\lambda t')\overline{F_k}(\mathbf{y}) \right) , \qquad (2.65)$$

which, evaluating explicitly the integral in λ,[h] reduces to

$$G(t, \mathbf{x}; t', \mathbf{y}) = \int_{\mathcal{K}} dk\, \frac{e^{-\omega_k|t-t'|} - e^{-\omega_k(t+t')}}{2\omega_k} F_k(\mathbf{x})\, \overline{F_k}(\mathbf{y}) . \qquad (2.66)$$

---

[h]It suffices to observe that, by symmetry arguments and the residue theorem, there holds

$$\int_0^{+\infty} d\lambda\, \frac{\sin(\lambda t)\sin(\lambda t')}{\lambda^2 + \omega_k^2} = \frac{1}{2} \int_{-\infty}^{+\infty} d\lambda\, \frac{\sin(\lambda t)\sin(\lambda t')}{\lambda^2 + \omega_k^2} = \frac{\pi}{4\omega_k} \left(e^{-\omega_k|t-t'|} - e^{-\omega_k(t+t')} \right) .$$

To go on, let us differentiate both sides of Eq. (2.66) with respect to t'; this gives

$$\partial_{t'} G(t, \mathbf{x}; t', \mathbf{y}) = \int_K dk \left[\frac{\text{sgn}(t - t')}{2} e^{-\omega_k |t - t'|} + \frac{1}{2} e^{-\omega_k (t + t')} \right] F_k(\mathbf{x}) \overline{F_k}(\mathbf{y}) .$$
(2.67)

Setting $t' = 0$ (and recalling that $t > 0$), the last equation yields

$$\partial_{t'} G(t, \mathbf{x}; t', \mathbf{y})\big|_{t'=0} = \int_K dk \, e^{-\omega_k t} F_k(\mathbf{x}) \overline{F_k}(\mathbf{y}) .$$
(2.68)

The right-hand side of the above equality is just the representation of the cylinder kernel T given by Eq. (2.45); thus we have proved Eq. (2.62). The previous considerations are readily generalized to the case when $\mathcal{A}$ is just non-negative; then, $\mathcal{P}$ is non-negative as well.

Let us conclude this subsection describing, as an example, how the approach based on (2.62) can be used to derive the expression (2.57) for the cylinder kernel T in the case where

$$\Omega := \mathbf{R}^d , \qquad \mathcal{A} := -\Delta .$$
(2.69)

In this case, the associated $(d + 1)$ dimensional domain and the operator $\mathcal{P}$ are

$$\mathcal{O} := (0, +\infty) \times \mathbf{R}^d , \qquad \mathcal{P} := -\partial_{tt} - \Delta = -\Delta_{d+1} ,$$
(2.70)

with Dirichlet boundary conditions on $\partial \mathcal{O}$, which coincides with the hyperplane $\{0\} \times \mathbf{R}^d$; of course, in the above Δ_{d+1} indicates the $(d + 1)$-dimensional Laplacian. The Green function $\mathcal{G}$ of $-\Delta_{d+1}$ on the half-space $\mathcal{O}$ can be obtained by the familiar method of images from the Green function $\mathcal{G}_0$ of $-\Delta_{d+1}$ on the full space $\mathbf{R}^{d+1}$; thus

$$\mathcal{G}(t, \mathbf{x}; t', \mathbf{y}) = \mathcal{G}_0(t, \mathbf{x}; t', \mathbf{y}) - \mathcal{G}_0(t, \mathbf{x}; -t', \mathbf{y}) ,$$
(2.71)

$$\mathcal{G}_0(t, \mathbf{x}; t', \mathbf{y}) := \begin{cases} \frac{1}{2\pi} \ln((t - t')^2 + (\mathbf{x} - \mathbf{y})^2) & \text{if } d = 1, \\ \dfrac{\Gamma(\frac{d+1}{2})}{(d-1)2\pi^{\frac{d+1}{2}} ((t - t')^2 + |\mathbf{x} - \mathbf{y}|^2)^{\frac{d-1}{2}}} & \text{if } d \geqslant 2. \end{cases}$$
(2.72)

Inserting Eqs. (2.71), (2.72) into Eq. (2.62) we obtain for the cylinder kernel T the expression (2.57)

$$T(t;\mathbf{x},\mathbf{y}) = \frac{\Gamma(\frac{d+1}{2})\, t}{\pi^{\frac{d+1}{2}}(t^2 + |\mathbf{x}-\mathbf{y}|^2)^{\frac{d+1}{2}}}$$

(in all dimensions, including $d = 1$).

2.6.5   *Behaviour of the heat and cylinder kernels for small and large* t

The small t limit. The asymptotic expansion of the heat kernel of an operator $\mathcal{A} = -\Delta + V$ on an open set $\Omega \subset \mathbf{R}^d$ has been extensively studied (see, e.g. the work of Minakshisundaram and Pleijel [107], or the already cited monographs [17, 32, 36, 44, 75, 80] on the heat kernel). From here to the end of this paragraph, we discuss the $t \to 0^+$ behavior of the heat and cylinder kernels of $\mathcal{A} = -\Delta + V$ under the assumptions (2.9) (these could be generalized to get the same results, but we are not going to discuss this subject). As known from the previously cited references, there is a unique sequence of real functions $a_n : \Omega \times \Omega \to \mathbf{R}$ ($n = 1, 2, 3....$), usually referred to as HMDS (Hadamard–Minakshisundaram–DeWitt–Seeley) coefficients, such that for any $N \in \{1, 2, 3, ...\}$ one has

$$K(t;\mathbf{x},\mathbf{y}) = \frac{1}{(4\pi t)^{d/2}} e^{-\frac{|\mathbf{x}-\mathbf{y}|^2}{4t}} \left[1 + \sum_{n=1}^{N} a_n(\mathbf{x},\mathbf{y}) t^n + O(t^{N+1}) \right] \text{ for } t \to 0^+.$$
$$(2.73)$$

In the above equation notice the factor $K_0(t;\mathbf{x},\mathbf{y}) := \frac{1}{(4\pi t)^{d/2}} e^{-\frac{|\mathbf{x}-\mathbf{y}|^2}{4t}}$, which is just the heat kernel associated to $-\Delta$ on $\mathbf{R}^d$. In the case $V = 0$, we have $a_n = 0$ for all n; thus, $K(t;\mathbf{x},\mathbf{y}) = K_0(t;\mathbf{x},\mathbf{y})[1 + O(t^\infty)]$ (where the last term indicates a remainder which is $O(t^N)$ for each $N \in \{1, 2, 3, ...\}$).

Along the diagonal $\mathbf{y} = \mathbf{x}$ (for any V) Eq. (2.73) reduces to

$$K(t;\mathbf{x},\mathbf{x}) = \frac{1}{(4\pi t)^{d/2}} \left[1 + \sum_{n=1}^{N} a_n(\mathbf{x})\, t^n + O(t^{N+1}) \right] \quad \text{for } t \to 0^+$$
$$(2.74)$$

where $a_n(\mathbf{x})$ is shorthand for $a_n(\mathbf{x}, \mathbf{x})$. The small t analysis of the cylinder kernel is more involved; however, Fulling proved (see, e.g. [68]) that its asymptotic behavior along the diagonal $\mathbf{y} = \mathbf{x}$ is as follows: there exist functions $e_n, f_n : \Omega \to \mathbf{R}$ $(n = 0, 1, 2, ...)$ such that, for any $N \in \{0, 1, 2, ...\}$,

$$T(t; \mathbf{x}, \mathbf{x}) = \frac{1}{t^d} \left[\sum_{n=0}^{N} e_n(\mathbf{x}) \, t^n + \sum_{\substack{n=d+1 \\ n-d \text{ odd}}}^{N} f_n(\mathbf{x}) \, t^n \ln t + O(t^{N+1} \ln t) \right] \quad \text{for } t \to 0^+.$$

$$(2.75)$$

As pointed out in [68], some of the functions e_n, f_n (but not all of them) can be expressed in terms of the diagonal HMDS coefficients $\mathbf{x} \mapsto a_n(\mathbf{x})$ mentioned previously.

Before proceeding, let us remark that the heat and cylinder traces $K(t), T(t)$ (see Eqs. (2.50), (2.51)) are well-known to admit small t expansions analogous to those in Eqs. (2.74), (2.75); see once more the references cited above. In particular, assuming again Ω to be compact and V to be smooth and bounded below, for $t \to 0^+$ there hold the following identities:

$$K(t) = \frac{1}{(4\pi t)^{d/2}} \left[\text{Vol}(\Omega) + \sum_{n=1}^{N} A_n \, t^{n/2} + O(t^{\frac{N+1}{2}}) \right], \qquad (2.76)$$

$$T(t) = \frac{1}{t^d} \left[\sum_{n=0}^{N} E_n \, t^n + \sum_{\substack{n=d+1 \\ n-d \text{ odd}}}^{N} F_n \, t^n \ln t + O(t^{N+1} \ln t) \right] \qquad (2.77)$$

($\text{Vol}(\Omega)$ denotes the volume of the spatial domain Ω). Notice, in particular, that the expansion (2.76) for $K(t)$ involves half-integer powers of t, whereas in expansions (2.73) and (2.74) for the local heat kernel $K(t; \mathbf{x}, \mathbf{y})$ only integer powers of t appear; besides, let us stress that the real coefficients A_n, E_n, F_n in Eqs. (2.76), (2.77) are not just the integrals over the spatial domain Ω of the functions $a_n(\mathbf{x}), e_n(\mathbf{x}), f_n(\mathbf{x})$ of Eqs. (2.74), (2.75), since boundary contributions arise as well.

The large t behavior. Let us move on and note that, as anticipated in Subsection 2.6.1, both the heat and the cylinder kernel (along with their traces) vanish exponentially for large t, under minimal regularity conditions. As an example, we will prove the exponential decay of the cylinder kernel making the following assumptions:

(a) $\mathcal{A}$ is a strictly positive, self-adjoint operator in $L^2(\Omega)$, possessing a generalized complete orthonormal set $(F_k)_{k \in \mathcal{K}}$ of eigenfunctions with eigenvalues ω_k^2 ($\omega_k \geqslant \varepsilon > 0$); the F_k's are continuous functions on Ω.

(b) for all points $\mathbf{x}, \mathbf{y} \in \Omega$ and $t > 0$, one has

$$\hat{T}(t; \mathbf{x}, \mathbf{y}) := \int_{\mathcal{K}} dk \, e^{-\omega_k t} |F_k(\mathbf{x})| \, |F_k(\mathbf{y})| < +\infty \ . \tag{2.78}$$

In this case, starting from the eigenfunction expansion (2.15) we find

$$|T(t; \mathbf{x}, \mathbf{y})| \leqslant \hat{T}(t; \mathbf{x}, \mathbf{y}) \tag{2.79}$$

at all points $\mathbf{x}, \mathbf{y} \in \Omega$. On the other hand, after fixing $\tau > 0$ we see that

$$\hat{T}(t; \mathbf{x}, \mathbf{y}) \leqslant e^{-\varepsilon(t-\tau)} \, \hat{T}(\tau; \mathbf{x}, \mathbf{y}) \quad \text{for all } t \geqslant \tau \tag{2.80}$$

(this follows from the definition (2.78) of $\hat{T}$, noting that the inequalities $\omega_k \geqslant \varepsilon > 0$ and $t - \tau \geqslant 0$ imply $e^{-\omega_k t} = e^{-\omega_k(t-\tau)} e^{-\omega_k \tau} \leqslant e^{-\varepsilon(t-\tau)} e^{-\omega_k \tau}$). The relations (2.79) and (2.80) give the desired result of exponential decay for T, since at all points $\mathbf{x}, \mathbf{y} \in \Omega$ they imply

$$|T(t; \mathbf{x}, \mathbf{y})| \leqslant e^{-\varepsilon(t-\tau)} \, \hat{T}(\tau; \mathbf{x}, \mathbf{y}) \qquad \text{for all } t \geqslant \tau \ . \tag{2.81}$$

In the above argument, condition (2.78) plays a crucial role; this condition is fulfilled, for example, if $\mathcal{A} = -\Delta + V$ and the assumptions (2.9) are satisfied. In this case we have a uniform bound $\hat{T}(t; \mathbf{x}, \mathbf{y}) \leqslant \check{T}(t) < +\infty$ for all $\mathbf{x}, \mathbf{y} \in \overline{\Omega}$ and $t > 0$ (see Appendix B, Eq. (B.37)), so that Eq. (2.81) implies

$$|T(t; \mathbf{x}, \mathbf{y})| \leqslant e^{-\varepsilon(t-\tau)} \, \check{T}(\tau) \qquad \text{for all } \mathbf{x}, \mathbf{y} \in \overline{\Omega} \text{ and } t \geqslant \tau \ . \tag{2.82}$$

The exponential decay of the heat kernel and of the traces of both the heat and cylinder kernels can be inferred by similar considerations. For alternative approaches, see [44] or [80].

2.7   The Dirichlet kernel as Mellin transform of the heat or cylinder kernel

The results reported in this section are well-known; they were derived by Dowker and Critchley [46], Hawking [84] and Wald [146] and later reconsidered by Moretti *et al.* (see [30, 110] and citations therein).

Let us first recall the definition of the Mellin transform. If $\mathcal{F} : (0, +\infty) \to \mathbf{C}$ is locally integrable, exponentially vanishing for $\mathfrak{t} \to +\infty^{i}$ and $\mathcal{F}(\mathfrak{t}) = O(1/\mathfrak{t}^{\rho})$ for $\mathfrak{t} \to 0^{+}$ with $\rho \in \mathbf{R}$, the Mellin transform of $\mathcal{F}$ is the function

$$\sigma \mapsto \mathfrak{M}(\sigma) := \int_{0}^{+\infty} d\mathfrak{t} \, \mathfrak{t}^{\sigma-1} \, \mathcal{F}(\mathfrak{t}) \qquad (\sigma \in \mathbf{C}, \, \Re\sigma > \rho) \, . \qquad (2.83)$$

As well-known, the map $\sigma \mapsto \mathfrak{M}(\sigma)$ is analytic; moreover, the integral representation in Eq. (2.83) can be used as a starting point to construct its analytic continuation with respect to σ to wider regions of the complex plane (a topic to be discussed in detail in the subsequent Section 2.8).

To proceed, let us pass to the representations of the Dirichlet kernel mentioned in the title of the present section; these can be derived starting from the well-known relation (see [117], page 139, Eq. (5.9.1))

$$\frac{1}{z^{s}} = \frac{1}{\Gamma(s)} \int_{0}^{+\infty} d\mathfrak{t} \, \mathfrak{t}^{s-1} \, e^{-z\mathfrak{t}} \quad \text{for all } z \in (0, +\infty), \, s \in \mathbf{C} \text{ with } \Re s > 0 \, .$$
$$(2.84)$$

The above identity, along with the eigenfunction expansions (2.15), (2.44) and (2.45) of the Dirichlet, heat and cylinder kernels, yields the identities

$$D_{s}(\mathbf{x}, \mathbf{y}) = \frac{1}{\Gamma(s)} \int_{0}^{+\infty} d\mathfrak{t} \, \mathfrak{t}^{s-1} \, K(\mathfrak{t}; \mathbf{x}, \mathbf{y}) \, , \qquad (2.85)$$

$$D_{s}(\mathbf{x}, \mathbf{y}) = \frac{1}{\Gamma(2s)} \int_{0}^{+\infty} d\mathfrak{t} \, \mathfrak{t}^{2s-1} \, T(\mathfrak{t}; \mathbf{x}, \mathbf{y}) \, . \qquad (2.86)$$

---

^{i}In fact, the assumption of exponential decay at infinity is not strictly necessary to define the Mellin transform of a function; however, for the applications to be considered in this book, this hypothesis implies no loss of generality. For a more general analysis of Mellin transforms, we refer to [62] and to the literature cited therein (see, in particular, [28, 150]).

For example, Eq. (2.85) is derived via the following chain of equalities:

$$D_s(\mathbf{x}, \mathbf{y}) = \int_{\mathcal{K}} \frac{dk}{\omega_k^{2s}} F_k(\mathbf{x}) \overline{F_k}(\mathbf{y})$$

$$= \int_{\mathcal{K}} dk \frac{1}{\Gamma(s)} \int_0^{+\infty} dt \, t^{s-1} \, e^{-\omega_k^2 t} F_k(\mathbf{x}) \overline{F_k}(\mathbf{y})$$

$$= \frac{1}{\Gamma(s)} \int_0^{+\infty} dt \, t^{s-1} \int_{\mathcal{K}} dk \, e^{-\omega_k^2 t} F_k(\mathbf{x}) \overline{F_k}(\mathbf{y})$$

$$= \frac{1}{\Gamma(s)} \int_0^{+\infty} dt \, t^{s-1} K(t; \mathbf{x}, \mathbf{y}) \, . \tag{2.87}$$

In the second passage above, we used Eq. (2.84) with $z = \omega_k^2$; in the third passage, the exchange in the order of integration is justifed with arguments similar to those in [106, 107]. The derivation of Eq. (2.86) is similar; in this case one has to resort to Eq. (2.84) with $z = \omega_k$ and s replaced by $2s$.

Equations (2.85), (2.86) state that the Dirichlet kernel can be represented as the *Mellin transform* of either the heat or the cylinder kernel; there are analogous relations for the derivatives of the Dirichlet kernel, involving the corresponding derivatives of the heat and cylinder kernels.

In the sequel we are especially interested in the case in which the integral representations (2.85) and (2.86) hold *pointwisely*, at a given pair of points $\mathbf{x}, \mathbf{y} \in \Omega$ (including the case $\mathbf{y} = \mathbf{x}$). This occurs under the following three conditions:

(a) $D_s(\mathbf{x}, \mathbf{y})$ is well defined at the given points $\mathbf{x}, \mathbf{y}$;

(b) $K(t; \mathbf{x}, \mathbf{y})$ and $T(t; \mathbf{x}, \mathbf{y})$ are also well-defined at these points, for all $t > 0$.

(c) the integrals in Eqs. (2.85), (2.86) converge.

With the assumptions (2.9), a) b) and c) hold at all points $\mathbf{x}, \mathbf{y} \in \Omega$ if $\Re s > d/2$. In fact: for $\Re s > d/2$, D_s is continuous on $\overline{\Omega} \times \overline{\Omega}$ (see Eq. (2.17)), so there is no problem with its pointwise evaluation; for each $t > 0$, the functions $K(t; \, , \,)$ and $T(t; \, , \,)$ are continuous (in fact C^∞) on $\overline{\Omega} \times \overline{\Omega}$, and again their pointwise evaluation is not a problem; due to the exponential decay of $K(t; \mathbf{x}, \mathbf{y})$ and $T(t; \mathbf{x}, \mathbf{y})$, the integrals (2.85) and (2.86) have no convergence problems for large t; for any $\mathbf{x}, \mathbf{y} \in \Omega$, due to the $t \to 0^+$ asymptotic expansions (2.73), (2.75), the same integrals are both convergent for t close to zero if $\Re s > d/2$.[j]

The pointwise representations (2.85), (2.86) can be used as a starting point to build the analytic continuation of the function $s \mapsto D_s(\mathbf{x}, \mathbf{y})$ to a larger domain than the one where they are initially granted to hold (e.g. larger than the half-plane $\{s \in \mathbf{C} \mid \Re s > d/2\}$ of case (2.9)); we return to this point in the next two sections.

One could make similar considerations for the derivatives of the Dirichlet kernel. For example, for any pair of multi-indices α, β, Eq. (2.85) gives

$$\partial_{\mathbf{x}}^{\alpha} \partial_{\mathbf{y}}^{\beta} D_s(\mathbf{x}, \mathbf{y}) = \frac{1}{\Gamma(s)} \int_0^{+\infty} dt \, t^{s-1} \, \partial_{\mathbf{x}}^{\alpha} \partial_{\mathbf{y}}^{\beta} K(t; \mathbf{x}, \mathbf{y}) . \qquad (2.88)$$

If we make the assumptions (2.9) on $\mathcal{A} = -\Delta + V$ and admit that the $t \to 0^+$ asymptotic expansion (2.73) can be differentiated term by term, Eq. (2.88) holds pointwisely at all $\mathbf{x}, \mathbf{y} \in \Omega$ if $\Re s > d/2 + j/2$, where $j := |\alpha| + |\beta|$. In fact, for $\Re s > d/2 + j/2$, D_s is C^j on $\overline{\Omega} \times \overline{\Omega}$ (see Eq. (2.17)); so, there is no problem with pointwise evaluation of its derivatives up to order j. Moreover, for each $t > 0$, the function $K(t; \, , \,)$ is C^∞ on $\overline{\Omega} \times \overline{\Omega}$, and again the pointwise evaluation of its derivatives is well defined; besides, the derivatives $\partial_{\mathbf{x}}^{\alpha} \partial_{\mathbf{y}}^{\beta} K(t; \mathbf{x}, \mathbf{y})$ are easily seen to decay exponentially, so the integral (2.88) has no convergence problem for large t. Therefore, deriving term by term the asymptotic expansion (2.73) we see that, for $t \to 0^+$, $\partial_{\mathbf{x}}^{\alpha} \partial_{\mathbf{y}}^{\beta} K(t; \mathbf{x}, \mathbf{y}) = O(1)$ for $\mathbf{y} \neq \mathbf{x}$ and $\partial_{\mathbf{x}}^{\alpha} \partial_{\mathbf{y}}^{\beta} K(t; \mathbf{x}, \mathbf{x}) = O(1/t^{d/2 + \lceil j/2 \rceil})$; these facts imply that, for $\Re s > d/2 + j/2$, the integral (2.88) is in any case convergent.

To conclude this section let us notice that, setting $\mathbf{y} = \mathbf{x}$ in Eqs. (2.85), (2.86) and integrating over the spatial domain Ω,[k] the relations (2.19) for the trace $\operatorname{Tr} \mathcal{A}^{-s}$ and (2.50), (2.51) for the heat

---

[j] For example, from (2.73) it is clear that, in the limit $t \to 0^+$, one has $K(t, \mathbf{x}, \mathbf{y}) = O(1)$ for $\mathbf{y} \neq \mathbf{x}$ and $K(t, \mathbf{x}, \mathbf{x}) = O(1/t^{d/2})$; so the integral in (2.85) converges for all $\mathbf{x}, \mathbf{y}$ if $\Re s > d/2$.

[k] Assuming that the order of integration can be interchanged for s is a suitable complex domain.

and cylinder traces $K(\mathfrak{t}), T(\mathfrak{t})$ allow us to infer

$$\operatorname{Tr} \mathcal{A}^{-s} = \frac{1}{\Gamma(s)} \int_0^{+\infty} d\mathfrak{t} \; \mathfrak{t}^{s-1} \, K(\mathfrak{t}) \; ; \qquad (2.89)$$

$$\operatorname{Tr} \mathcal{A}^{-s} = \frac{1}{\Gamma(2s)} \int_0^{+\infty} d\mathfrak{t} \; \mathfrak{t}^{2s-1} \, T(\mathfrak{t}) \; . \qquad (2.90)$$

Equations (2.89), (2.90) can be used to continue analytically the function $s \mapsto \operatorname{Tr}(\mathcal{A}^{-s})$; the situation is similar to the one outlined previously for the local counterparts of these equations, and will be reconsidered in the next two sections.

2.8   Analytic continuation of Mellin transforms

In the forthcoming subsections we are going to describe three different methods allowing to construct the analytic continuation of Mellin transforms, under suitable assumptions for the integrand functions; in fact, we present these methods with increasingly stronger hypotheses on the integrands.

In each subsection, we first describe the general continuation techniques for an arbitrary Mellin transform function; next, we specialize these techniques to the computation of the analytic continuation of the Dirichlet kernel, which was proven in the previous section to coincide with the Mellin transforms of the heat and cylinder kernels. Analogous results for the corresponding trace $\operatorname{Tr} \mathcal{A}^{-s}$ (in terms of the heat and cylinder traces) are also discussed. For more information and references on these topics, see e.g. [62], which also includes a bibliography on the subject.

2.8.1   *Continuation via asymptotic expansions*

Let $\mathcal{F} : (0, +\infty) \to \mathbf{C}$ be any locally integrable, exponentially vanishing function; moreover, assume there exist $\rho \in \mathbf{C}$, $N, P \in \mathbf{N}$ and two families of coefficients $\mathfrak{a}_n, \mathfrak{f}_{np}$ ($n \in \{0, ..., N+1\}$, $p \in \{0, ..., P\}$) with

$$\mathfrak{a}_n \in \mathbf{R} , \quad 0 \leqslant \mathfrak{a}_0 < \mathfrak{a}_1 < ... < \mathfrak{a}_N < \mathfrak{a}_{N+1} \quad \text{and} \quad \mathfrak{f}_{np} \in \mathbf{C} , \qquad (2.91)$$

such that there holds the asymptotic expansion

$$\mathcal{F}(t) = \frac{1}{t^\rho} \sum_{n=0}^{N} \sum_{p=0}^{P} f_{np}\, t^{\mathfrak{a}_n} (\ln t)^p + O(t^{\mathfrak{a}_{N+1}-\rho}) \quad \text{for } t \to 0^+. \quad (2.92)$$

Then, the Mellin transform of $\mathcal{F}$, i.e. the map $\sigma \mapsto \mathfrak{M}(\sigma)$ of Eq. (2.83), is well-defined and analytic for any $\sigma \in \mathbf{C}$ with $\Re\sigma > \Re\rho - \mathfrak{a}_0$. On the other hand, for any such σ and for any fixed $T > 0$ the integral in the right-hand side of Eq. (2.83) can be re-expressed as follows:[1]

$$\mathfrak{M}(\sigma) = \sum_{n=0}^{N} \sum_{p=0}^{P} \sum_{j=0}^{p} \left(\frac{(-1)^j p!}{(p-j)!} \right) \frac{f_{np}\, T^{\sigma-\rho+\mathfrak{a}_n} (\ln T)^{p-j}}{(\sigma-\rho+\mathfrak{a}_n)^j}$$

$$+ \int_0^T dt\, t^{\sigma-1} \left(\mathcal{F}(t) - \frac{1}{t^\rho} \sum_{n=0}^{N} \sum_{p=0}^{P} f_{np}\, t^{\mathfrak{a}_n} (\ln t)^p \right)$$

$$+ \int_T^{+\infty} dt\, t^{\sigma-1} \mathcal{F}(t). \quad (2.93)$$

In fact, under the assumptions on $\mathcal{F}$ made in the present section, the above representation analytically continues the Mellin transform $\mathfrak{M}(\sigma)$ to a function which is meromorphic in the region

$$\{\sigma \in \mathbf{C} \mid \Re\sigma > \Re\rho - \mathfrak{a}_{N+1}\}, \quad (2.94)$$

with possible pole singularities at the points

$$\sigma \in \{-\mathfrak{a}_0, \ldots, -\mathfrak{a}_N\}. \quad (2.95)$$

Now, let us show how the above results can be used to construct the analytic continuation of the Dirichlet kernel $D_s(\mathbf{x}, \mathbf{y})$ along the diagonal $\mathbf{y} = \mathbf{x}$, for fixed $\mathbf{x} \in \Omega$, starting from its representations (2.85), (2.86) in terms of the heat and cylinder kernel, respectively.

---

[1] To derive Eq. (2.93) one must first consider the integral in Eq. (2.83) and write it as $\mathfrak{M}(\sigma) = \int_0^T dt\, \mathcal{F}(t) + \int_T^{+\infty} dt\, \mathcal{F}(t)$. Next, one should add and subtract the asymptotic expansion of $\mathcal{F}$ in the first integral. Then, Eq. (2.93) follows noting that $t^{s-1} (\ln t)^p = (d^p/ds^p) t^{s-1}$, so that

$$\int_0^T dt\, t^{s-1} (\ln t)^p = \frac{d^p}{ds^p} \int_0^T dt\, t^{s-1} = \frac{d^p}{ds^p} \left(\frac{T^s}{s} \right) \qquad \text{for all } s \in \mathbf{C} \text{ with } \Re s > 0.$$

To this purpose, let us first recall that for $t \to 0^+$ the latter kernels possess asymptotic expansions of the forms (2.74) (2.75), which we report here for reading convenience (setting $a_0 := 1$):

$$K(t;\mathbf{x},\mathbf{x}) = \frac{1}{(4\pi t)^{d/2}} \left[\sum_{n=0}^{N} a_n(\mathbf{x})\, t^n + O(t^{N+1}) \right],$$

$$T(t;\mathbf{x},\mathbf{x}) = \frac{1}{t^d} \left[\sum_{n=0}^{N} e_n(\mathbf{x})\, t^n + \sum_{\substack{n=d+1 \\ n-d \text{ odd}}}^{N} f_n(\mathbf{x})\, t^n \ln t + O(t^{N+1}\ln t) \right],$$

for any $N \in \mathbf{N}$ and suitable functions $a_n, e_n, f_n : \Omega \to \mathbf{R}$ ($n = 0, ..., N$). In view of this fact, the basic integral relations (2.85) and (2.86), along with the general result (2.93), can be used to derive the following identities:

$$D_s(\mathbf{x},\mathbf{x}) = \frac{1}{\Gamma(s)} \left[\frac{1}{(4\pi)^{d/2}} \sum_{n=1}^{N} \frac{a_n(\mathbf{x})\, T^{s-d/2+n}}{s - d/2 + n} \right.$$

$$+ \int_0^T dt\, t^{s-1} \left(K(t;\mathbf{x},\mathbf{x}) - \frac{1}{(4\pi t)^{d/2}} \sum_{n=1}^{N} a_n(\mathbf{x})\, t^n \right)$$

$$\left. + \int_T^{+\infty} dt\, t^{s-1} K(t;\mathbf{x},\mathbf{x}) \right]; \qquad (2.96)$$

$$D_s(\mathbf{x},\mathbf{x}) = \frac{1}{\Gamma(2s)} \left[\sum_{n=0}^{N} \frac{T^{2s-d+n}\, e_n(\mathbf{x})}{2s - d + n} \right.$$

$$+ \sum_{\substack{n=d+1 \\ n-d \text{ odd}}}^{N} \frac{f_n(\mathbf{x})\, T^{2s-d+n}((2s-d+n)\log T + 1)}{(2s - d + n)^2}$$

$$+ \int_0^T dt\, t^{2s-1} \left(T(t;\mathbf{x},\mathbf{x}) - \frac{1}{t^d} \left[\sum_{n=0}^{N} e_n(\mathbf{x})\, t^n \right. \right.$$

$$\left. \left. + \sum_{\substack{n=d+1 \\ n-d \text{ odd}}}^{N} f_n(\mathbf{x})\, t^n \ln t \right] \right) + \left. \int_T^{+\infty} dt\, t^{2s-1} T(t;\mathbf{x},\mathbf{x}) \right].$$

$$(2.97)$$

For any fixed $N \in \mathbf{N}$, Eq. (2.96) determines the analytic continuation of the diagonal Dirichlet kernel $D_s(\mathbf{x}, \mathbf{x})$ to the region $\{s \in \mathbf{C} \,|\, \Re s > d/2 - N - 1\}$, except at the points $s \in \{d/2, d/2 - 1, ..., d/2 - N\}$ where simple poles may arise. Similarly, Eq. (2.97) gives the analytic continuation of $D_s(\mathbf{x}, \mathbf{x})$ on $\{s \in \mathbf{C} \,|\, \Re s > (d - N - 1)/2\}$, to a meromorphic function with possible poles at $s \in \{d/2, (d - 1)/2, ..., (d - N)/2\}$.

Let us mention that similar results can be derived for the derivatives of the Dirichlet kernel evaluated along the diagonal, assuming the heat and cylinder kernel derivatives to possess asymptotic expansions analogous to those in Eqs. (2.74), (2.75) when evaluated along the diagonal.

Finally, let us point out that the analytic continuation of the trace $\operatorname{Tr} \mathcal{A}^{-s}$ can be determined by means of the same technique whenever the heat and cylinder traces $\operatorname{Tr} e^{-t\mathcal{A}}$, $\operatorname{Tr} e^{-t\sqrt{\mathcal{A}}}$ have suitable asymptotic expansions for $t \to 0^+$. For example, let us assume Ω to be bounded and the heat trace $\operatorname{Tr} e^{-t\mathcal{A}}$ to decay exponentially at infinity and to fulfill Eq. (2.76), so that

$$K(t) = \frac{1}{(4\pi t)^{d/2}} \sum_{n=0}^{N} A_n \, t^{n/2} + O(t^{\frac{N+1}{2}}),$$

for suitable $A_n \in \mathbf{R}$ $(n = 0, ..., N)$ with $A_0 := \operatorname{Vol}(\Omega)$. Then, the integral representation (2.89) and the general rule (2.93) allow to infer that

$$\operatorname{Tr} \mathcal{A}^{-s} = \frac{1}{\Gamma(s)} \left[\frac{1}{(4\pi)^{d/2}} \sum_{n=0}^{N} \frac{A_n \, T^{s-(d-n)/2}}{s - (d - n)/2} \right.$$

$$+ \int_0^T dt \, t^{s-1} \left(K(t) - \frac{1}{(4\pi t)^{d/2}} \sum_{n=0}^{N} A_n \, t^{n/2} \right)$$

$$\left. + \int_T^{+\infty} dt \, t^{s-1} \, K(t) \right]. \tag{2.98}$$

The above integral representation analytically continues the map $s \mapsto \operatorname{Tr} \mathcal{A}^{-s}$ to a function which is meromorphic in the region $\{s \in \mathbf{C} \,|\, \Re s > (d - N - 1)/2\}$, with possible simple poles at the points $s \in \{d/2, (d - 1)/2, ..., (d - N)/2\}$.

2.8.2   *Continuation via integration by parts*

Let $\mathcal{F} : (0, +\infty) \to \mathbf{C}$ be a function of the form

$$\mathcal{F}(t) = \frac{1}{t^\rho} \mathcal{H}(t) \tag{2.99}$$

for some $\rho \in \mathbf{C}$ and some smooth function $\mathcal{H} : [0, +\infty) \to \mathbf{C}$, vanishing exponentially for $t \to +\infty$. Due to the hypotheses on $\mathcal{F}$, its Mellin transform $\mathfrak{M}(\sigma)$ (see Eq. (2.83)) is well defined only for $\sigma \in \mathbf{C}$ with $\Re\sigma > \Re\rho$, and gives an analytic function of σ in this region. However, integrating by parts n times (for any $n \in \{1, 2, 3, ...\}$) the integral in Eq. (2.83) and noting that the boundary terms vanish (for $\Re\sigma > \Re\rho$), we obtain

$$\mathfrak{M}(\sigma) = \frac{(-1)^n}{(\sigma-\rho)...(\sigma-\rho+n-1)} \int_0^{+\infty} dt \; t^{\sigma-\rho+n-1} \frac{d^n \mathcal{H}}{dt^n}(t) \; . \tag{2.100}$$

In consequence of the features of the function $\mathcal{H}$, the above integral converges for $\Re\sigma > \Re\rho - n$; thus, Eq. (2.100) yields the analytic continuation of the Mellin transform $\mathfrak{M}(\sigma)$ to the region

$$\{\sigma \in \mathbf{C} \mid \Re\sigma > \Re\rho - n\} \tag{2.101}$$

from which the zeros of the denominator in (2.100) must be removed. This gives a meromorphic function with possible simple poles at the above zeros, which are the points

$$\sigma \in \{\rho, \, \rho-1, \, ..., \, \rho-n+1\} \; . \tag{2.102}$$

Moreover, since the above results hold for any given $n \in \{1, 2, 3, ...\}$, they actually allow to determine the analytic continuation of $\mathfrak{M}(\sigma)$ to the whole complex plane with simple poles at the points $\sigma \in \{\rho, \rho-1, \rho-2, ...\}$.

As mentioned before, the above results can be employed to obtain the sought-for analytic continuation of the Dirichlet kernel $D_s(\mathbf{x}, \mathbf{y})$ (treating $\mathbf{x}, \mathbf{y} \in \Omega$ as fixed parameters) starting from its representations (2.85), (2.86) in terms of the heat and cylinder kernel, respectively.

More precisely, consider the case in which the heat or the cylinder kernel is given by a smooth function of t rapidly vanishing at infinity, divided by a power of t; by this we mean that

$$K(t; \mathbf{x}, \mathbf{y}) = \frac{1}{t^p} H(t; \mathbf{x}, \mathbf{y}) \quad \text{or} \quad T(t; \mathbf{x}, \mathbf{y}) = \frac{1}{t^q} J(t; \mathbf{x}, \mathbf{y}) \; , \tag{2.103}$$

where $p, q \in \mathbf{R}$, $H, J : [0, +\infty) \times \Omega \times \Omega \to \mathbf{R}$, and it is assumed that (for fixed $\mathbf{x}, \mathbf{y} \in \Omega$) the function $t \in [0, +\infty) \mapsto H(t; \mathbf{x}, \mathbf{y})$ or $J(t; \mathbf{x}, \mathbf{y})$ is smooth and rapidly vanishing for $t \to +\infty$. In these cases the integrals in the right-hand sides of Eqs. (2.85) and (2.86) converge for $\Re s > p$ and $\Re s > q/2$, respectively. Let us mention that, with the assumptions (2.9), the heat kernel of $\mathcal{A} = -\Delta + V$ is as in Eq. (2.103) with $p = d/2$ (recall the asymptotics (2.73)); again with the assumptions (2.9), the cylinder kernel of $\mathcal{A}$ is as in (2.103) with $q = d$, provided that no logarithmic terms appear in the asymptotic expansion (2.75).

Under the previous assumptions, Eq. (2.100), along with Eqs. (2.85), (2.86), gives the following for any $n \in \{1, 2, 3, ...\}$:[m]

$$D_s(\mathbf{x}, \mathbf{y}) = \frac{(-1)^n}{\Gamma(s)(s-p)...(s-p+n-1)} \int_0^{+\infty} dt \; t^{s-p+n-1} \partial_t^n H(t; \mathbf{x}, \mathbf{y}) \; ;$$
(2.104)

$$D_s(\mathbf{x}, \mathbf{y}) = \frac{(-1)^n}{\Gamma(2s)(2s-q)...(2s-q+n-1)} \int_0^{+\infty} dt \; t^{2s-q+n-1} \partial_t^n J(t; \mathbf{x}, \mathbf{y}) .$$
(2.105)

Comments analogous to the ones below Eq. (2.100) can be done for the above representations. More in detail, on the one hand Eq. (2.104) gives the analytic continuation of the Dirichlet kernel D_s in the region $\{s \in \mathbf{C} \mid \Re s > p - n\}$ to a meromorphic function with simple poles at $s \in \{p, p-1, ..., p-n+1\}$; on the other hand, Eq. (2.105) gives the analytic continuation of D_s to the region $\{s \in \mathbf{C} \mid \Re s > (q-n)/2\}$, with possible simple poles at $s \in \{q/2, (q-1)/2, ..., (q-n+1)/2\}$.

Of course, relations analogous to (2.104) and (2.105) hold as well for the spatial derivatives of the Dirichlet kernel.

In conclusion, let us stress that similar results can be deduced for the trace $\mathrm{Tr}\,\mathcal{A}^{-s}$ (see Eq. (2.19)), determining its analytic continuation to wider regions in the complex plane. For example, assume the heat trace has the form (compare with the first relation in

---

[m]To obtain Eq. (2.104) one uses Eqs. (2.99)–(2.100) with $\mathcal{F}(t) = K(t; \mathbf{x}, \mathbf{y})$, $\rho = p$, $\mathcal{H}(t) = H(t; \mathbf{x}, \mathbf{y})$) and $\sigma = s$. To obtain Eq. (2.105) one uses Eqs. (2.99)–(2.100) with $\mathcal{F}(t) = T(t; \mathbf{x}, \mathbf{y})$, $\rho = q$, $\mathcal{H}(t) = J(t; \mathbf{x}, \mathbf{y})$ and $\sigma = 2s$.

Eq. (2.103))

$$K(t) = \frac{1}{t^p} H(t) ,$$
(2.106)

for some $p \in \mathbf{R}$ and some smooth function $H : [0, +\infty) \to \mathbf{R}$, rapidly vanishing for $t \to +\infty$; then, starting with Eq. (2.89) and using the relations (2.99)–(2.100), we obtain the following, for $n \in \{1, 2, 3, ...\}$:

$$\operatorname{Tr} \mathcal{A}^{-s} = \frac{(-1)^n}{\Gamma(s)(s-p)...(s-p+n-1)} \int_0^{+\infty} dt\, t^{s-p+n-1} \frac{d^n H}{dt^n}(t) .$$
(2.107)

The above relation gives the analytic continuation of $\operatorname{Tr} \mathcal{A}^{-s}$ in the region $\{s \in \mathbf{C} \mid \Re s > p - n\}$ to a meromorphic function with simple poles at $s \in \{p, p-1, ..., p-n+1\}$. A similar result can be derived using the cylinder trace $T(t)$.

2.8.3   *Continuation via complex integration*

Another way to obtain the analytic continuation of the Mellin transform of a given function is available (assuming the latter to fulfill suitable conditions). Consider again the Mellin transform in Eq. (2.83); this time the idea is to re-express the integral appearing in the cited equation as an integral along a suitable path in the complex plane.

To this purpose, we first consider the following identity concerning Mellin transforms. Let $t \mapsto h(t)$ be a complex-valued function, analytic in a complex neighborhood of $[0, +\infty)$ and exponentially vanishing for $\Re t \to +\infty$ in this neighborhood; then

$$\int_0^{+\infty} dt\, t^{s-1} h(t)$$

$$= \frac{e^{-i\pi s}}{2i \sin(\pi s)} \int_{\mathfrak{H}} dt\, t^{s-1} h(t) \quad \text{for } s \in \mathbf{C} \backslash \{1, 2, 3, ...\},\ \Re s > 0,$$
(2.108)

where $\mathfrak{H}$ denotes the *Hankel contour*, that is a simple path in the complex plane that starts in the upper half-plane near $+\infty$, encircles the origin counterclockwise and returns to $+\infty$ in the lower half-plane (see Fig. 2.1 below).

In the right-hand side of Eq. (2.108), the complex power t^{s-1} is defined making reference to Eqs. (1.5), (1.6); in the left-hand side,

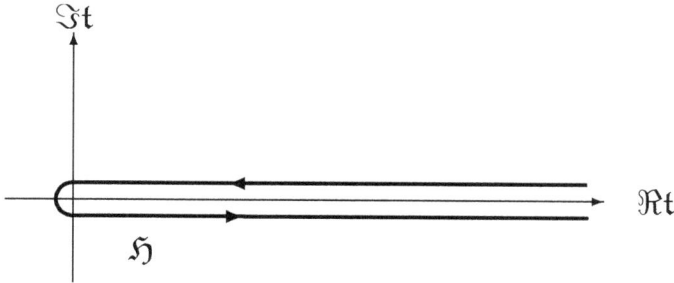

Fig. 2.1   The Hankel contour $\mathfrak{H}$.

since $t \in (0, +\infty)$, we define t^{s-1} according to the standard convention (1.4). See Appendix C for the derivation of Eq. (2.108).

Assume now $\mathcal{F}$ to be as in Eq. (2.99) with $\mathcal{H}$ a complex function, analytic in a neighborhood of $[0, +\infty)$ and exponentially vanishing for $\Re t \to +\infty$; then, considering the Mellin transform $\mathfrak{M}(\sigma)$ in Eq. (2.83) and using Eq. (2.108) with $s = \sigma - \rho$ and $h = \mathcal{H}$, we obtain

$$\mathfrak{M}(\sigma) = \frac{e^{-i\pi(\sigma-\rho)}}{2i \sin(\pi(\sigma-\rho))} \int_{\mathfrak{H}} dt \, t^{\sigma-1} \mathcal{F}(t) \ . \qquad (2.109)$$

In principle, for Eq. (2.109) to hold, certain conditions must be fulfilled: to ensure existence of $\mathfrak{M}(\sigma)$ as defined by Eq. (2.83) we must require $\Re\sigma > \Re\rho$, and the denominator $\sin(\pi(\sigma-\rho))$ must be nonzero. Nevertheless, the integral in Eq. (2.109) converges for any $\sigma \in \mathbf{C}$, so this equation yields the analytic continuation of the Mellin transform $\mathfrak{M}(\sigma)$ to the whole complex plane, possibly with simple poles for

$$\sigma \in \{\rho, \, \rho - 1, \, \rho - 2, \, ...\} \ , \qquad (2.110)$$

due to the vanishing of the sine function in the denominator.[n]

Recall once more that, according to Eqs. (2.85), (2.86), the Dirichlet kernel can be expressed as the Mellin transform of either the heat or the cylinder kernel; so, the above results on the analytic continuation via contour integration can be applied to $D_s(\mathbf{x}, \mathbf{y})$ (for fixed

---

[n]By inspection of the denominator, it would seem that also the points $\sigma \in \{\rho+1, \, \rho+2, \, \rho+3, \, ...\}$ are singular, but we know this is not the case since the original expression (2.83) for $\mathfrak{M}(\sigma)$ is regular at these points. The reason for the apparent contradiction lies in the fact that the integral over the Hankel contour in Eq. (2.109) vanishes for the above mentioned values of σ (as can be easily checked via the residue theorem recalling the properties of $\mathcal{F}$), thus yielding an indeterminate form $\infty \cdot 0$.

$\mathbf{x}, \mathbf{y} \in \Omega$). More precisely, suppose that either the heat or the cylinder kernel has the form (2.103):

$$K(\mathfrak{t};\mathbf{x},\mathbf{y}) = \frac{1}{\mathfrak{t}^p} H(\mathfrak{t};\mathbf{x},\mathbf{y}) \quad \text{or} \quad T(\mathfrak{t};\mathbf{x},\mathbf{y}) = \frac{1}{\mathfrak{t}^q} J(\mathfrak{t};\mathbf{x},\mathbf{y}) ,$$

with $p, q \in \mathbf{R}$ and suitable functions $H, J : [0, +\infty) \times \Omega \times \Omega \to \mathbf{R}$. Assume these functions to possess extensions $H, J : \mathcal{U}([0, +\infty)) \times \Omega \times \Omega \to \mathbf{C}$, where $\mathcal{U}([0, +\infty)) \subset \mathbf{C}$ is an open neighborhood of the interval $[0, +\infty)$ and, for fixed $\mathbf{x}, \mathbf{y} \in \Omega$, the function $\mathfrak{t} \in \mathcal{U}([0, +\infty)) \mapsto H(\mathfrak{t};\mathbf{x},\mathbf{y})$ or $J(\mathfrak{t};\mathbf{x},\mathbf{y})$ is analytic and exponentially vanishing for $\Re\mathfrak{t} \to +\infty$. Making these hypotheses and using Eq. (2.109) along with Eqs. (2.85), (2.86), we obtain[o]

$$D_s(\mathbf{x},\mathbf{y}) = \frac{e^{-i\pi(s-p)}}{2i\,\Gamma(s)\,\sin(\pi(s-p))} \int_{\mathfrak{H}} d\mathfrak{t}\; \mathfrak{t}^{s-1}\, K(\mathfrak{t};\mathbf{x},\mathbf{y}) ; \qquad (2.111)$$

$$D_s(\mathbf{x},\mathbf{y}) = \frac{e^{-i\pi(2s-q)}}{2i\,\Gamma(2s)\,\sin(\pi(2s-q))} \int_{\mathfrak{H}} d\mathfrak{t}\; \mathfrak{t}^{2s-1}\, T(\mathfrak{t};\mathbf{x},\mathbf{y}) . \qquad (2.112)$$

Due to the comments reported below Eq. (2.109), both the above identities yield the analytic continuation of the Dirichlet kernel to a meromorphic function on the whole complex plane; more precisely, the analytic continuations obtained via Eqs. (2.111) and (2.112) may have simple poles respectively for $s \in \{p, p-1, p-2, ...\} \setminus \{0, -1, -2, ...\}$ and $s \in \{q/2, (q-1)/2, (q-2)/2, ...\} \setminus \{0, -1/2, -1, -3/2, ...\}$ (as readily understood analyzing the denominators in the right-hand sides of the cited equations).

In the subcases where $p, q \in \mathbf{Z} = \{0, \pm1, \pm2, ...\}$, using trivial trigonometric identities and recalling that $\Gamma(s)\Gamma(1-s)\sin(\pi s) = \pi$ for any $s \in \mathbf{C}$ (see [117], page 138, Eq. (5.5.3)), Eqs. (2.111), (2.112) can be rephrased as follows:

$$D_s(\mathbf{x},\mathbf{y}) = \frac{e^{-i\pi s}\,\Gamma(1-s)}{2\pi i} \int_{\mathfrak{H}} d\mathfrak{t}\; \mathfrak{t}^{s-1} K(\mathfrak{t};\mathbf{x},\mathbf{y}) , \qquad (2.113)$$

$$D_s(\mathbf{x},\mathbf{y}) = \frac{e^{-2i\pi s}\,\Gamma(1-2s)}{2\pi i} \int_{\mathfrak{H}} d\mathfrak{t}\; \mathfrak{t}^{2s-1} T(\mathfrak{t};\mathbf{x},\mathbf{y}) . \qquad (2.114)$$

---

[o]One proceeds as in footnote *m*, using Eq. (2.109) in place of Eq. (2.100).

In these subcases the integrals along the Hankel contour can be computed straightforwardly for integer and half-integer values of s, respectively, by means of the residue theorem; for example, for $s = -n/2$ and $n \in \{0, 1, 2, ...\}$, Eq. (2.114) yields

$$D_{-\frac{n}{2}}(\mathbf{x}, \mathbf{y}) = (-1)^n \, \Gamma(n+1) \, \mathrm{Res}\left(\mathsf{t}^{-(n+1)} \, T(\mathsf{t}\,;\mathbf{x}, \mathbf{y})\,;0\right) . \qquad (2.115)$$

We can obtain a variant of Eq. (2.114), giving the analytic continuation of the Dirichlet kernel in terms of the modified cylinder kernel $\tilde{T}$ (recall Eqs. (2.53), (2.55)). To this purpose, we assume $\tilde{T}$ to admit a meromorphic extension in t to a neighborhood of $[0, +\infty)$, having a pole in $\mathsf{t} = 0$ and rapidly vanishing for $\Re\mathsf{t} \to +\infty$; then, expressing the cylinder kernel $T(\mathsf{t}\,;\mathbf{x}, \mathbf{y})$ in Eq. (2.114) as $-\partial_{\mathsf{t}}\tilde{T}(\mathsf{t}\,;\mathbf{x}, \mathbf{y})$ and integrating by parts, we obtain

$$D_s(\mathbf{x}, \mathbf{y}) = -\,\frac{e^{-2i\pi s}\,\Gamma(2-2s)}{2\pi i} \int_{\mathfrak{H}} d\mathsf{t}\; \mathsf{t}^{2s-2} \, \tilde{T}(\mathsf{t}\,;\mathbf{x}, \mathbf{y}) \qquad (2.116)$$

(note that no boundary contribution arises, due to the rapid vanishing of $\tilde{T}(\mathsf{t}\,;\mathbf{x}, \mathbf{y})$ for $\Re\mathsf{t} \to +\infty$). Again, for half-integer values of s we can compute explicitly the analytic continuation (2.116) by means of the residue theorem; to be more precise, for $s = -n/2$ and $n \in \{-1, 0, 1, 2, ...\}$ we have

$$D_{-\frac{n}{2}}(\mathbf{x}, \mathbf{y}) = (-1)^{n+1} \, \Gamma(n+2) \, \mathrm{Res}\left(\mathsf{t}^{-(n+2)} \, \tilde{T}(\mathsf{t}\,;\mathbf{x}, \mathbf{y})\,;0\right) . \quad (2.117)$$

Relations similar to the ones obtained above hold for the spatial derivatives of the Dirichlet kernel, allowing in turn to determine their analytic continuations.

To conclude, let us mention that similar results hold as well for the trace $\mathrm{Tr}\, \mathcal{A}^{-s}$; these are obtained using the representations (2.89), (2.90) in terms of the heat and cylinder trace $K(\mathsf{t}), T(\mathsf{t})$, and assuming suitable features for the latter traces. In particular, if the map $\mathsf{t} \mapsto T(\mathsf{t})$ admits a meromorphic extension to a neighborhood of $[0, +\infty)$ which only has a pole singularity at $\mathsf{t} = 0$ and vanishes exponentially for $\Re\mathsf{t} \to +\infty$, for any $n \in \{0, 1, 2, ...\}$ we have

$$\mathrm{Tr}\, \mathcal{A}^{n/2} = (-1)^n \, \Gamma(n+1) \, \mathrm{Res}\left(\mathsf{t}^{-(n+1)} \, T(\mathsf{t})\,;0\right) . \qquad (2.118)$$

2.9   Product configurations: factorization of the heat kernel

Let us consider the case where $\mathcal{A} = -\Delta + V$ and the spatial domain $\Omega \subset \mathbf{R}^d$ has the form

$$\Omega = \Omega_1 \times \Omega_2 \qquad (2.119)$$

with Ω_a $(a \in \{1, 2\})$ indicating an open subset of $\mathbf{R}^{d_a}$ $(d_1 + d_2 = d)$. In this case, points of Ω will be indicated with

$$\mathbf{x} = (\mathbf{x}_1, \mathbf{x}_2) , \qquad \mathbf{y} = (\mathbf{y}_1, \mathbf{y}_2) \qquad (2.120)$$

etc. where $\mathbf{x}_a, \mathbf{y}_a \in \Omega_a$ $(a \in \{1, 2\})$. In addition to Eq. (2.119), we assume that the external potential has the form

$$V(\mathbf{x}) = V_1(\mathbf{x}_1) + V_2(\mathbf{x}_2) \qquad (2.121)$$

and that the boundary conditions specified on $\partial\Omega = (\partial\Omega_1 \times \Omega_2) \cup (\Omega_1 \times \partial\Omega_2)$ arise from suitable boundary conditions prescribed separately on $\partial\Omega_1$ and $\partial\Omega_2$ in such a way that, for $a \in \{1, 2\}$, the operator

$$\mathcal{A}_a := -\Delta_a + V(\mathbf{x}_a) \qquad (2.122)$$

(Δ_a the Laplacian on Ω_a) is self-adjoint in $L^2(\Omega_a)$. Moreover, each one of these operators is assumed to be strictly positive or, at least, non-negative.

In the situation described above, the Hilbert space $L^2(\Omega)$ and the fundamental operator $\mathcal{A} := -\Delta + V(\mathbf{x})$ acting therein can be represented, respectively, as

$$L^2(\Omega) = L^2(\Omega_1) \otimes L^2(\Omega_2) , \qquad \mathcal{A} = \mathcal{A}_1 \otimes 1 + 1 \otimes \mathcal{A}_2 . \qquad (2.123)$$

Because of the assumptions we have made, each of the operators $\mathcal{A}_a$ $(a \in \{1, 2\})$ possesses a complete orthonormal system of eigenfunctions $(F_{a,k_a})_{k_a \in \mathcal{K}_a}$ with eigenvalues ω_{a,k_a}^2; as for the fundamental operator $\mathcal{A}$, we see that it has a complete orthonormal set of eigenfunctions of the form

$$F_k(\mathbf{x}) := F_{1,k_1}(\mathbf{x}_1) F_{2,k_2}(\mathbf{x}_2) \qquad \text{for} \ \ k = (k_1, k_2) \in \mathcal{K}_1 \times \mathcal{K}_2 \quad (2.124)$$

and that

$$\mathcal{A} F_k = \omega_k^2 F_k , \qquad \omega_k^2 = \omega_{1,k_1}^2 + \omega_{2,k_2}^2 . \qquad (2.125)$$

Of course $\sigma(\mathcal{A}) = \sigma(\mathcal{A}_1) + \sigma(\mathcal{A}_2)$, so that $\mathcal{A}$ is non-negative; besides, $\mathcal{A}$ is strictly positive as soon as at least one between $\mathcal{A}_1$ and $\mathcal{A}_2$ is so.

In the product case under analysis, a number of interesting facts occurs for the integral kernels associated to $\mathcal{A}$ and $\mathcal{A}_1$, $\mathcal{A}_2$. The most elementary of these facts is the factorization of the heat kernel; by this we mean that the kernels

$$
\begin{aligned}
K(t; \mathbf{x}, \mathbf{y}) &:= (e^{-t\mathcal{A}})(\mathbf{x}, \mathbf{y}) , \\
K_a(t; \mathbf{x}_a, \mathbf{y}_a) &:= (e^{-t\mathcal{A}_a})(\mathbf{x}_a, \mathbf{y}_a) \quad (a \in \{1,2\})
\end{aligned}
\tag{2.126}
$$

are related by

$$
K(t; \mathbf{x}, \mathbf{y}) = K_1(t; \mathbf{x}_1, \mathbf{y}_1) \, K_2(t; \mathbf{x}_1, \mathbf{x}_2) ,
\tag{2.127}
$$

a fact that is made apparent by the eigenfunction expansion (2.44) and by Eqs. (2.124), (2.125).

In passing, let also mention that an analogous relation can be easily derived for the heat trace; writing $K(t)$, $K_a(t)$ for the heat traces of $\mathcal{A}$ and $\mathcal{A}_a$ ($a \in \{1,2\}$), respectively, we obtain[p]

$$
K(t) = K_1(t) \, K_2(t) .
\tag{2.128}
$$

In the present section we have analyzed the case of a product configuration with two factors; as a straightforward generalization, one can consider a product with an arbitrary number of factors. Examples of such multiple products will appear in Chapters 9 and 10.

2.10   Slab configurations: reduction to a lower-dimensional problem

By definition, we have a slab if

$$
\Omega = \Omega_1 \times \mathbf{R}^{d_2} , \qquad V = V(\mathbf{x}_1)
\tag{2.129}
$$

with Ω_1 a domain in $\mathbf{R}^{d_1}$ ($d_1 + d_2 = d$), and if the boundary conditions prescribed for the field refer to its behavior on $\partial\Omega_1 \times \mathbf{R}^{d_2}$. Clearly, a

---

[p]In fact, Eqs. (2.119), (2.127) and the relations (2.50), (2.51) allow us to infer the following chain of equalities proving Eq. (2.128):

$$
K(t) = \int_\Omega d\mathbf{x} \, K(t; \mathbf{x}, \mathbf{x}) = \int_{\Omega_1} d\mathbf{x}_1 \, K_1(t; \mathbf{x}_1, \mathbf{x}_1) \int_{\Omega_2} d\mathbf{x}_2 \, K_2(t; \mathbf{x}_2, \mathbf{x}_2) = K_1(t) \, K_2(t) .
$$

slab is a subcase of the general product case discussed in the previous section, with $\Omega_2 = \mathbf{R}^{d_2}$ and $V_2 = 0$. In this subcase the relevant operators are $\mathcal{A} = -\Delta + V(\mathbf{x}_1)$ acting in $L^2(\Omega)$,

$$\mathcal{A}_1 := -\Delta_1 + V(\mathbf{x}_1) \tag{2.130}$$

acting in $L^2(\Omega_1)$, and $\mathcal{A}_2 := -\Delta_2$ acting in $L^2(\mathbf{R}^{d_2})$.

The operator $\mathcal{A}_1$ has its own eigenfunctions $F_{1,k_1}(\mathbf{x}_1) \equiv \mathfrak{F}_{k_1}(\mathbf{x}_1)$ and eigenvalues $\omega_{1,k_1}^2 \equiv \varpi_{k_1}^2$ $(k_1 \in \mathcal{K}_1)$; we assume $\mathcal{A}_1$ to be strictly positive, so that $\varpi_{k_1} \geqslant \varepsilon$ for some $\varepsilon > 0$. Of course, $-\Delta_2$ is non-negative with eigenfunctions $F_{2,k_2}(\mathbf{x}_2) = (2\pi)^{-d_2/2} e^{i\mathbf{k}_2 \cdot \mathbf{x}_2}$ and eigenvalues $\omega_{2,\mathbf{k}_2}^2 = |\mathbf{k}_2|^2$, for $\mathbf{k}_2 \in \mathbf{R}^{d_2}$.

In the sequel we write $D_s(\mathbf{x}_1, \mathbf{x}_2; \mathbf{y}_1, \mathbf{y}_2)$ for the Dirichlet kernel of $\mathcal{A}$ at the points $\mathbf{x} = (\mathbf{x}_1, \mathbf{x}_2)$ and $\mathbf{y} = (\mathbf{y}_1, \mathbf{y}_2)$; $D_s^{(1)}(\mathbf{x}_1, \mathbf{y}_1)$ will be the Dirichlet kernel of $\mathcal{A}_1$.

According to the general results (2.28)–(2.30), the regularized VEV $\langle 0|\widehat{T}_{\mu\nu}^u(\mathbf{x})|0\rangle$ is determined by the Dirichlet kernel $D_s(\mathbf{x}, \mathbf{y})$ and its derivatives evaluated on the diagonal $\mathbf{y} = \mathbf{x}$. The main intent of this section is to express D_s and its derivatives in terms of the reduced kernel $D_s^{(1)}$ at points on the diagonal $\mathbf{y}_1 = \mathbf{x}_1$. The starting point towards this goal is the identity

$$D_s(\mathbf{x}_1, \mathbf{x}_2; \mathbf{y}_1, \mathbf{y}_2) = \hat{D}_s(\mathbf{x}_1, \mathbf{y}_1; |\mathbf{x}_2 - \mathbf{y}_2|^2) , \tag{2.131}$$

involving a function $\hat{D}_s : \Omega_1 \times \Omega_1 \times [0, +\infty) \to \mathbf{C}$, $(\mathbf{x}_1, \mathbf{y}_1, q) \mapsto \hat{D}_s(\mathbf{x}_1, \mathbf{y}_1, q)$. This function and its partial derivatives with respect to q, at $q = 0$, are completely determined by the kernel $D_s^{(1)}$, according to the following rules:

$$\hat{D}_s(\mathbf{x}_1, \mathbf{y}_1; 0) = \frac{\Gamma(s - \frac{d_1}{2})}{(4\pi)^{d_1/2}\, \Gamma(s)}\, D_{s-\frac{d_1}{2}}^{(1)}(\mathbf{x}_1, \mathbf{y}_1) \quad \text{for } s \in \mathbf{C},\ \Re s > \frac{d_1}{2}\,;$$

$$\tag{2.132}$$

$$\frac{\partial^n \hat{D}_s}{\partial q^n}(\mathbf{x}_1, \mathbf{y}_1; 0) = \frac{(-1)^n \Gamma(s - \frac{d_1}{2} - n)}{(4\pi)^{d_1/2}\, 4^n\, \Gamma(s)}\, D_{s-\frac{d_1}{2}-n}^{(1)}(\mathbf{x}_1, \mathbf{y}_1)$$

$$\text{for } s \in \mathbf{C},\ \Re s > \frac{d_1}{2} + n. \tag{2.133}$$

We defer to Appendix D the derivation of Eqs. (2.131)–(2.133). In particular, for the derivatives involved in Eqs. (2.28)–(2.30) on

$\langle 0|\widehat{T}^u_{\mu\nu}(\mathbf{x})|0\rangle$ we obtain the following expressions (where, for simplicity of notation, we write $(\mathbf{x}, \mathbf{y})$ for $(\mathbf{x}_1, \mathbf{x}_2; \mathbf{y}_1, \mathbf{y}_2)$):

$$D_{\frac{u\pm1}{2}}(\mathbf{x}, \mathbf{y})\Big|_{\mathbf{y}=\mathbf{x}} = \frac{\Gamma(\frac{u-d_2\pm1}{2})}{(4\pi)^{d_2/2}\,\Gamma(\frac{u\pm1}{2})}\, D^{(1)}_{\frac{u-d_2\pm1}{2}}(\mathbf{x}_1, \mathbf{y}_1)\Big|_{\mathbf{y}_1=\mathbf{x}_1}\,;$$

(2.134)

$$\partial_{x_a^i y_b^j} D_{\frac{u+1}{2}}(\mathbf{x}, \mathbf{y})\Big|_{\mathbf{y}=\mathbf{x}} = \partial_{x_a^i x_b^j} D_{\frac{u+1}{2}}(\mathbf{x}, \mathbf{y})\Big|_{\mathbf{y}=\mathbf{x}} = \partial_{y_a^i y_b^j} D_{\frac{u+1}{2}}(\mathbf{x}, \mathbf{y})\Big|_{\mathbf{y}=\mathbf{x}} = 0$$

for $(a, b) = (1, 2)$ or $(a, b) = (2, 1)$ and $i \in \{1, ...d_a\}$, $j \in \{1, ..., d_b\}$;

(2.135)

$$\partial_{z_1^i w_1^j} D_{\frac{u+1}{2}}(\mathbf{x}, \mathbf{y})\Big|_{\mathbf{y}=\mathbf{x}} = \frac{\Gamma(\frac{u-d_2+1}{2})}{(4\pi)^{d_2/2}\,\Gamma(\frac{u+1}{2})}\, \partial_{z_1^i w_1^j} D^{(1)}_{\frac{u-d_2+1}{2}}(\mathbf{x}_1, \mathbf{y}_1)\Big|_{\mathbf{y}_1=\mathbf{x}_1}$$

for $z, w \in \{x, y\}$ and $i, j \in \{1, ..., d_1\}$;

(2.136)

$$\partial_{x_2^i y_2^j} D_{\frac{u+1}{2}}(\mathbf{x}, \mathbf{y})\Big|_{\mathbf{y}=\mathbf{x}}$$

$$= -\,\partial_{x_2^i x_2^j} D_{\frac{u+1}{2}}(\mathbf{x}, \mathbf{y})\Big|_{\mathbf{y}=\mathbf{x}} = -\,\partial_{y_2^i y_2^j} D_{\frac{u+1}{2}}(\mathbf{x}, \mathbf{y})\Big|_{\mathbf{y}=\mathbf{x}}$$

$$= \delta_{ij}\, \frac{\Gamma(\frac{u-d_2-1}{2})}{(4\pi)^{d_2/2}\, 2\,\Gamma(\frac{u+1}{2})}\, D^{(1)}_{\frac{u-d_2-1}{2}}(\mathbf{x}_1, \mathbf{y}_1)\Big|_{\mathbf{y}_1=\mathbf{x}_1} \qquad \text{for } i, j \in \{1, ..., d_2\}\,.$$

(2.137)

The identities in Eqs. (2.134)–(2.137) are derived assuming $\Re u > d_2+1$, but it follows from them that the analytic continuations in u of the Dirichlet kernel, of its reduced analogue and of their derivatives fulfill the very same relations.

Let us remark that the left-hand sides of the above equations depend in principle on $\mathbf{x} = (\mathbf{x}_1, \mathbf{x}_2)$, while the right-hand sides only contain $\mathbf{x}_1$; this confirms the expectation that the stress-energy VEV ought to be independent of the variable $\mathbf{x}_2$, due to the symmetry of the slab configuration under translations regarding this variable alone. Finally, let us remark that using Eqs. (2.134)–(2.137) and

(2.28)–(2.30), it can be easily checked that the components of the regularized stress-energy tensor VEV fulfill

$$\langle 0|\widehat{T}^u_{ij}(\mathbf{x})|0\rangle = 0 \quad \text{for } i,j \in \{d_1+1,...,d\},\, i \neq j \ ;$$

$$\langle 0|\widehat{T}^u_{ij}(\mathbf{x})|0\rangle = \langle 0|\widehat{T}^u_{ji}(\mathbf{x})|0\rangle = 0 \quad \text{for } i \in \{1,...,d_1\},\, j \in \{d_1+1,...,d\} \ .$$

$$(2.138)$$

Chapter 3

Total energy and forces on the boundary

Let us refer to the general framework of Chapter 1, where a quantized scalar field on a spatial domain Ω and the associated stress-energy VEV are considered; as usual, $\mathcal{A}$ indicates the fundamental operator $-\Delta + V$ acting in $L^2(\Omega)$. In this chapter and in the associated Appendix E we relate to the previous framework the total energy, the pressure and the total forces on the boundary. We also take the occasion to prove (at the regularized level) the equivalence between two alternative definitions of the pressure, often assumed without proof in the literature: pressure as the action of the stress-energy VEV on the normal to the boundary, and pressure as the functional derivative of the bulk energy with respect to deformations of the spatial domain Ω.

3.1 The total energy

As anticipated in Subsection 1.4.3, the zeta-regularized total energy can be defined as

$$\mathcal{E}^u := \int_\Omega d\mathbf{x} \, \langle 0|\widehat{T}_{00}^u(\mathbf{x})|0\rangle \, , \qquad (3.1)$$

provided that the above integral converges for u in a suitable complex domain; using Eq. (2.28) for the regularized VEV $\langle 0|\widehat{T}_{00}^u(\mathbf{x})|0\rangle$, the

55

above definition yields

$$\mathcal{E}^u = \kappa^u \left(\frac{1}{4} + \xi\right) \int_\Omega d\mathbf{x}\, D_{\frac{u-1}{2}}(\mathbf{x},\mathbf{y})\Big|_{\mathbf{y}=\mathbf{x}}$$

$$+ \kappa^u \left(\frac{1}{4} - \xi\right) \int_\Omega d\mathbf{x}\, \left[(\partial^{x^\ell}\partial_{y^\ell} + V(\mathbf{x}))D_{\frac{u+1}{2}}(\mathbf{x},\mathbf{y})\right]_{\mathbf{y}=\mathbf{x}} . \qquad (3.2)$$

On the other hand, the eigenfunction expansion (2.15) for the Dirichlet kernel (here used with $s = \frac{u+1}{2}$) gives

$$\int_\Omega d\mathbf{x}\, \left[(\partial^{x^\ell}\partial_{y^\ell} + V(\mathbf{x}))D_{\frac{u+1}{2}}(\mathbf{x},\mathbf{y})\right]_{\mathbf{y}=\mathbf{x}}$$

$$= \int_K \frac{dk}{\omega_k^{u+1}} \int_\Omega d\mathbf{x}\, \left(\partial^\ell F_k(\mathbf{x})\partial_\ell \overline{F_k}(\mathbf{x}) + V(\mathbf{x})F_k(\mathbf{x})\overline{F_k}(\mathbf{x})\right)$$

$$= \int_K \frac{dk}{\omega_k^{u+1}} \left(\int_\Omega d\mathbf{x}\, \left(F_k(\mathbf{x})(-\partial^\ell\partial_\ell + V(\mathbf{x}))\overline{F_k}(\mathbf{x})\right)\right.$$

$$\left. + \int_{\partial\Omega} da(\mathbf{x})\, F_k(\mathbf{x})\, n^\ell(\mathbf{x})\partial_\ell \overline{F_k}(\mathbf{x})\right) \qquad (3.3)$$

where, in the last step, we have integrated by parts.[a]

To go on, we note that $(-\partial^\ell\partial_\ell + V)\overline{F_k} = \overline{AF_k} = \omega_k^2\,\overline{F_k}$ and $F_k(\mathbf{x})n^\ell(\mathbf{x})\partial_\ell\overline{F_k}(\mathbf{x}) = F_k(\mathbf{x})\frac{\partial\overline{F_k}}{\partial n}(\mathbf{x}) = F_k(\mathbf{x})\frac{\partial\overline{F_k}(\mathbf{y})}{\partial n_\mathbf{y}}\big|_{\mathbf{y}=\mathbf{x}}$, where we have introduced the normal derivative $\partial/\partial n := n^\ell\partial_{x^\ell}$; substituting these identities into Eq. (3.3) and summing over $k \in K$, we obtain

$$\int_\Omega d\mathbf{x}\, \left[(\partial^{x^\ell}\partial_{y^\ell} + V(\mathbf{x}))D_{\frac{u+1}{2}}(\mathbf{x},\mathbf{y})\right]_{\mathbf{y}=\mathbf{x}}$$

$$= \int_\Omega d\mathbf{x}\, D_{\frac{u-1}{2}}(\mathbf{x},\mathbf{y})\Big|_{\mathbf{y}=\mathbf{x}} + \int_{\partial\Omega} da(\mathbf{x})\, \frac{\partial}{\partial n_\mathbf{y}}\, D_{\frac{u+1}{2}}(\mathbf{x},\mathbf{y})\Big|_{\mathbf{y}=\mathbf{x}} . \qquad (3.4)$$

Inserting this result into Eq. (3.2), we conclude

$$\mathcal{E}^u = E^u + B^u , \qquad (3.5)$$

---

[a] Here and in similar situations, whenever we speak of an integration by parts we refer to the identity

$$\int_\Omega d\mathbf{x}\, (\partial_\ell f)g = \int_{\partial\Omega} da\, f\, g\, n_\ell - \int_\Omega d\mathbf{x}\, f\, \partial_\ell g ,$$

holding for all sufficiently smooth functions f, g.

where we have introduced the *regularized bulk* and *boundary* energies

$$E^u := \frac{\kappa^u}{2} \int_\Omega d\mathbf{x} \, D_{\frac{u-1}{2}}(\mathbf{x}, \mathbf{x}) = \frac{\kappa^u}{2} \operatorname{Tr} \mathcal{A}^{\frac{1-u}{2}} , \tag{3.6}$$

$$B^u := \kappa^u \left(\frac{1}{4} - \xi \right) \int_{\partial\Omega} da(\mathbf{x}) \, \frac{\partial}{\partial n_\mathbf{y}} D_{\frac{u+1}{2}}(\mathbf{x}, \mathbf{y}) \Big|_{\mathbf{y}=\mathbf{x}} . \tag{3.7}$$

The derivation of the above result is a bit formal since, in general, one cannot grant convergence of the integrals defining E^u and B^u, for suitable values $u \in \mathbf{C}$.

As for the bulk energy, it is easy to give an example in which finiteness is granted for appropriate u. To this purpose let us consider the case (2.9) (involving a bounded domain with Dirichlet boundary conditions). In this case, recalling the Weyl estimates (2.10), we infer the following:

$$E^u \text{ is finite if } \Re u > d{+}1 . \tag{3.8}$$

Concerning the regularized boundary energy B^u let us mention that, for Ω bounded,

$$B^u = 0 \quad \text{under Dirichlet or Neumann boundary conditions on } \partial\Omega \tag{3.9}$$

(since in the Dirichlet case we have $D_s(\mathbf{x}, \mathbf{y}) = 0$ for $\mathbf{x} \in \partial\Omega$ and all $\mathbf{y}$, while in the Neumann case $\frac{\partial}{\partial n_\mathbf{y}} D_s(\mathbf{x}, \mathbf{y}) = 0$ for $\mathbf{y} \in \partial\Omega$ and all $\mathbf{x}$).

Applying the above considerations in the case of an unbounded domain requires much caution. On the one hand, E^u can be infinite for all $u \in \mathbf{C}$; on the other hand, in the definition (3.7) of B^u it might be necessary to intend the integral $\int_{\partial\Omega} da$ as a limit $\lim_{\ell \to +\infty} \int_{\partial\Omega_\ell} da$, where $(\Omega_\ell)_{\ell=0,1,2,...}$ is a sequence of bounded subdomains such that $\Omega_\ell \subset \Omega_{\ell+1}$ (for any $\ell \in \{0, 1, 2, ..\}$) and $\cup_{\ell=0}^{+\infty} \Omega_\ell = \Omega$ (note that this limit could either be infinite or even fail to exist).

If E^u exists finite for u belonging to a suitable open subset of $\mathbf{C}$ and it is an analytic function of u on this domain, a renormalization by analytic continuation can be implemented. In general, following the extended version of the zeta approach, we define the renormalized bulk energy as

$$E^{\text{ren}} := RP\Big|_{u=0} E^u = RP\Big|_{u=0} \left(\frac{\kappa^u}{2} \operatorname{Tr} \mathcal{A}^{\frac{1-u}{2}} \right). \qquad (3.10)$$

When the analytic continuation of $\operatorname{Tr} \mathcal{A}^{\frac{1-u}{2}}$ is regular up to $u = 0$ the above prescription reduces to

$$E^{\text{ren}} := E^u\Big|_{u=0} = \frac{1}{2} \operatorname{Tr} \mathcal{A}^{1/2} \qquad (3.11)$$

(of course $\operatorname{Tr} \mathcal{A}^{1/2}$ indicates the analytic continuation of $\operatorname{Tr} \mathcal{A}^{\frac{1-u}{2}}$ at $u = 0$).

In a similar way one can define the renormalized boundary and total energies as

$$B^{\text{ren}} := RP\Big|_{u=0} B^u \; ; \qquad (3.12)$$

$$\mathcal{E}^{\text{ren}} := RP\Big|_{u=0} \mathcal{E}^u \; . \qquad (3.13)$$

An alternative definition of the renormalized total energy could be

$$\mathcal{E}^{\text{ren}} := \int_{\Omega} d\mathbf{x} \, \langle 0 | \widehat{T}_{00}(\mathbf{x}) | 0 \rangle_{\text{ren}} \; . \qquad (3.14)$$

This possibility, which is considered rarely in this book, is not granted to be equivalent to (3.13); for example, it may happen that the integral in the right-hand side of Eq. (3.14) diverges, while the prescription (3.13) always gives a finite result by construction. For a comparison between the alternatives (3.13), (3.14), see the final lines of Section 5.4 (dealing with a field on a segment, for several types of boundary conditions).

3.1.1 *The reduced energy for a slab configuration*

Let us consider the slab configuration introduced in Section 2.10, so that $\Omega := \Omega_1 \times \mathbf{R}^{d_2}$, the potential V depends only on $\mathbf{x}_1 \in \Omega_1$, and the boundary conditions regard only $\partial\Omega_1 \times \mathbf{R}^{d_2}$. We already observed in the above mentioned section that the regularized VEV $\langle 0 | \widehat{T}^u_{\mu\nu} | 0 \rangle$ depends only on $\mathbf{x}_1 \in \Omega_1$ (and not on $\mathbf{x}_2 \in \mathbf{R}^{d_2}$); so, the integral in

Eq. (3.1) defining the total energy diverges due to an infinite volume factor.

As a matter of fact, when dealing with a slab configuration one usually considers in place of the total energy $\mathcal{E}^u$ the *reduced total energy* $\mathcal{E}_1^u$; this is the total energy per unit volume in the "free" dimensions, i.e.

$$\mathcal{E}_1^u := \int_{\Omega_1} d\mathbf{x}_1 \, \langle 0|\widehat{T}_{00}^u|0\rangle \; . \tag{3.15}$$

Recalling Eqs. (2.134)–(2.137) and using some well-known identities regarding the Euler gamma function, we infer

$$\mathcal{E}_1^u = \frac{\kappa^u \, \Gamma(\frac{u-d_2+1}{2})}{(4\pi)^{d_2/2} \, \Gamma(\frac{u+1}{2})} \left\{ \left(\frac{u-1+d_2}{4(u-1-d_2)} + \xi \right) \int_{\Omega_1} d\mathbf{x}_1 D_{\frac{u-d_2-1}{2}}^{(1)}(\mathbf{x}_1, \mathbf{y}_1) \Big|_{\mathbf{y}_1=\mathbf{x}_1} \right.$$

$$\left. + \left(\frac{1}{4} - \xi \right) \int_{\Omega_1} d\mathbf{x}_1 \left[(\partial^{x_1^\ell} \partial_{y_1^\ell} + V(\mathbf{x}_1)) D_{\frac{u-d_2+1}{2}}^{(1)}(\mathbf{x}_1, \mathbf{y}_1) \right]_{\mathbf{y}_1=\mathbf{x}_1} \right\} . \tag{3.16}$$

Concerning the second term above, we can express the reduced Dirichlet kernel $D_{\frac{u-d_2+1}{2}}^{(1)}$ in terms of the eigenfunctions $(\mathfrak{F}_{k_1}(\mathbf{x}_1))_{k_1 \in \mathcal{K}_1}$ and the eigenvalues $(\varpi_{k_1})_{k_1 \in \mathcal{K}_1}$ of the reduced operator $\mathcal{A}_1 = -\Delta_1 + V(\mathbf{x}_1)$ and integrate by parts as in the general setting; working as in the derivation of Eqs. (3.3), (3.4) and keeping in mind that $\mathcal{A}_1 \mathfrak{F}_{k_1} = \varpi_{k_1}^2 \mathfrak{F}_{k_1}$, we obtain

$$\int_{\Omega_1} d\mathbf{x}_1 \left[(\partial^{x_1^\ell} \partial_{y_1^\ell} + V(\mathbf{x}_1)) D_{\frac{u-d_2+1}{2}}^{(1)}(\mathbf{x}_1, \mathbf{y}_1) \right]_{\mathbf{y}_1=\mathbf{x}_1}$$

$$= \int_{\Omega_1} d\mathbf{x}_1 \, D_{\frac{u-d_2-1}{2}}^{(1)}(\mathbf{x}_1, \mathbf{y}_1) \Big|_{\mathbf{y}_1=\mathbf{x}_1}$$

$$+ \int_{\partial\Omega_1} da(\mathbf{x}_1) \frac{\partial}{\partial n_{\mathbf{y}_1}} D_{\frac{u-d_2+1}{2}}^{(1)}(\mathbf{x}_1, \mathbf{y}_1) \Big|_{\mathbf{y}_1=\mathbf{x}_1} . \tag{3.17}$$

In conclusion, we have a result similar to Eq. (3.5):

$$\mathcal{E}_1^u = E_1^u + B_1^u \; , \tag{3.18}$$

where we have introduced the *regularized reduced bulk* and *boundary energies*

$$E_1^u := \frac{\kappa^u \, \Gamma(\frac{u-d_2-1}{2})}{2 \, (4\pi)^{d_2/2} \, \Gamma(\frac{u-1}{2})} \int_{\Omega_1} d\mathbf{x}_1 \, D^{(1)}_{\frac{u-d_2-1}{2}}(\mathbf{x}_1, \mathbf{x}_1)$$

$$= \frac{\kappa^u \, \Gamma(\frac{u-d_2-1}{2})}{2 \, (4\pi)^{d_2/2} \, \Gamma(\frac{u-1}{2})} \, \mathrm{Tr} \, \mathcal{A}_1^{\frac{d_2+1-u}{2}} , \tag{3.19}$$

$$B_1^u := \frac{\kappa^u \, \Gamma(\frac{u-d_2+1}{2})}{(4\pi)^{d_2/2} \, \Gamma(\frac{u+1}{2})} \left(\frac{1}{4} - \xi\right) \int_{\partial\Omega_1} da(\mathbf{x}_1) \, \frac{\partial}{\partial n_{\mathbf{y}_1}} D^{(1)}_{\frac{u-d_2+1}{2}}(\mathbf{x}_1, \mathbf{y}_1) \Bigg|_{\mathbf{y}_1 = \mathbf{x}_1} . \tag{3.20}$$

The considerations of the previous section about the convergence of the bulk and boundary energies E^u, B^u have obvious analogues for the reduced energies E_1^u, B_1^u. Of course, the reduced bulk and boundary energies are renormalized in terms of the corresponding analytic continuations (or, possibly, of their regular parts) at $u = 0$.

3.2   The pressure on the boundary

In this section we are interested in the *pressure* $\mathbf{p}(\mathbf{x}) \equiv (p_i(\mathbf{x}))_{i=1,...,d}$, i.e. the force per unit area produced by the field inside Ω at a point $\mathbf{x}$ on the boundary $\partial\Omega$. A possible characterization is the one described hereafter. We first introduce, for $\Re u$ large enough, the *regularized pressure* $\mathbf{p}^u(\mathbf{x})$ of components

$$p_i^u(\mathbf{x}) := \langle 0|\widehat{T}_{ij}^u(\mathbf{x})|0\rangle \, n^j(\mathbf{x}) \qquad \text{for } i \in \{1, ..., d\} ; \tag{3.21}$$

then, we define the *renormalized pressure* at $\mathbf{x}$ setting

$$p_i^{\mathrm{ren}}(\mathbf{x}) := RP\Big|_{u=0} p_i^u(\mathbf{x}) , \tag{3.22}$$

where $RP|_{u=0}$ indicates the regular part of the analytic continuation at $u = 0$ (of course, if the mentioned continuation is regular up to $u = 0$, the above prescription reduces to $p_i^{\mathrm{ren}}(\mathbf{x}) := p_i^u(\mathbf{x})|_{u=0}$, meaning that the analytic continuation at $u = 0$ has to be considered).

It is important to point out that the one introduced above is not the only reasonable definition for the renormalized pressure at a point $\mathbf{x} \in \Omega$; another possibility is to put

$$p_i^{\text{ren}}(\mathbf{x}) := \left(\lim_{\mathbf{x}' \in \Omega, \mathbf{x}' \to \mathbf{x}} \langle 0 | \widehat{T}_{ij}(\mathbf{x}') | 0 \rangle_{\text{ren}} \right) n^j(\mathbf{x}) . \tag{3.23}$$

In few words: in the approach (3.21)–(3.22), one *stays at a point on the boundary*, and performs therein the renormalization; in the approach (3.23), one renormalizes *at points inside* Ω, and then moves towards the boundary. Notice that both approaches require the existence of the normal $\mathbf{n}(\mathbf{x})$ (and thus lose meaning on edges and corner points of $\partial\Omega$).

As a matter of fact, the prescriptions (3.21)–(3.22) and (3.23) do not always agree. The approach (3.22) (possibly, in the restricted version, see (1.22)) gives by construction a finite pressure; on the contrary, this is not granted for the alternative prescription (3.23). As an example, in Chapter 8 we discuss the case where the spatial domain Ω is a wedge; in this case at all boundary points not in the edge, where the normal is well defined, the pressure defined according to Eq. (3.23) diverges. We conjecture that, in general, at points $\mathbf{x} \in \partial\Omega$ where the normal is well defined and the approach (3.23) gives a finite pressure, the result obtained according to the latter prescription agrees with the renormalized pressure defined according to Eq. (3.22); in fact, this happens in all the examples analyzed in this book.

In the rest of the present chapter, we will mainly refer to the approach (3.21)–(3.22) for the analysis of boundary forces.

In applications, one often considers a situation where a quantized field is present both inside Ω and in the complementary region $\Omega^c := \mathbf{R}^d \setminus \Omega$. In this setting the force per unit area acting on the boundary is the resultant of the pressure produced by the field inside Ω, on the one hand, and by the field inside Ω^c, on the other; the renormalized versions of both these observables can be computed separately, using either one of the two approaches mentioned before.

3.2.1 *An explicit expression for the regularized pressure*

Let us stick to the viewpoint (3.21)–(3.22); in order to implement it, we use Eq. (2.30) for the regularized stress-energy tensor that gives the following, for any $\mathbf{x} \in \partial\Omega$ where $\mathbf{n}(\mathbf{x})$ is well defined:

$$p_i^u(\mathbf{x}) = \kappa^u \left[\left(\frac{1}{4} - \xi \right) \delta_{ij} \left(D_{\frac{u-1}{2}}(\mathbf{x}, \mathbf{y}) - (\partial^{x^\ell} \partial_{y^\ell} + V(\mathbf{x})) D_{\frac{u+1}{2}}(\mathbf{x}, \mathbf{y}) \right) \right.$$

$$\left. + \left(\left(\frac{1}{2} - \xi \right) \partial_{x^i y^j} - \xi \, \partial_{x^i x^j} \right) D_{\frac{u+1}{2}}(\mathbf{x}, \mathbf{y}) \right]_{\mathbf{y}=\mathbf{x}} n^j(\mathbf{x}) . \quad (3.24)$$

To go on, let us restrict the attention to the case of *Dirichlet boundary conditions*; then, only the terms involving mixed derivatives (both with respect to $\mathbf{x}$ and $\mathbf{y}$) of the Dirichlet kernel yield non-vanishing contributions on the boundary $\partial\Omega$. Furthermore, the terms proportional to ξ in Eq. (3.24) can be shown to vanish, so that

$$p_i^u(\mathbf{x}) = \kappa^u \left[\left(-\frac{1}{4} \delta_{ij} \, \partial^{x^\ell} \partial_{y^\ell} + \frac{1}{2} \partial_{x^i y^j} \right) D_{\frac{u+1}{2}}(\mathbf{x}, \mathbf{y}) \right]_{\mathbf{y}=\mathbf{x}} n^j(\mathbf{x}) ;$$
$$(3.25)$$

see Appendix E for the proof. As a final step, the analytic continuation of $\boldsymbol{p}^u$ at $u = 0$ must be considered.

3.3   An equivalent characterization of boundary forces

In the literature, forces on $\partial\Omega$ are often characterized by a different approach, which does not require the knowledge of the full stress-energy tensor; see, e.g. the monographs by Bordag *et al.* [21, 22], Milton [105] and Plunien *et al.* [120]. In this approach one considers a variation of the spatial domain Ω controlled by a real parameter, and defines the pressure in terms of the derivative of the bulk energy with respect to the mentioned parameter. For example, if Ω is a parallelepiped $(0, a) \times (0, b) \times (0, c)$ one could consider the variation of the length of any one of its sides, say a; it is customary to define the total force on the face $\{x^1 = a\}$ as the derivative of the bulk energy with respect to a, with the sign changed.

The idea that boundary forces are related to the variation of Ω has been typically presented in simple examples like the previous one; it can be of interest to propose a general formulation of this idea, and to compare it with the characterization of boundary forces given in Section 3.2 via the stress-energy tensor.

For the sake of definiteness, let us consider the case where Ω is a bounded domain in $\mathbf{R}^d$ and Dirichlet boundary conditions are prescribed on $\partial\Omega$; besides, let $\mathfrak{S} : \mathbf{R}^d \to \mathbf{R}^d$ be a vector field. We assume Ω, its boundary $\partial\Omega$ and $\mathfrak{S}$ are regular enough to permit the forthcoming computations.

First of all, consider the family of diffeomorphism

$$\boldsymbol{S}_\epsilon : \mathbf{R}^d \to \mathbf{R}^d , \quad \mathbf{x} \mapsto \boldsymbol{S}_\epsilon(\mathbf{x}) := \mathbf{x} + \epsilon \, \mathfrak{S}(\mathbf{x}) , \qquad (3.26)$$

labeled by a small parameter $\epsilon > 0$, that will be ultimately sent to zero. For any ϵ, the spatial domain

$$\Omega_\epsilon := \boldsymbol{S}_\epsilon(\Omega) \quad (\subset \mathbf{R}^d) \qquad (3.27)$$

can be regarded as a deformation of the initial domain Ω.

Of course, a relation analogous to (3.6) holds for the regularized bulk energy E_ϵ^u associated to the deformed domain Ω_ϵ, i.e.

$$E_\epsilon^u = \frac{\kappa^u}{2} \sum_{k \in \mathcal{K}} (\omega_{\epsilon,k}^2)^{\frac{1-u}{2}} ; \qquad (3.28)$$

in the above $(\omega_{\epsilon,k}^2)_{k \in \mathcal{K}}$ denote the eigenvalues of the fundamental operator $\mathcal{A}_\epsilon$, that is the operator $-\Delta + V$ acting on the Hilbert space $L^2(\Omega_\epsilon)$ (with Dirichlet boundary conditions on $\partial\Omega_\epsilon$) instead of $L^2(\Omega)$.

Let us now consider the expansion of E_ϵ^u to the first order in ϵ, which describes the variation of the regularized bulk energy under the deformation (3.26), (3.27) of the space domain. Due to Eq. (3.28), the calculation of this expansion can be reduced to the first order expansion of the eigenvalues $\omega_{\epsilon,k}$; this can be done by standard perturbation techniques, as well known from the classic work of Rellich [124]. As illustrated in Appendix E, the conclusion of this analysis is

$$E_\epsilon^u = E^u - \epsilon \, (1-u) \int_{\partial\Omega} da(\mathbf{x}) \, \mathfrak{S}^i(\mathbf{x}) \, p_i^u(\mathbf{x}) + O(\epsilon^2) \qquad (3.29)$$

where $\boldsymbol{p}^u \equiv (p_i^u)$ is the regularized pressure, given by Eq. (E.11). Equation (3.29) can be used for an alternative, but equivalent definition of the regularized pressure; from this viewpoint, the regularized pressure field $\boldsymbol{p}^u$ is the unique vector field on $\partial\Omega$ such that (3.29) holds for each one-parameter deformation of the domain Ω of the form (3.26)–(3.27).

Let us now perform the analytic continuation up to $u = 0$, assuming *that no pole occurs at this point*; $E_\epsilon^u\big|_{u=0}$ and $\boldsymbol{p}^u\big|_{u=0}$ are the renormalized bulk energy and pressure, and Eq. (3.29) yields the relation

$$E_\epsilon^{\text{ren}} = E^{\text{ren}} - \epsilon \int_{\partial\Omega} da(\mathbf{x})\, \mathfrak{S}^i(\mathbf{x})\, p_i^{\text{ren}}(\mathbf{x}) + O(\epsilon^2)\,. \qquad (3.30)$$

This is the result anticipated at the beginning of this section: a characterization of boundary forces in terms of the bulk energy variation under deformations of the domain. We already mentioned the frequent use of this idea in the literature, for particular choices of the spatial domain Ω.

3.4 Integrated forces on the boundary

Let us now discuss the evaluation of the integrated force $\mathfrak{F}_{\mathfrak{O}}$ acting on an arbitrary subset $\mathfrak{O}$ of the spatial boundary ($\mathfrak{O} \subset \partial\Omega$; possibly, $\mathfrak{O} = \partial\Omega$). As in the case of the pressure considered in Section 3.2, we can give several alternative definitions of this quantity. As a first possibility, we introduce the *regularized total force* on $\mathfrak{O}$ (for large $\Re u$)

$$\mathfrak{F}_{\mathfrak{O}}^u := \int_{\mathfrak{O}} da(\mathbf{x})\, \mathbf{p}^u(\mathbf{x})\,, \qquad (3.31)$$

where $\mathbf{p}^u \equiv (p_i^u(\mathbf{x}))$ indicates the regularized pressure (see Eq. (3.21)); then, we define the *renormalized total force* on $\mathfrak{O}$ as

$$\mathfrak{F}_{\mathfrak{O}}^{\text{ren}} := RP\big|_{u=0} \mathfrak{F}_{\mathfrak{O}}^u \qquad (3.32)$$

(clearly, when there is no pole in $u = 0$, the above prescription reduces to $\mathfrak{F}_{\mathfrak{O}}^{\text{ren}} := \mathfrak{F}_{\mathfrak{O}}^u\big|_{u=0}$, meaning that the analytic continuation in $u = 0$ has to be considered).

Another possibility is to put

$$\mathfrak{F}_{\mathfrak{O}}^{\text{ren}} := \int_{\mathfrak{O}} da(\mathbf{x})\, \mathbf{p}^{\text{ren}}(\mathbf{x})\,, \qquad (3.33)$$

where $\mathbf{p}^{\text{ren}}(\mathbf{x}) = (p_i^{\text{ren}}(\mathbf{x}))$ is the renormalized pressure, defined according to either Eq. (3.22) or Eq. (3.23).

Similarly to what we pointed out in Section 3.2 for the pressure, in general the two alternatives (3.32), (3.33) give different results; in fact, the prescription (3.32) always gives a finite result for $\mathfrak{F}_{\mathfrak{O}}^{\mathrm{ren}}$, while (3.33) can give an infinite result.

In conclusion, let us stress that for the integrated force there hold comments analogous to the ones at the end of Section 3.2, concerning the case where a quantized field is present both inside Ω and in the complementary region $\Omega^c := \mathbf{R}^d \setminus \Omega$. In this case the total force on any subset $\mathfrak{O} \subset \partial\Omega$ is given by the resultant of the forces corresponding, respectively, to the field inside and outside the spatial domain Ω.

3.5   A comment on some previous "anomalies"

In the previous sections we have pointed out that the renormalized versions of the total energy, of the pressure and of the integrated forces on the boundary can be defined according to different prescriptions, which in general are not equivalent (see Eqs. (3.13), (3.14), (3.22), (3.23), (3.32), (3.33) and the considerations in the corresponding sections).

In consequence of this, there arise unavoidable ambiguities, or *anomalies*, when talking about the renormalized observables mentioned above. For example, we have mentioned previously the possible non-equivalence of the alternatives (3.13), (3.14) for the total energy $\mathcal{E}^{\mathrm{ren}}$ and (3.22), (3.23) for the pressure $\mathbf{p}^{\mathrm{ren}}$; recall that it may happen that the prescriptions (3.14) and (3.23) give infinite results for $\mathcal{E}^{\mathrm{ren}}$ and $\mathbf{p}^{\mathrm{ren}}$, due to boundary singularities of the renormalized stress-energy VEV which make divergent the integral $\int_\Omega d\mathbf{x} \, \langle 0|\widehat{T}_{00}(\mathbf{x})|0\rangle_{\mathrm{ren}}$ or the limit $\lim_{\mathbf{x}' \in \Omega, \, \mathbf{x}' \to \mathbf{x}} \langle 0|\widehat{T}_{ij}(\mathbf{x}')|0\rangle_{\mathrm{ren}} \, n^j(\mathbf{x})$ ($\mathbf{x} \in \partial\Omega$, $i \in \{1, ..., d\}$).

On the other hand, such boundary singularities of the renormalized stress-energy VEV are not a specific consequence of the zeta regularization; indeed, they also appear if one uses point-splitting, as indicated by the very systematic analysis of Deutsch and Candelas [45]. For the moment, the above mentioned anomalies must be accepted as a problematic aspect of the main regularization schemes;

what we can do is just to record them when they appear, and hope that in the future they can be better understood.

Perhaps, one could look for the origin of such anomalies in some excessive idealization of the physical model (for example, one could try to describe in a more realistic manner the boundaries of the spatial domain; these are "hard" and "deterministic" in the present formulation, but could perhaps be replaced with "soft" or "stochastic" [12, 64] boundaries). For a different type of boundary anomaly, related to vacuum effects for charged fermions, and for its cure taking into account back reaction effects, see [134].

Chapter 4

Some variations of the previous schemes

In this chapter and in the related Appendix F we consider some variations of the general framework of Chapters 1–3 concerning the following situations: (i) the case where 0 is either an isolated or non-isolated point of the spectrum of the fundamental operator $\mathcal{A} := -\Delta + V$; (ii) the case where the flat spatial domain Ω is described via curvilinear coordinates or, more generally, the case where Ω is a (possibly non-flat) Riemannian manifold equipped with arbitrary coordinates (or an open subset of it, with prescribed boundary conditions). Some of these variations will be used in the applications of Part 2.

4.1 The basic Hilbert space when 0 is a proper eigenvalue of $\mathcal{A}$; the case of Neumann and periodic boundary conditions

In this section we are going to consider a variation of the framework developed in Chapters 1 and 2 to deal with the cases where the fundamental operator $\mathcal{A} = -\Delta + V$ acting on $L^2(\Omega)$ has spectrum contained in $[0, +\infty)$, with 0 an isolated point; in other terms, $0 \in \sigma(\mathcal{A}) \subset \{0\} \cup [\varepsilon^2, +\infty)$, for some $\varepsilon > 0$. In this case 0 is a proper eigenvalue, as it always occurs for isolated points of the spectrum.

A standard approach employed in the physical literature to treat problems of this kind is to simply neglect the states of "zero energy"; see, e.g. [84, 86, 146]. According to the formulation considered in this book, this amounts to the following procedure: in place of $L^2(\Omega)$, we define the basic Hilbert space as the orthogonal complement in $L^2(\Omega)$ of the null eigenspace, that is

$$L_0^2(\Omega) := (\ker \mathcal{A})^\perp \quad (\subset L^2(\Omega)) . \tag{4.1}$$

It should be noted that the restriction of $\mathcal{A}$ to $L_0^2(\Omega)$ is a self-adjoint, strictly positive operator in $L_0^2(\Omega)$ with spectrum contained in $[\varepsilon^2, +\infty)$. In this situation, $L_0^2(\Omega)$ is the basic Hilbert space even from the viewpoint of field quantization.[a]

Let us recall the definition (2.1) for the integral kernel associated to a given operator on $L^2(\Omega)$ and consider Eqs. (2.14), (2.43) and (2.53) for the Dirichlet, heat, cylinder and modified cylinder kernels associated to $\mathcal{A}$; if the latter is redefined as the restriction of $-\Delta + V$ to $L_0^2(\Omega)$, in the cited equations we should formally replace $\delta_{\mathbf{x}}, \delta_{\mathbf{y}}$ with $E_0\delta_{\mathbf{x}}, E_0\delta_{\mathbf{y}}$ where E_0 is the orthogonal projection onto $L_0^2(\Omega)$ (suitably extended to distributions, so that it can be applied to $\delta_{\mathbf{x}}, \delta_{\mathbf{y}}$). With this modification, the expansions (2.15), (2.44), (2.45) and (2.53) for the kernels mentioned above hold again, using the eigenfunctions of $\mathcal{A}$ in $L_0^2(\Omega)$.[b]

Typical configurations of the above type are those where $\mathcal{A} = -\Delta$, the spatial domain Ω is bounded and the field fulfills either Neumann or periodic boundary conditions on $\partial\Omega$;[c] indeed, in such cases the spectrum of $\mathcal{A}$ in $L^2(\Omega)$ is purely pointwise, 0 is an eigenvalue and $\ker\mathcal{A}$ is formed by the constant functions. Therefore $L_0^2(\Omega)$, defined via Eq. (4.1), is formed by the functions with mean zero:

$$L_0^2(\Omega) = \left\{ f \in L^2(\Omega) \;\middle|\; \int_\Omega d\mathbf{x}\, f(\mathbf{x}) = 0 \right\} . \tag{4.2}$$

Let us mention that an analogous framework can be considered for slab configurations where $\Omega = \Omega_1 \times \mathbf{R}^{d_2}$ and Neumann or periodic boundary conditions are prescribed on $\partial\Omega_1 \times \mathbf{R}^{d_2}$. In these cases one

---

[a]By this, we mean that the Fock space $\mathfrak{F}$ of the quantized scalar field living in Ω is the direct sum of all symmetrized tensor powers of $L_0^2(\Omega)$.

[b]As an example, in the case described by Eq. (4.2) the projection E_0 onto $L_0^2(\Omega)$ is given by $E_0 f = f - \frac{1}{\mathrm{Vol}(\Omega)} \int_\Omega d\mathbf{x}\, f(\mathbf{x})$ ($\mathrm{Vol}(\Omega)$ is the volume of Ω); the previous prescription makes sense as well for $f = \delta_{\mathbf{x}}$ and gives $E_0\delta_{\mathbf{x}} = \delta_{\mathbf{x}} - \frac{1}{\mathrm{Vol}(\Omega)}$.

[c]As will be observed in Section 4.3, the case of periodic boundaries would be more properly formulated in terms of a free field on a torus, but this is cause of no concern for the present considerations.

works with the reduced operator $\mathcal{A}_1$ acting in $L^2(\Omega_1)$; the latter must then be replaced with the Hilbert space

$$L_0^2(\Omega_1) := (\ker \mathcal{A}_1)^\perp = \left\{ f \in L^2(\Omega_1) \,\Big|\, \int_{\Omega_1} d\mathbf{x}_1 \, f(\mathbf{x}_1) = 0 \right\} \quad (4.3)$$

and the basic Hilbert space for the full theory on Ω is $L_0^2(\Omega_1) \otimes L^2(\mathbf{R}^{d_2})$.

In the applications to be considered in the following, whenever 0 is an isolated point of the spectrum we will always assume that the fundamental operator $\mathcal{A}$ (resp. $\mathcal{A}_1$) has been redefined so that it acts on the Hilbert space $L_0^2(\Omega)$ of Eq. (4.1) (resp. $L_0^2(\Omega_1)$ of Eq. (4.3)).

4.2 The case where 0 belongs to the continuous spectrum of $\mathcal{A}$

Let us pass to the case where the fundamental operator $\mathcal{A} = -\Delta + V$ is non-negative ($\sigma(\mathcal{A}) \subset [0, +\infty)$), and 0 is in the continuous spectrum of $\mathcal{A}$; then 0 has zero spectral measure and it is a non isolated point of the spectrum (otherwise, it would be a proper eigenvalue).

Hereafter we describe two examples of this kind. To obtain them we consider the operator $\mathcal{A} := -\Delta$ in $L^2(\mathbf{R}^d)$, or the operator $\mathcal{A} := -\Delta$ in $L^2(\Omega)$ where Ω is the half-space $\{\mathbf{x} \in \mathbf{R}^d \mid x^1 > 0\}$, and suitable boundary conditions, say Dirichlet, are specified on $\partial\Omega = \{x^1 = 0\}$. In these cases $\mathcal{A}$ has a complete orthonormal system of (improper) eigenfunctions $(F_\mathbf{k})_{\mathbf{k} \in \mathcal{K}}$ with corresponding eigenvalues $\omega_\mathbf{k}^2$, where: in the first case, $\mathcal{K} = \mathbf{R}^d$ (with the Lebesgue measure $d\mathbf{k}$), $\mathcal{F}_\mathbf{k}(\mathbf{x}) := (2\pi)^{-d/2} e^{i\mathbf{k} \cdot \mathbf{x}}$, $\omega_\mathbf{k} := |\mathbf{k}|$; in the second case, $\mathcal{K} = (0, +\infty) \times \mathbf{R}^{d-1}$ (again, with the Lebesgue measure $d\mathbf{k}$), $\mathcal{F}_\mathbf{k}(\mathbf{x}) := \sqrt{2}(2\pi)^{-d/2} \sin(k^1 x^1) e^{ik^2 x^2 + \ldots + k^d x^d}$ and, again, $\omega_\mathbf{k} := |\mathbf{k}|$. In both cases $\sigma(\mathcal{A}) = [0, +\infty)$ and the spectrum is purely continuous.

The case of $\mathcal{A}$ non-negative, with 0 in the continuous spectrum, cannot be treated with the approach of the previous section: there is no way to obtain a strictly positive operator by simply removing 0 from the spectrum. In a more physical language, infrared divergences cannot be simply ignored and we must devise a more sophisticated way to treat them, as we are currently doing for ultraviolet divergences. A natural approach to the problem is to represent $\mathcal{A}$ as a limit

$$\mathcal{A} := \text{``}\lim_{\varepsilon \to 0^+}\text{''} \mathcal{A}_\varepsilon \qquad (4.4)$$

where $\mathcal{A}_\varepsilon$ is a self-adjoint operator, depending on a parameter $\varepsilon \in (0, \varepsilon_0)$ in such a way that its spectrum is contained in $[\varepsilon^2, +\infty)$; the deformed operator $\mathcal{A}_\varepsilon$ is used everywhere in place of $\mathcal{A}$, and the limit $\varepsilon \to 0^+$ is performed only at the end, after zeta renormalization has been carried out. In particular, we define the *deformed, regularized field operator*

$$\widehat{\phi}^{\varepsilon u} := (\kappa^{-2}\mathcal{A}_\varepsilon)^{-u/4}\widehat{\phi} \qquad (4.5)$$

and the corresponding *deformed, regularized stress-energy tensor* $\widehat{T}^{\varepsilon u}_{\mu\nu}(x)$ whose VEV is given by

$$\langle 0|\widehat{T}^{\varepsilon u}_{\mu\nu}(x)|0\rangle = \left(\frac{1}{2}-\xi\right)(\partial_{x^\mu y^\nu}+\partial_{x^\nu y^\mu}) - \left(\frac{1}{2}-2\xi\right)\eta_{\mu\nu}\left(\partial^{x^\lambda}\partial_{y^\lambda} + V\right)$$

$$-\xi(\partial_{x^\mu x^\nu}+\partial_{y^\mu y^\nu})|_{y=x} \cdot \langle 0|\widehat{\phi}^{\varepsilon u}(x)\widehat{\phi}^{\varepsilon u}(y)|0\rangle \ . \qquad (4.6)$$

For the above VEV we have expression analogous to (2.28)–(2.30) in terms of the *deformed Dirichlet kernel*

$$D^\varepsilon_s(\mathbf{x},\mathbf{y}) := \mathcal{A}^{-s}_\varepsilon(\mathbf{x},\mathbf{y}) = \langle\delta_\mathbf{x}|\mathcal{A}^{-s}_\varepsilon\,\delta_\mathbf{y}\rangle \ , \qquad (4.7)$$

with $s = (u \pm 1)/2$.

As mentioned above, in this generalized version of the local zeta regularization the limit $\varepsilon \to 0^+$ must be considered only after the analytic continuation has been performed; in particular, we define

$$\langle 0|\widehat{T}_{\mu\nu}(x)|0\rangle_{\mathrm{ren}} := \lim_{\varepsilon \to 0^+} RP\Big|_{u=0} \langle 0|\widehat{T}^{\varepsilon u}_{\mu\nu}(x)|0\rangle \ . \qquad (4.8)$$

The above renormalized VEV can be expressed in terms of the renormalized kernels $D^{(\kappa)}_{\pm 1/2}(\mathbf{x},\mathbf{y})$ and $\partial_{zw}D^{(\kappa)}_{1/2}(\mathbf{x},\mathbf{y})$, where

$$D^{(\kappa)}_{\pm\frac{1}{2}}(\mathbf{x},\mathbf{y}) := \lim_{\varepsilon\to 0^+} RP\Big|_{u=0}\left(\kappa^u D^\varepsilon_{\frac{u\pm 1}{2}}(\mathbf{x},\mathbf{y})\right)$$

$$= \lim_{\varepsilon\to 0^+} RP\Big|_{s=\pm\frac{1}{2}}\left(\kappa^{2s\mp 1}D^\varepsilon_s(\mathbf{x},\mathbf{y})\right) ,$$

$$\partial_{zw}D^{(\kappa)}_{\frac{1}{2}}(\mathbf{x},\mathbf{y}) := \lim_{\varepsilon\to 0^+} RP\Big|_{u=0}\left(\kappa^u \partial_{zw}D^\varepsilon_{\frac{u+1}{2}}(\mathbf{x},\mathbf{y})\right)$$

$$= \lim_{\varepsilon\to 0^+} RP\Big|_{s=\frac{1}{2}}\left(\kappa^{2s-1}\partial_{zw}D^\varepsilon_s(\mathbf{x},\mathbf{y})\right) ; \qquad (4.9)$$

these functions play a role very similar to the ones introduced in
Eq. (2.34) for a strictly positive $\mathcal{A}$, and allow to express the renormalized VEV (4.8) as in Eqs. (2.36)–(2.38).

In the sequel we will write K^ε and T^ε (or $\tilde{T}^\varepsilon$), respectively, for
the heat and cylinder (or modified cylinder) kernel associated to $\mathcal{A}_\varepsilon$.
Proceeding as in Chapter 2 we obtain, for $\Re s$ sufficiently large,

$$D_s^\varepsilon(\mathbf{x},\mathbf{y}) = \frac{1}{\Gamma(s)} \int_0^{+\infty} dt\, t^{s-1}\, K^\varepsilon(t;\mathbf{x},\mathbf{y}) \; ; \qquad (4.10)$$

$$D_s^\varepsilon(\mathbf{x},\mathbf{y}) = \frac{1}{\Gamma(2s)} \int_0^{+\infty} dt\, t^{2s-1}\, T^\varepsilon(t;\mathbf{x},\mathbf{y}) \;. \qquad (4.11)$$

These identities are the starting point to discuss the analytic contin-
uation in s of the Dirichlet kernel D_s^ε, for any fixed $\varepsilon \in (0,\varepsilon_0)$.

Let us associate to the "undeformed" fundamental operator $\mathcal{A}$ the
heat and cylinder kernels

$$K(t;\mathbf{x},\mathbf{y}) := e^{-t\mathcal{A}}(\mathbf{x},\mathbf{y}) \;, \qquad T(t;\mathbf{x},\mathbf{y}) := e^{-t\sqrt{\mathcal{A}}}(\mathbf{x},\mathbf{y}) \;, \qquad (4.12)$$

as well as the modified cylinder kernel

$$\tilde{T}(t;\mathbf{x},\mathbf{y}) := (\sqrt{\mathcal{A}}^{\,-1} e^{-t\sqrt{\mathcal{A}}})(\mathbf{x},\mathbf{y}) \qquad (4.13)$$

(see subsection 2.6.3); these kernels can be represented as in
Eqs. (2.44), (2.45) and (2.53) in terms of the eigenfunctions $(F_k)_{k\in\mathcal{K}}$
and eigenvalues $(\omega_k)_{k\in\mathcal{K}}$ of $\mathcal{A}$. We stress that the functions $D_{\pm 1/2}^{(\kappa)}$ of
Eq. (4.9) do *not* possess integral representations of the form (4.10)
(4.11) with $K^\varepsilon, T^\varepsilon$ replaced by K, T; in fact, the corresponding integrals for K, T are typically divergent. In the sequel, we will present
a more subtle way to obtain $D_{\pm 1/2}^{(\kappa)}$ from K or T (and $\tilde{T}$).

Up to now we have not specified any particular form for $\mathcal{A}_\varepsilon$. The
following two choices seem to be natural and particulary useful in
applications:

$$\mathcal{A}_\varepsilon := \mathcal{A} + \varepsilon^2 \;, \qquad (4.14)$$

$$\mathcal{A}_\varepsilon := (\sqrt{\mathcal{A}} + \varepsilon)^2 \;. \qquad (4.15)$$

The first one corresponds to the idea, widespread in the physical literature, to treat infrared divergences adding a small mass ε [101,136];

the second one is less familiar and is justified by the considerations that follow.

Assuming $\mathcal{A}_\varepsilon$ to have either the form (4.14) or (4.15), we readily obtain the following relations allowing to express the deformed kernels $K^\varepsilon, T^\varepsilon$ in terms of the analogous undeformed kernels K, T:

$$\mathcal{A}_\varepsilon := \mathcal{A} + \varepsilon^2 \Rightarrow K^\varepsilon(t;\mathbf{x},\mathbf{y}) = e^{-\varepsilon^2 t}\, K(t;\mathbf{x},\mathbf{y}) \; ; \qquad (4.16)$$

$$\mathcal{A}_\varepsilon := (\sqrt{\mathcal{A}} + \varepsilon)^2 \Rightarrow T^\varepsilon(t;\mathbf{x},\mathbf{y}) = e^{-\varepsilon t}\, T(t;\mathbf{x},\mathbf{y}) \; . \qquad (4.17)$$

In particular, assume the kernels K, T to have analytic extensions in t to a neighbourhood of $[0, +\infty)$, except a possible pole at $t = 0$, and (at most) polynomial growth for $\Re t \to +\infty$; in these cases, the deformed Dirichlet kernel D_s^ε admits integral representations analogous to (2.113), (2.114), involving the Hankel contour $\mathfrak{H}$. For example, one can write

$$D_s^\varepsilon(\mathbf{x},\mathbf{y}) = \frac{e^{-2i\pi s}\,\Gamma(1-2s)}{2\pi i} \int_{\mathfrak{H}} dt\; t^{2s-1}\, T^\varepsilon(t;\mathbf{x},\mathbf{y}) \; . \qquad (4.18)$$

Starting from the above representation we can derive explicit expressions for the renormalized functions $D_{\pm 1/2}^{(\kappa)}(\mathbf{x},\mathbf{y})$, and, more generally, for

$$D_{-\frac{n}{2}}^{(\kappa)}(\mathbf{x},\mathbf{y}) := \lim_{\varepsilon \to 0} RP\Big|_{s=-\frac{n}{2}} \left(\kappa^{2s+n} D_s^\varepsilon(\mathbf{x},\mathbf{y})\right) \quad (n \in \{-1,0,1,2,...\}) \; , \qquad (4.19)$$

that could be called "renormalized Dirichlet kernels" of order $-n/2$. More precisely, let us assume the modified cylinder kernel $\tilde{T}(t;\mathbf{x},\mathbf{y})$ to be a meromorphic function of t in the neighbourhood of the positive real semiaxis (like T). Let $n = -1, 0, 1, 2, ...$; if $n = -1$, assume in addition

$$|T(t;\mathbf{x},\mathbf{y})| \leqslant C\,|t|^{-a-1} \quad \text{for } \Re t \to +\infty \text{ and some } C, a > 0 \; . \qquad (4.20)$$

Then, we obtain the following result (see Appendix F):

$$D_{-\frac{n}{2}}^{(\kappa)}(\mathbf{x},\mathbf{y}) = (-1)^{n+1}\,\Gamma(n+2)\,\mathrm{Res}\left(t^{-(n+2)}\,\tilde{T}(t;\mathbf{x},\mathbf{y})\,;0\right) \; . \qquad (4.21)$$

Let us remark that Eq. (4.21) has the same structure of Eq. (2.117), dealing with the Dirichlet kernel when $\mathcal{A}$ is strictly positive. In the strictly positive case, the cited result was derived rigorously (from Eq. (2.116)), with no need to introduce a regulating parameter ε; in the present framework, instead, it would be impossible to establish (4.21) without using the regulator ε.

One could derive results similar to Eq. (4.21), involving the "renormalized derivatives", e.g.

$$\partial_{zw} D^{(\kappa)}_{-\frac{n}{2}}(\mathbf{x}, \mathbf{y}) := \lim_{\varepsilon \to 0} RP\Big|_{s=-\frac{n}{2}} \left(\kappa^{2s+n} \, \partial_{zw} D^\varepsilon_s(\mathbf{x}, \mathbf{y}) \right), \qquad (4.22)$$

where z, w are spatial variables; indeed, assuming $\partial_{zw}\tilde{T}(\mathfrak{t}; \mathbf{x}, \mathbf{y})$ to fulfill conditions of the form stipulated previously for $\tilde{T}(\mathfrak{t}; \mathbf{x}, \mathbf{y})$ (see, in particular, Eq. (4.20)), for $n = -1, 0, 1, 2, \ldots$ we have

$$\partial_{zw} D^{(\kappa)}_{-\frac{n}{2}}(\mathbf{x}, \mathbf{y}) = (-1)^{n+1} \, \Gamma(n+2) \operatorname{Res}\left(\mathfrak{t}^{-(n+2)} \, \partial_{zw}\tilde{T}(\mathfrak{t}; \mathbf{x}, \mathbf{y}); 0 \right).$$
$$(4.23)$$

To conclude, let us discuss the pressure on the boundary in the present framework; the main point is the fact that, as in the case of strictly positive $\mathcal{A}$, there are two possible prescriptions for the renormalized pressure. The first alternative is to introduce, at each point $\mathbf{x} \in \partial\Omega$, a *deformed, regularized pressure* with components

$$p^{\varepsilon u}_i(\mathbf{x}) := \langle 0|\widehat{T}^{\varepsilon u}_{ij}(\mathbf{x})|0\rangle n^j(\mathbf{x}) \qquad (i \in \{1, ..., d\}), \qquad (4.24)$$

where $\mathbf{n}(\mathbf{x}) \equiv (n^j(\mathbf{x}))$ is the outer unit normal to the boundary; we then define the renormalized pressure at $\mathbf{x}$ as

$$p^{\mathrm{ren}}_i(\mathbf{x}) := \lim_{\varepsilon \to 0+} RP\Big|_{u=0} p^{\varepsilon u}_i(\mathbf{x}). \qquad (4.25)$$

The second alternative is to put

$$p^{\mathrm{ren}}_i(\mathbf{x}) := \left(\lim_{\mathbf{x}' \in \Omega, \mathbf{x}' \to \mathbf{x}} \langle 0|\widehat{T}_{ij}(\mathbf{x}')|0\rangle_{\mathrm{ren}} \right) n^j(\mathbf{x}) \qquad (4.26)$$

where $\langle 0|\widehat{T}_{ij}(\mathbf{x}')|0\rangle_{\mathrm{ren}}$ is defined according to Eq. (4.8) at all interior points $\mathbf{x}'$ of the domain. The prescriptions (4.25), (4.26) can, in general, give different results: this happens, for example, in the case of a massless field on a wedge-shaped domain, to be discussed in Chapter 8 of Part 2.

4.3   Variations involving the spatial domain

In the literature, a scalar field fulfilling periodic boundary conditions is often considered. To give a rigorous description of this configuration, one should better give up to viewing Ω as an open subset of $\mathbf{R}^d$ and pass to a description in terms of tori. For example, it is customary to speak of a field on the hypercube $(0, a)^d$ with periodic boundary conditions, where $a > 0$ is some given length. In a more precise description of this configuration, Ω is not $(0, a)^d$ but rather the d-dimensional torus $\mathbf{T}_a^d := \mathbf{R}^d/(a\mathbf{Z})^d \simeq (\mathbf{R}/a\mathbf{Z})^d$ (where $\mathbf{Z}$ is the set of integers, so that $a\mathbf{Z} = \{..., -2a, -a, 0, a, 2a, ...\}$).[d]

In some applications to be considered in the subsequent Part 2, the space domain Ω is an open subset of $\mathbf{R}^d$ but, in place of the Cartesian coordinates $\mathbf{x} \equiv (x^i)$, it is natural to use for it some curvilinear coordinates $(q^i)_{i=1,...,d} \equiv \mathbf{q}$; in these coordinates, the line element of $\mathbf{R}^d$ has the form

$$d\ell^2 = a_{ij}(\mathbf{q})\, dq^i dq^j \ . \tag{4.27}$$

The above spatial coordinates induce a set of spacetime coordinates $(q^\mu)_{\mu=0,...,d} \equiv q$ on $\mathbf{R} \times \Omega$ where $q^0 := t$ and the q^i's are as before; clearly, the spacetime line element $ds^2 = -dt^2 + d\ell^2$ has the form

$$ds^2 = g_{\mu\nu}(q)\, dq^\mu dq^\nu \ ,$$

$$g_{00} := -1 \ , \qquad g_{i0} = g_{0i} := 0 \ , \tag{4.28}$$

$$g_{ij}(q) := a_{ij}(\mathbf{q}) \quad \text{for } i, j \in \{1, ..., d\} \ .$$

The analogue of Eq. (1.20) in the coordinate system (q^μ) is

$$\widehat{T}_{\mu\nu}^u := (1 - 2\xi)\partial_\mu \widehat{\phi}^u \circ \partial_\nu \widehat{\phi}^u - \left(\frac{1}{2} - 2\xi\right) g_{\mu\nu}\left(\partial^\lambda \widehat{\phi}^u \partial_\lambda \widehat{\phi}^u + V(\widehat{\phi}^u)^2\right)$$

$$- 2\xi\, \widehat{\phi}^u \circ \nabla_{\mu\nu}\widehat{\phi}^u \ , \tag{4.29}$$

---

[d]The considerations of Section 4.1 for the periodic case are easily rephrased in terms of the torus $\mathbf{T}_a^d$. The operator $\mathcal{A} := -\Delta$ acting in $L^2(\mathbf{T}_a^d)$ has 0 as an eigenvalue, with $\ker \mathcal{A}$ formed by the constant functions; again, 0 is eliminated viewing $\mathcal{A}$ as an operator acting in

$$L_0^2(\mathbf{T}_a^d) := (\ker \mathcal{A})^\perp = \left\{ f \in L^2(\mathbf{T}_a^d) \ \Big| \ \int_{\mathbf{T}_a^d} d\mathbf{x}\, f(\mathbf{x}) = 0 \right\} \ .$$

where ∇_μ is the covariant derivative induced by the (flat) spacetime metric (4.28).

In principle, the covariant derivative ∇_μ should appear in place of any derivative ∂_μ; however we are working with a *scalar* field and it is well-known that

$$\nabla_\mu f = \partial_\mu f \quad \text{if } f \text{ is a scalar function .} \tag{4.30}$$

The situation is different when we consider second order derivatives, which explains the appearing of $\nabla_{\mu\nu}$ in (4.29). Let us recall that

$$\nabla_{\mu\nu} f = \partial_{\mu\nu} f - \Gamma^\lambda_{\mu\nu} \partial_\lambda f \quad (= \nabla_{\nu\mu} f) \quad \text{if } f \text{ is a scalar function ,} \tag{4.31}$$

where we are using the spacetime Christoffel symbols $\Gamma^\lambda_{\mu\nu} := \frac{1}{2} g^{\lambda\rho}(\partial_\mu g_{\rho\nu} + \partial_\nu g_{\mu\rho} - \partial_\rho g_{\mu\nu})$. The above computational rule is more efficiently implemented recalling that $q^0 = t$ and using the space covariant derivatives D_i corresponding to the line element (4.27); these rely on the Christoffel symbols $\gamma^k_{ij} := \frac{1}{2} a^{kh}(\partial_i a_{hj} + \partial_j a_{ih} - \partial_h a_{ij})$. From Eq. (4.28) one easily infers that $\Gamma^k_{ij} = \gamma^k_{ij}$ (for $i, j, k \in \{1, ..., d\}$) are the only non-vanishing coefficients; so, Eq. (4.31) for a scalar function f on the spacetime $\mathbf{R} \times \Omega$ implies

$$\nabla_{ij} f = D_{ij} f = \partial_{ij} f - \gamma^k_{ij} \partial_k f ,$$
$$\nabla_{0i} f = \partial_0(\partial_i f) = \partial_i(\partial_0 f) = \nabla_{i0} f , \quad \nabla_{00} f = \partial_{00} f . \tag{4.32}$$

As a further variation of the general schemes presented in the preceding chapters, we can stipulate the spatial domain Ω to be an arbitrary d-dimensional Riemannian manifold, possibly non flat; in any coordinate system $(q^i)_{i=1,...,d} \equiv \mathbf{q}$ of Ω, the Riemannian line element $d\ell^2$ will have a representation of the form (4.27). (Of course the position $\Omega = \mathbf{T}^d_a$, considered at the beginning of this paragraph in relation to periodic boundary conditions, amounts to choosing for Ω a very simple, flat Riemannian manifold). Given any Riemannian manifold Ω, we can associate to it the (ultrastatic) spacetime $\mathbf{R} \times \Omega$ equipped with the line element $ds^2 = -dt^2 + d\ell^2$; this takes the form (4.28) using the coordinates $(t, q^i) \equiv (q^0, q^i) \equiv (q^\mu)_{\mu=0,...,d}$.

As a final variation, we can assume the space domain Ω to be an open subset of a Riemannian manifold and prescribe boundary conditions on $\partial\Omega$.

Many results of Chapters 1, 2 and 3 can be readily adapted to the case where Ω is a subset of $\mathbf{R}^d$ with curvilinear coordinates, or a flat Riemannian manifold, or an open subset of it with boundary conditions. An essential point in making these adaptations is to remember that, when an arbitrary coordinate system is employed, the second order derivatives of scalar functions must be intended in a covariant sense and the computational rules (4.32) must be applied.

The extension of our formalism to the case where Ω is a curved Riemannian manifold (or an open subset of it) is a bit more engaging: in particular, the stress energy tensor of the scalar field contains an additional term of the form [118]

$$-\xi \left(R_{\mu\nu} - \frac{1}{2} g_{\mu\nu} R \right) \widehat{\phi}^2 \qquad (4.33)$$

where $g_{\mu\nu}, R_{\mu\nu}, R$ are the metric tensor, the Ricci tensor and the scalar curvature of the spacetime $\mathbf{R} \times \Omega$ with line element $ds^2 = -dt^2 + d\ell^2$; of course, this must be regularized replacing $\widehat{\phi}$ with $\widehat{\phi}^u$.

PART 2
Applications

Chapter 5

A massless field on the segment

Part 2 of this book shows in action the mechanical procedures for analytic continuation presented in the preceding Part 1. In the present Chapter 5 we start from a very simple case, namely, a one-dimensional model with a massless scalar field living on a segment, with no background potential. The formal description of this setting is presented in Section 5.1 (see, in particular, Eq. (5.1)); in the subsequent Sections 5.2–5.5, we describe the general methods that can be used to compute the renormalized VEVs of the stress-energy tensor, of the total energy and of the force on the end-points. Sections 5.6–5.9 take into account several types of boundary conditions, deriving for each one of them explicit expressions for the above mentioned VEVs. The final Section 5.10 gives some details about the comparison with some previous works which are related to the setting under analysis here.

Let us anticipate some comparison with the previous literature about the Casimir effect for a scalar field on a segment. Concerning global aspects, we wish to mention the book of Bordag *et al.* [22] (see Chapter 2 therein) and the work by Fulling *et al.* [71]; these authors derive the total bulk energy, for several boundary conditions, using regularization methods different from the zeta approach. More precisely, [22] uses an exponential cut-off type regularization followed by Abel–Plana resummation, while [71] employs essentially a point-splitting procedure. These authors also obtain the force acting on the end-points of the segment by differentiating the expression for the total energy with respect to the length of the segment (see the comments at the beginning of Section 3.3 of this book). As for local computations, some important references are a classical paper by Mamaev and Trunov [97], and two more recent ones by Romeo and Saharian [128], and by Mera and Fulling [103]. The work [97] derives the stress-energy VEV for a massless scalar field on a segment in the case of periodic boundary conditions, via point-splitting regularization. The same technique is used in [128] to compute the renormalized VEV of the stress-energy and of the total energy in the case of a massless scalar field confined between two parallel planes with Robin boundary conditions; of course, when specialized to the case with spatial dimension $d = 1$, this setting coincides with that of a segment, considered here. Reference [103] deals with a *massive* field with Dirichlet or Neumann boundary conditions, regularized via an exponential cutoff, and computes the stress-energy VEV by the method of images, i.e. as a sum over infinitely many optical paths.[a] The authors of [103] consider the zero cutoff limit, whose treatment can be understood as a renormalization and is connected with the Pauli–Villars regularization; they also present some results on the zero-mass limit.

In all cases where the present chapter has an intersection with [22, 71, 97, 103, 128], the results that we present here are in agreement with the cited references (see, in particular, Section 5.10 for a detailed comparison with [97, 103, 128]).

5.1   Introducing the problem for arbitrary boundary conditions

Throughout this chapter we consider the following setting:

$$d = 1 \,, \qquad \Omega = (0, a) \quad (a > 0) \,,$$
$$\mathcal{A} = -\partial_{x^1 x^1} \quad (V = 0) \tag{5.1}$$

---

[a]Even though different from zeta regularization, this approach is related to the cylinder kernel; in particular, the cutoff can be identified with the "time" variable t of this kernel.

where $x^1 \in (0, a)$ is a standard Cartesian coordinate.[b] The field is assumed to fulfill Dirichlet, Neumann or periodic boundary conditions at the boundary $\partial\Omega = \{0\} \cup \{a\}$; we will deal with each one of these alternatives separately, in the subsequent Sections 5.2–5.9. In passing, we note that the setting described above is the $d = 1$ case both for the configuration with two parallel hyperplanes and for the d-dimensional box, to be considered in the subsequent Chapters 6 and 10, respectively.

5.2   The cylinder and Dirichlet kernels

For any one of the previously mentioned boundary conditions, we proceed in the manner explained hereafter. First of all, we determine explicitly the cylinder kernel $T(\mathfrak{t}\,; x^1, y^1)$ associated to the fundamental operator $\mathcal{A}$; to this purpose we consider a complete orthonormal set of eigenfunctions $(F_k)_{k \in \mathcal{K}}$ for $\mathcal{A}$ with eigenvalues $(\omega_k^2)_{k \in \mathcal{K}}$. The label set $\mathcal{K}$ is countable and $\int_{\mathcal{K}} dk$ means $\sum_{k \in \mathcal{K}}$; so, the eigenfunction expansion (2.45) for the cylinder kernel reads

$$T(\mathfrak{t}\,; x^1, y^1) = \sum_{k \in \mathcal{K}} e^{-\omega_k \mathfrak{t}} F_k(x^1) \overline{F_k}(y^1) \,. \tag{5.2}$$

Once the cylinder kernel has been computed explicitly by evaluating the above sum, we can proceed to determine the modified cylinder kernel $\tilde{T}(\mathfrak{t}\,; x^1, y^1)$ as the primitive of $-T(\mathfrak{t}\,; x^1, y^1)$ vanishing for $\mathfrak{t} \to +\infty$ (see Eq. (2.55)).

In the next sections T and $\tilde{T}$ will be computed explicitly, for several types of boundary conditions. In all cases, the cylinder kernel T and the *spatial derivatives* of the modified cylinder kernel $\tilde{T}$ will be found to have meromorphic extensions in the $\mathfrak{t}$ variable to an open complex neighborhood of $[0, +\infty)$, with possible poles only at $\mathfrak{t} = 0$, vanishing exponentially for $\Re \mathfrak{t} \to +\infty$. Due to these facts, the framework of Subsection 2.8.3 can be applied straightforwardly.

---

[b]We could have used, in place of x^1, the rescaled spatial variable

$$x_\star^1 := x^1/a \in (0, 1) \,,$$

in terms of which, we would have obtained simpler expressions for the results reported in sequel. Yet, we choose not to employ this rescaled coordinate in order to make the comparison with known results more straightforward.

To be more precise: for $y^1 \neq x^1$, $T, \tilde{T}$ and their derivatives have analytic extensions in t to a neighborhood of $[0, +\infty)$. When evaluated on the diagonal $y^1 = x^1$, the modified cylinder kernel $\tilde{T}$ is found to have a logarithmic singularity in t = 0, while the cylinder kernel T and the spatial derivatives of both T and $\tilde{T}$ are meromorphic in a neighborhood of $[0, +\infty)$ with only a pole singularity in t = 0. Because of this, one can resort to Eqs. (2.114), (2.116) to obtain the analytic continuation of the Dirichlet kernel and of its derivatives, required in order to determine the regularized VEV of the stress-energy tensor; explicitly, we have

$$D_{\frac{u-1}{2}}(x^1, y^1)\Big|_{y^1 = x^1} = \frac{e^{-i\pi(u-1)}\,\Gamma(2-u)}{2\pi i}\int_{\mathfrak{H}} dt\; t^{u-2}\, T(t; x^1, y^1)\Big|_{y^1 = x^1}\,,$$

$$(5.3)$$

$$\partial_{zw} D_{\frac{u+1}{2}}(x^1, y^1)\Big|_{y^1 = x^1}$$

$$= -\frac{e^{-i\pi(u+1)}\,\Gamma(1-u)}{2\pi i}\int_{\mathfrak{H}} dt\; t^{u-1}\, \partial_{zw}\tilde{T}(t; x^1, y^1)\Big|_{y^1 = x^1} \qquad \text{for } z, w \in \{x^1, y^1\}\,.$$

$$(5.4)$$

In order to obtain the analytic continuations at u = 0 of the above functions, one can simply set u = 0 in the expressions on the right-hand sides of Eqs. (5.3), (5.4) and explicitly evaluate the remaining integrals along the Hankel contour via the residue theorem.[c]   As stated in Subsection 2.8.3, this gives

$$D_{-\frac{1}{2}}(x^1, y^1)\Big|_{y^1 = x^1} = -\,\mathrm{Res}\Big(t^{-2}\, T(t; x^1, y^1)\Big|_{y^1 = x^1}; 0\Big)\,, \qquad (5.5)$$

$$\partial_{zw} D_{\frac{1}{2}}(x^1, y^1)\Big|_{y^1 = x^1}$$

$$= \mathrm{Res}\Big(t^{-1}\partial_{zw}\tilde{T}(t; x^1, y^1)\Big|_{y^1 = x^1}; 0\Big) \qquad \text{for } z, w \in \{x^1, y^1\} \quad (5.6)$$

---

[c]Notice that the analogue of Eq. (5.3) in terms of $\tilde{T}$ is not so simple and straightforward to employ, due to the logarithmic behavior of the modified cylinder kernel near t = 0; on the other hand, the analogue of (5.4) in terms of T has a singularity in the gamma function for u = 0. Thus, for the computations in which we are interested, there is no better strategy than using Eq. (5.3) with T and Eq. (5.4) with $\tilde{T}$; this also explains why, in the sequel, we will frequently refer to both kernels.

(use Eq. (2.115) and the analogue of Eq. (2.117) for the derivatives of D_s).

Before moving on, let us mention that analogous considerations can be made concerning the traces $\operatorname{Tr}\mathcal{A}^{-s}$, $T(\mathfrak{t}) = \operatorname{Tr} e^{-\mathfrak{t}\sqrt{\mathcal{A}}}$ (see Eq. (2.19) and Eqs. (2.50), (2.51), respectively). Indeed, we can compute the cylinder trace according to Eq. (2.52), which in the present case reads

$$T(\mathfrak{t}) = \sum_{k\in\mathcal{K}} e^{-\omega_k \mathfrak{t}} \ . \tag{5.7}$$

By explicit evaluation of the above sum for the boundary conditions to be considered in the sequel, it becomes apparent that $T(\mathfrak{t})$ possesses the same features of its local counterpart. Thus, we can resort again to the general framework of Subsection 2.8.3 to obtain the analytic continuation of $\operatorname{Tr}\mathcal{A}^{-s}$; in particular, due to Eq. (2.118), the continuation at $s = -1/2$ is given by

$$\operatorname{Tr}\mathcal{A}^{1/2} = -\operatorname{Res}\left(\mathfrak{t}^{-2}\,T(\mathfrak{t})\,;0\right) \ . \tag{5.8}$$

5.3 The stress-energy tensor

We can now determine explicitly the renormalized VEV of the stress-energy tensor; in fact, since no singularity arises, Eqs. (2.36)–(2.38) and (2.39) imply

$$\langle 0|\widehat{T}_{00}(x^1)|0\rangle_{\text{ren}} = \left[\left(\frac{1}{4}+\xi\right)D_{-\frac{1}{2}}(x^1,y^1)\right.$$
$$\left.+\left(\frac{1}{4}-\xi\right)\partial_{x^1y^1}D_{\frac{1}{2}}(x^1,y^1)\right]_{y^1=x^1}, \tag{5.9}$$

$$\langle 0|\widehat{T}_{01}(x^1)|0\rangle_{\text{ren}} = \langle 0|\widehat{T}_{10}(x^1)|0\rangle_{\text{ren}} = 0 \ , \tag{5.10}$$

$$\langle 0|\widehat{T}_{11}(x^1)|0\rangle_{\text{ren}} = \left[\left(\frac{1}{4}-\xi\right)D_{-\frac{1}{2}}(x^1,y^1)+\frac{1}{4}\,\partial_{x^1y^1}D_{\frac{1}{2}}(x^1,y^1)\right.$$
$$\left.-\xi\,\partial_{x^1x^1}D_{\frac{1}{2}}(x^1,y^1)\right]_{y^1=x^1}. \tag{5.11}$$

In the following, the scheme outlined above will be illustrated in detail, as an example, for the case of Dirichlet boundary conditions. For the other boundary conditions we will be more synthetic but, in any case, we will always report the expressions for $T(\mathsf{t};\mathbf{x},\mathbf{y})$, $\widetilde{T}(\mathsf{t};\mathbf{x},\mathbf{y})$ and $\langle 0|\widehat{T}_{\mu\nu}|0\rangle_{\mathrm{ren}}$ ($\mu,\nu \in \{0,1\}$).

Making reference to the considerations of Subsection 1.4.2, we will always present the final result for the renormalized stress-energy VEV in the form (1.32); to this purpose, let us remark that in the one-dimensional case under analysis here Eq. (1.28) gives

$$\xi_1 = 0 \ . \tag{5.12}$$

5.4   The total energy

Since no singularity appears, we can use the general prescription (3.11) to express the renormalized bulk energy E^{ren} in terms of the trace $\mathrm{Tr}\,\mathcal{A}^{1/2}$ (to be understood as the analytic continuation of the regularized trace $\mathrm{Tr}\,\mathcal{A}^{\frac{1-u}{2}}$ at $u = 0$); the said prescription, along with Eq. (5.8), allows to derive an explicit expression for the bulk energy, for any one of the boundary conditions to be considered in the following. More precisely, we have

$$E^{\mathrm{ren}} = -\frac{1}{2}\,\mathrm{Res}\!\left(\mathsf{t}^{-2}\,T(\mathsf{t})\,;0\right) \tag{5.13}$$

where $T(\mathsf{t})$ is the cylinder trace determined according to Eq. (5.7).

In passing, let us remark that the renormalized boundary energy B^{ren} always vanishes identically in all the cases we consider, due to the prescribed boundary conditions; indeed, the same statement can be made for the regularized version B^u (see Eq. (3.9)).

We also mention the following fact: by direct comparison of the results reported in Sections 5.2–5.9 it appears that the results derived using Eq. (5.13) could as well be deduced integrating over $(0,a)$ the conformal part of the renormalized energy density $\langle 0|\widehat{T}_{00}|0\rangle_{\mathrm{ren}}$. On the contrary, the non-conformal part of the latter appears to diverge in a non-integrable manner near the end-points $x = 0$ and $x = a$. This is one of the anomalies related to the boundary behavior of the renormalized stress-energy VEV, a general discussion of which was given in Section 3.5.

5.5 The boundary forces

Let us remark that, since the boundary is zero-dimensional, the nominal "pressure" on the boundary points $x^1 = 0$, $x^1 = a$ does in fact coincide with the force on these points; because of this, we adopt the notation

$$F_{\text{ren}}(x^1) \equiv p^{\text{ren}}(x^1) \qquad \text{for } x^1 \in \{0, a\} \,. \tag{5.14}$$

For all the (non periodic) boundary conditions to be analyzed in the following sections, there are in principle two definitions of the renormalized boundary forces; these descend from the two alternatives pointed out in the general discussion on pressure of Section 3.2.

Let us indicate with $n^1(x^1)$ the outer unit normal at the points on the boundary, so that $n^1(0) = -1$ and $n^1(a) = 1$.

The first definition reads (see Eq. (3.22) and notice that the prescription of taking the regular part is superfluous, since no singularity arises)

$$F_{\text{ren}}(x^1) := \langle 0 | \widehat{T}^u_{11}(x^1) | 0 \rangle \Big|_{u=0} n^1(x^1) \qquad (x^1 \in \{0, a\}) \,; \tag{5.15}$$

namely, we first compute the regularized stress-energy tensor at the boundary point $x^1 \in \{0, a\}$ and then we analytically continue at $u = 0$.

The second alternative is to define (see Eq. (3.23))

$$F_{\text{ren}}(x^1) := \left(\lim_{x'^1 \in (0,a),\, x'^1 \to x^1} \langle 0 | \widehat{T}_{11}(x'^1) | 0 \rangle_{\text{ren}} \right) n^1(x^1) \,, \tag{5.16}$$

i.e. we first renormalize at inner points of the interval $(0, a)$ and then move towards the boundary.

For all boundary conditions considered in the next sections, the equivalence between (5.15) and (5.16) will be checked by direct computation.

5.6 Dirichlet boundary conditions

As a first example, let us consider the case where the field fulfills Dirichlet conditions at both the end points of the segment $(0, a)$, that is

$$\widehat{\phi}(t, x^1) = 0 \qquad \text{for any } t \in \mathbf{R}, \text{ and } x^1 = 0 \text{ or } x^1 = a \,. \tag{5.17}$$

A complete orthonormal set of eigenfunctions $(F_k)_{k\in\mathcal{K}}$ for $\mathcal{A}$ and the related eigenvalues $(\omega_k^2)_{k\in\mathcal{K}}$ are given by

$$F_k(x^1) := \sqrt{\frac{2}{a}}\,\sin(k\,x^1)\,,$$

$$\omega_k^2 := k^2 \quad \text{for } k \in \mathcal{K} \equiv \left\{ \frac{n\pi}{a} \,\middle|\, n = 1,2,3,... \right\}. \tag{5.18}$$

The expansion (5.2) for the cylinder kernel associated to $\mathcal{A}$ reads

$$T(\mathfrak{t};x^1,y^1) = \frac{2}{a} \sum_{n=1}^{+\infty} e^{-\frac{n\pi}{a}\mathfrak{t}} \sin\left(\frac{n\pi}{a}x^1\right)\sin\left(\frac{n\pi}{a}y^1\right). \tag{5.19}$$

Re-writing the trigonometric functions in terms of complex exponentials, the right-hand side of the above equation reduces to a sum of four geometric series, which can be explicitly evaluated; the final result is

$$T(\mathfrak{t};x^1,y^1) = \frac{1}{2a}\left[\frac{\cos(\frac{\pi}{a}(x^1-y^1)) - e^{-\frac{\pi}{a}\mathfrak{t}}}{\cosh(\frac{\pi}{a}\mathfrak{t}) - \cos(\frac{\pi}{a}(x^1-y^1))}\right.$$

$$\left. - \frac{\cos(\frac{\pi}{a}(x^1+y^1)) - e^{-\frac{\pi}{a}\mathfrak{t}}}{\cosh(\frac{\pi}{a}\mathfrak{t}) - \cos(\frac{\pi}{a}(x^1+y^1))}\right]. \tag{5.20}$$

The same expression is also reported, e.g. in [68, 71], but therein it is not used to compute the full, renormalized stress-energy VEV.

Expressing the hyperbolic functions appearing in Eq. (5.20) in terms of exponentials, we can easily obtain the primitive of T which vanishes exponentially for $\mathfrak{t} \to +\infty$, that is $-\tilde{T}$; explicitly, we have

$$\tilde{T}(\mathfrak{t};x^1,y^1) = -\frac{1}{2\pi}\left[\ln\left(1 - 2e^{-\frac{\pi}{a}\mathfrak{t}}\cos\left(\frac{\pi}{a}(x^1-y^1)\right) + e^{-\frac{2\pi}{a}\mathfrak{t}}\right)\right.$$

$$\left. - \ln\left(1 - 2e^{-\frac{\pi}{a}\mathfrak{t}}\cos\left(\frac{\pi}{a}(x^1+y^1)\right) + e^{-\frac{2\pi}{a}\mathfrak{t}}\right)\right]. \tag{5.21}$$

Both T and the space derivatives of $\tilde{T}$ have meromorphic extensions in $\mathfrak{t}$ to a complex neighborhood of $[0,+\infty)$, with poles only at $\mathfrak{t} = 0$; so, we can employ Eqs. (5.5), (5.6) to obtain the renormalized

Dirichlet kernel and its spatial derivatives. For example, noting that

$$T(t;x^1,y^1)\Big|_{y^1=x^1} = \frac{1}{\pi t} - \frac{\pi(3-\sin^2(\frac{\pi}{a}x^1))}{12a^2\sin^2(\frac{\pi}{a}x^1)}\,t$$

$$+ \frac{\pi^3(15(2+\cos(\frac{2\pi}{a}x^1)) - \sin^4(\frac{\pi}{a}x^1))}{720a^4\sin^4(\frac{\pi}{a}x^1)}\,t^3$$

$$+ O(t^5)\,, \tag{5.22}$$

for $t \to 0$ and evaluating explicitly the residue in Eq. (5.5), we infer

$$D_{-\frac{1}{2}}(x^1,y^1)\Big|_{y^1=x^1} = \frac{\pi}{12a^2}\frac{3-\sin^2(\frac{\pi}{a}x^1)}{\sin^2(\frac{\pi}{a}x^1)}\,. \tag{5.23}$$

Proceeding similarly for the derivatives of the Dirichlet kernel, and then using Eqs. (5.9)–(5.11), one obtains the following expression for the renormalized VEV of the stress-energy tensor:

$$\langle 0|\widehat{T}_{\mu\nu}(x^1)|0\rangle_{\text{ren}}\Big|_{\mu,\nu=0,1} = A\begin{pmatrix}-1 & 0\\ 0 & -1\end{pmatrix} + \xi\,B(x^1)\begin{pmatrix}1 & 0\\ 0 & 0\end{pmatrix},$$
$$\tag{5.24}$$

$$A := \frac{\pi}{24a^2}\,,\quad B(x^1) := \frac{\pi}{2a^2}\frac{1}{\sin^2(\frac{\pi}{a}x^1)}\quad\text{for } x^1 \in (0,a)\,.$$

Let us now discuss the renormalized bulk energy. To this purpose, we first note that the expansion (2.52) for the cylinder trace gives

$$T(t) = \sum_{n=1}^{+\infty} e^{-\frac{n\pi}{a}t} = \frac{1}{e^{\frac{\pi}{a}t}-1}\,; \tag{5.25}$$

then, using prescription (5.13), we readily infer

$$E^{\text{ren}} = -\frac{\pi}{24a}\,. \tag{5.26}$$

In conclusion, let us consider the boundary forces; it is easily seen that both definitions (5.15) and (5.16) give

$$F_{\text{ren}}(0) = \frac{\pi}{24a^2}\,,\quad F_{\text{ren}}(a) = -\frac{\pi}{24a^2}\,. \tag{5.27}$$

5.7   Dirichlet–Neumann boundary conditions

Let us now consider the case where Dirichlet and Neumann boundary conditions are respectively prescribed at the two end points of the segment $(0, a)$: we assume

$$\widehat{\phi}(t, 0) = 0 , \qquad \partial_{x^1}\widehat{\phi}(t, a) = 0 \quad \text{for } t \in \mathbf{R} . \tag{5.28}$$

In this case, a complete orthonormal set of eigenfunctions $(F_k)_{k \in \mathcal{K}}$ for $\mathcal{A}$ and the related eigenvalues $(\omega_k^2)_{k \in \mathcal{K}}$ are

$$F_k(x^1) := \sqrt{\frac{2}{a}} \sin(k\, x^1),$$

$$\omega_k^2 := k^2 \quad \text{for } k \in \mathcal{K} \equiv \left\{ \left(n + \frac{1}{2} \right) \frac{\pi}{a} \,\middle|\, n = 0, 1, 2, ... \right\}. \tag{5.29}$$

Using the expansion (5.2), we can determine the cylinder kernel T and then obtain the modified kernel $\tilde{T}$ as minus the primitive of T, vanishing for $t \to +\infty$; the final results are

$$T(t; x^1, y^1) = \frac{1}{a} \left[\frac{\sinh(\frac{\pi}{2a}\, t)\cos(\frac{\pi}{2a}(x^1 - y^1))}{\cosh(\frac{\pi}{a}\, t) - \cos(\frac{\pi}{a}(x^1 - y^1))} \right.$$

$$\left. - \frac{\sinh(\frac{\pi}{2a}\, t)\cos(\frac{\pi}{2a}(x^1 + y^1))}{\cosh(\frac{\pi}{a}\, t) - \cos(\frac{\pi}{a}(x^1 + y^1))} \right], \tag{5.30}$$

$$\tilde{T}(t; x^1, y^1) = \frac{1}{2\pi} \left[\ln \left(\frac{\cos(\frac{\pi}{2a}(x^1 - y^1)) + \cosh(\frac{\pi}{2a}t)}{\cos(\frac{\pi}{2a}(x^1 - y^1)) - \cosh(\frac{\pi}{2a}t)} \right) \right.$$

$$\left. - \ln \left(\frac{\cos(\frac{\pi}{2a}(x^1 + y^1)) + \cosh(\frac{\pi}{2a}t)}{\cos(\frac{\pi}{2a}(x^1 + y^1)) - \cosh(\frac{\pi}{2a}t)} \right) \right]. \tag{5.31}$$

Using the above expressions along with Eqs. (5.5), (5.6) and (5.9)–(5.11), one obtains the renormalized VEV of the stress-energy tensor:

$$\langle 0|\widehat{T}_{\mu\nu}(x^1)|0\rangle_{\text{ren}}\Big|_{\mu,\nu=0,1} = A \begin{pmatrix} 1 & 0 \\ 0 & 1 \end{pmatrix} + \xi\, B(x^1) \begin{pmatrix} 1 & 0 \\ 0 & 0 \end{pmatrix}, \tag{5.32}$$

$$A := \frac{\pi}{48a^2} , \qquad B(x^1) := \frac{\pi}{2a^2} \frac{\cos(\frac{\pi}{a}\, x^1)}{\sin^2(\frac{\pi}{a}\, x^1)} \quad \text{for } x^1 \in (0, a) .$$

Next, we derive the cylinder trace using again the expansion (2.52):

$$T(t) = \sum_{n=0}^{+\infty} e^{-(n+\frac{1}{2})\frac{\pi}{a}t} = \frac{e^{\frac{\pi}{2a}t}}{e^{\frac{\pi}{a}t} - 1} . \tag{5.33}$$

Now prescription (5.13) allows us to obtain the renormalized bulk energy:

$$E^{\mathrm{ren}} = \frac{\pi}{48a} . \tag{5.34}$$

Concerning the boundary forces, also in this case definitions (5.15), (5.16) agree and give

$$F_{\mathrm{ren}}(0) = -\frac{\pi}{48a^2} , \quad F_{\mathrm{ren}}(a) = \frac{\pi}{48a^2} ; \tag{5.35}$$

we remark, in particular, that the above expressions have the opposite sign with respect to the ones of Eq. (5.27), corresponding the case of Dirichlet boundary conditions.

5.8 Neumann boundary conditions

We are now going to study the case where

$$\partial_{x^1}\widehat{\phi}(t, x^1) = 0 \quad \text{for } t \in \mathbf{R}, \text{ and } x^1 = 0 \text{ or } x^1 = a . \tag{5.36}$$

In this case, according to the considerations of Section 4.1, the Hilbert space $L^2(0, a)$ has to be replaced with the space $L_0^2(0, a)$ of square integrable functions on $(0, a)$ with mean zero (see Eq. (4.2)); in this space, a complete orthonormal set of eigenfunctions $(F_k)_{k \in \mathcal{K}}$ for $\mathcal{A} = -\partial_{x^1 x^1}$ and the corresponding eigenvalues $(\omega_k^2)_{k \in \mathcal{K}}$ are given by

$$F_k(x^1) := \sqrt{\frac{2}{a}} \cos(k \, x^1) ,$$

$$\omega_k^2 := k^2 \quad \text{for } k \in \mathcal{K} \equiv \left\{ \frac{n\pi}{a} \ \middle| \ n = 1, 2, 3, ... \right\}. \tag{5.37}$$

The cylinder kernel associated to $\mathcal{A}$ can be evaluated according to Eq. (5.2) to obtain

$$T(t ; x^1, y^1) = \frac{1}{2a} \left[\frac{\cos(\frac{\pi}{a}(x^1 - y^1)) - e^{-t}}{\cosh t - \cos(\frac{\pi}{a}(x^1 - y^1))} + \frac{\cos(\frac{\pi}{a}(x^1 + y^1)) - e^{-t}}{\cosh t - \cos(\frac{\pi}{a}(x^1 + y^1))} \right] ; \tag{5.38}$$

on the other hand, computing the modified cylinder kernel as minus the primitive of T, we obtain

$$\tilde{T}(t; x^1, y^1) = -\frac{1}{2\pi} \left[\ln\left(1 - 2e^{-\frac{\pi}{a}t} \cos\left(\frac{\pi}{a}(x^1 - y^1)\right) + e^{-\frac{2\pi}{a}t} \right) \right.$$

$$\left. + \ln\left(1 - 2e^{-\frac{\pi}{a}t} \cos\left(\frac{\pi}{a}(x^1 + y^1)\right) + e^{-\frac{2\pi}{a}t} \right) \right]. \quad (5.39)$$

Resorting once more to Eqs. (5.5), (5.6) and (5.9)–(5.11), the renormalized VEV of the stress-energy tensor is found to be

$$\langle 0|\widehat{T}_{\mu\nu}(x^1)|0\rangle_{\text{ren}}\Big|_{\mu,\nu=0,1} = A \begin{pmatrix} -1 & 0 \\ 0 & -1 \end{pmatrix} - \xi\, B(x^1) \begin{pmatrix} 1 & 0 \\ 0 & 0 \end{pmatrix}, \quad (5.40)$$

where A and $B(x^1)$ are defined as in Eq. (5.24), corresponding to the case of Dirichlet boundary conditions. In the case under analysis the spectrum of A coincides with the one obtained for Dirichlet boundary conditions (compare Eqs. (5.18), (5.37)); therefore, the cylinder trace is again given by Eq. (5.25), and we derive the same renormalized bulk energy as in Eq. (5.26):

$$E^{\text{ren}} = -\frac{\pi}{24a} \,.$$

Finally, both definitions (5.15), (5.16) give for the boundary forces the same results as in the case of Dirichlet boundary conditions (see Eq. (5.27)):

$$F_{\text{ren}}(0) = \frac{\pi}{24a^2} \,, \qquad F_{\text{ren}}(a) = -\frac{\pi}{24a^2} \,.$$

5.9   Periodic boundary conditions

The last case we consider for the segment configuration is the one where the field fulfills periodic boundary conditions:

$$\widehat{\phi}(t, 0) = \widehat{\phi}(t, a)\,, \qquad \partial_{x^1}\widehat{\phi}(t, 0) = \partial_{x^1}\widehat{\phi}(t, a) \qquad \text{for } t \in \mathbf{R}\,. \quad (5.41)$$

As explained in Section 4.3, this case would be more properly formulated in terms of a free scalar field on the one-dimensional torus $\mathbf{T}_a^1 := \mathbf{R}/(a\mathbf{Z})$. Besides, similarly to the case of Neumann boundary conditions, the basic Hilbert space that we consider is

$L_0^2(\mathbf{T}_a^1) = \{f \in L^2(\mathbf{T}_a^1) \mid \int_0^a dx^1 f(x^1) = 0\}$ (see Section 4.1 and footnote d in Chapter 4). In this space a complete orthonormal set of eigenfunctions $(F_k)_{k \in \mathcal{K}}$ for $\mathcal{A}$, with the corresponding eigenvalues $(\omega_k^2)_{k \in \mathcal{K}}$, is

$$F_k(x^1) := \sqrt{\frac{1}{a}}\, e^{ik\,x^1},$$

$$\omega_k^2 := k^2 \quad \text{for } k \in \mathcal{K} \equiv \left\{\pm \frac{2n\pi}{a} \;\middle|\; n = 1,2,3,...\right\}. \tag{5.42}$$

Let us pass to determine the cylinder and modified cylinder kernel associated to $\mathcal{A}$; using the same methods of the previous sections, we obtain

$$T(\mathfrak{t};x^1,y^1) = \frac{\cos(\frac{2\pi}{a}(x^1-y^1)) - e^{-\frac{2\pi}{a}\mathfrak{t}}}{a\left[\cosh(\frac{2\pi}{a}\mathfrak{t}) - \cos(\frac{2\pi}{a}(x^1-y^1))\right]}, \tag{5.43}$$

$$\tilde{T}(\mathfrak{t};x^1,y^1) = -\frac{1}{2\pi}\ln\left(1 - 2\,e^{-\frac{2\pi}{a}\mathfrak{t}}\cos\left(\frac{2\pi}{a}(x^1-y^1)\right) + e^{-\frac{4\pi}{a}\mathfrak{t}}\right) \tag{5.44}$$

(the same expression for T is also reported, e.g. in [68], again for other purposes).

Equations (5.5), (5.6) and (5.9)–(5.11) yield the following expression for the renormalized VEV of the stress-energy tensor:

$$\langle 0|\widehat{T}_{\mu\nu}(x^1)|0\rangle_{\text{ren}}\Big|_{\mu,\nu=0,1} = A\begin{pmatrix} -1 & 0 \\ 0 & -1 \end{pmatrix}, \qquad A := \frac{\pi}{6a^2}. \tag{5.45}$$

Let us stress that the above result respects the invariance under translations $x^1 \mapsto x^1 + \alpha$ (for any $\alpha \in \mathbf{R}$) of the given configuration, since it does not depend explicitly on the spatial coordinate x^1.

To conclude, we discuss the renormalized bulk energy. We first note that expansion (2.52) for the cylinder trace yields, in the present case,

$$T(\mathfrak{t}) = \left(\sum_{n=-\infty}^{-1} + \sum_{n=1}^{+\infty}\right)e^{-\frac{2|n|\pi}{a}\mathfrak{t}} = 2\sum_{n=1}^{+\infty}e^{-\frac{2n\pi}{a}\mathfrak{t}} = \frac{2}{e^{\frac{2\pi}{a}\mathfrak{t}} - 1}; \tag{5.46}$$

then, using once more the prescription (5.13), we obtain for the renormalized bulk energy

$$E^{\text{ren}} = -\frac{\pi}{6a}. \tag{5.47}$$

5.10   Comparison with the previous literature

As mentioned at the beginning of this chapter, the expressions for the renormalized stress-energy VEV presented in the preceding sections for several types of boundary conditions agree with the analogous objects computed in previous works by Mamaev and Trunov [97], Romeo and Saharian [128], and by Mera and Fulling [103], in all the cases where there is intersection. In this section we give more information on the comparison with the cited works.

Mamaev and Trunov [97] consider the case of a scalar field on a segment with periodic boundary conditions (i.e. on the torus $\mathbf{T}_a^1$). They use point-splitting regularization to compute the stress-energy VEV (see Eq. (5) on page 767 of [97]); in the renormalization limit, after subtraction of the divergent terms, their result appears to be equivalent to ours (see Eq. (5.45) of this book).

The comparison with [128] and [103] is less straightforward. As an example, we consider the $(0,0)$ component of the stress-energy VEV (i.e. the energy density) in the case of Neumann boundary conditions; let us recall that our result for this object reads (see Eq. (5.40), along with Eq. (5.24))

$$\langle 0|\widehat{T}_{00}(x^1)|0\rangle_{\mathrm{ren}} = -\frac{\pi}{24a^2} - \xi\,\frac{\pi}{2a^2}\,\frac{1}{\sin^2(\frac{\pi}{a}x^1)} \qquad \text{for all } x^1 \in (0,a)\ .$$

$$(5.48)$$

This should be kept in mind in the sequel.

Comparison with [128].   Romeo and Saharian consider the general setting of a massless scalar field in arbitrary spatial dimension d, fulfilling Robin boundary conditions on two parallel hyperplanes. The case of a field on a segment $(0,a)$ that we are considering in the present chapter corresponds to the choice $d = 1$; in this case, the boundary conditions considered in [128] read

$$(1 + \beta_1\partial_{x^1})\widehat{\phi}\Big|_{x^1=0} = (1 - \beta_2\partial_{x^1})\widehat{\phi}\Big|_{x^1=a} = 0 \qquad (5.49)$$

for some $\beta_1, \beta_2 \in \mathbf{R}$. Under these assumptions, the renormalized energy density VEV is determined by means of point-splitting; the explicit expression for this object is given by the sum of three terms, i.e. (see Eq. (4.22) on page 1309 of [128])

$$\langle 0|\widehat{T}_{00}(x^1)|0\rangle_{\text{ren}} = q(x^1;0) + q(x^1;a) + \triangle q(x^1) , \tag{5.50}$$

where (see Eqs. (3.21), (3.22) and (4.23) in [128])

$$q(x^1;0) = \frac{\xi}{2\pi(x^1)^2}\left[1 + \sum_{j=2}^{+\infty}\frac{j!}{y^{j-1}}\right]_{y=2x^1/\beta_1}, \tag{5.51}$$

$$q(x^1;a) = \frac{\xi}{2\pi(a-x^1)^2}\left[1 + \sum_{j=2}^{+\infty}\frac{j!}{y^{j-1}}\right]_{y=2(a-x^1)/\beta_2}, \tag{5.52}$$

$$\triangle q(x^1) = -\frac{1}{\pi a^2}\,\text{p.v.}\int_0^{+\infty}\frac{t\,dt}{\frac{(\frac{\beta_1}{a}t-1)(\frac{\beta_2}{a}t-1)}{(\frac{\beta_1}{a}t+1)(\frac{\beta_2}{a}t+1)}e^{2t}-1}$$

$$\cdot\left[1 + 2\xi\left(\frac{\frac{\beta_1}{a}t+1}{\frac{\beta_1}{a}t-1}e^{-\frac{2tx^1}{a}} + \frac{\frac{\beta_2}{a}t+1}{\frac{\beta_2}{a}t-1}e^{-\frac{2t(a-x^1)}{a}}\right)\right]. \tag{5.53}$$

The case of Neumann boundary conditions corresponds formally to the limit $\beta_1 = \beta_2 \to +\infty$; in this limit, it is shown in [128] that

$$q(x^1;0) \to -\frac{\xi}{2\pi(x^1)^2} , \qquad q(x^1;a) \to -\frac{\xi}{2\pi(a-x^1)^2} . \tag{5.54}$$

On the other hand, one has (see [117], page 140, Eq. (5.9.12))

$$\triangle q(x^1) \to -\frac{1}{\pi a^2}\int_0^{+\infty}\frac{t\,dt}{e^{2t}-1}\left[1 + 2\xi\left(e^{-\frac{2tx^1}{a}} + e^{-\frac{2t(a-x^1)}{a}}\right)\right]$$

$$= -\frac{\pi}{24a^2} - \frac{\xi}{2\pi a^2}\left[\psi^{(1)}\left(1+\frac{x^1}{a}\right) + \psi^{(1)}\left(1+\frac{a-x^1}{a}\right)\right], \tag{5.55}$$

where $\psi^{(1)}(z)$ is the polygamma function of first order (i.e. the first derivative of the digamma function $\psi(z) := \Gamma'(z)/\Gamma(z)$, where Γ is the Euler gamma function). Summing up, Eqs. (5.50)–(5.55) allow to infer that[d]

$$\lim_{\beta_1=\beta_2\to+\infty} \langle 0|\widehat{T}_{00}(x^1)|0\rangle_{\text{ren}}$$

$$= -\frac{\pi}{24a^2} - \frac{\xi}{2\pi a^2}\left[\psi^{(1)}\left(\frac{x^1}{a}\right) + \psi^{(1)}\left(1 - \frac{x^1}{a}\right)\right]. \quad (5.57)$$

The expression in the right-hand side above is found to coincide with the one in Eq. (5.48) of this book, using the identity (see [117], page 144, Eq. (5.15.6))

$$\psi^{(1)}\left(\frac{x^1}{a}\right) + \psi^{(1)}\left(1 - \frac{x^1}{a}\right) = \left[-\pi\frac{d}{dz}\cot(\pi z)\right]_{z=x^1/a} = \frac{\pi^2}{\sin^2(\pi\frac{x^1}{a})}.$$

$$(5.58)$$

Comparison with [103].   This paper deals with the case of a massive field on a segment with either Dirichlet or Neumann boundary conditions and it uses an exponential cut-off regularization. The stress-energy VEV is expressed as the sum of a term diverging quadratically in the cutoff, and of other terms which remain finite when the cutoff is removed. If one renormalizes subtracting the divergent term, in the case of Neumann boundary conditions one obtains

---

[d]To prove Eq. (5.57), one should first recall that the polygamma function admits the series representation (see [117], page 144, Eq. (5.15.1))

$$\psi^{(1)}(z) = \sum_{n=0}^{+\infty}\frac{1}{(n+z)^2} \qquad \text{(for all } z > 0\text{)}; \qquad (5.56)$$

keeping in mind this fact, by simple algebraic manipulations one infers, e.g.

$$\frac{1}{(x^1)^2} + \frac{1}{a^2}\psi^{(1)}\left(1 + \frac{x^1}{a}\right) = \frac{1}{a^2}\left[\frac{1}{(x^1/a)^2} + \sum_{n=0}^{+\infty}\frac{1}{(n+1+(x^1/a))^2}\right]$$

$$= \frac{1}{a^2}\sum_{n=0}^{+\infty}\frac{1}{(n+(x^1/a))^2} = \frac{1}{a^2}\psi^{(1)}\left(\frac{x^1}{a}\right),$$

where the second equality follows by a trivial relabeling of the summation index.

from [103] that

$$\langle 0|T_{00}(x^1)|0\rangle_{\text{ren}} = -\frac{m}{2\pi a}\sum_{n=1}^{+\infty}\frac{K_1(2mna)}{n} - \frac{m\xi}{\pi}\sum_{n=-\infty}^{+\infty}\frac{K_1(2m|x^1+na|)}{|x^1+na|}$$

$$-\frac{2m^2}{\pi}\left(\xi - \frac{1}{4}\right)\sum_{n=-\infty}^{+\infty}K_0(2m|x^1+na|) \qquad (5.59)$$

where m is the field mass and K_0, K_1 are modified Bessel functions of the second kind (see Eqs. (10), (15)–(17) and (51) of [103], here re-written with $L = a$, $\beta = \xi - 1/4$ and putting $r = l = 0$ to indicate the choice of Neumann boundary conditions). It is known that $K_0(z) = -\ln z + O(1)$ and $K_1(z) = 1/z + O(z\ln z)$ for $z \to 0^+$. Therefore, one expects that, for $m \to 0^+$,

$$m\sum_{n=1}^{+\infty}\frac{K_1(2mna)}{n} \to \frac{1}{2a}\sum_{n=1}^{+\infty}\frac{1}{n^2} = \frac{\pi^2}{12a} ,$$

$$m^2\sum_{n=-\infty}^{+\infty}K_0(2m|x^1+na|) \to 0 ; \qquad (5.60)$$

these facts are established in [103]. Let us add to these results the following remark:

$$m\sum_{n=-\infty}^{+\infty}\frac{K_1(2m|x^1+na|)}{|x^1+na|}$$

$$\to \frac{1}{2}\sum_{n=-\infty}^{+\infty}\frac{1}{(x^1+na)^2} = \frac{\pi^2}{2a^2\sin^2(\frac{\pi}{a}x^1)} \quad \text{for } m \to 0^+ \qquad (5.61)$$

(for the computation of the last series by contour integral methods see [85], page 268, Eq. (4.9-4)). In view of these facts, there holds

$$\lim_{m\to 0^+}\langle 0|T_{00}(x^1)|0\rangle_{\text{ren}} = -\frac{\pi}{24a^2} - \xi\frac{\pi}{2a^2\sin^2(\frac{\pi}{a}x^1)} , \qquad (5.62)$$

in agreement with Eq. (5.48) of this book. Let us emphasize that, differently from [103], the zeta approach gives the renormalized stress-energy VEV by mere analytic continuation, with no need to remove divergent terms.

Chapter 6

A massless field between parallel hyperplanes

We shall now pass to consider the case of a massless field confined between two parallel (hyper-)planes, for several kinds of boundary conditions; we restrict the attention to the case of odd spatial dimension, for technical reasons to be discussed later on. The formal description of this setting is given in Section 6.1; therein it is pointed out that the analysis of this problem can be reduced to that of a one-dimensional problem, analogous to the one already considered in the preceding Chapter 5. Sections 6.2–6.5 present the general techniques which allow to determine the analytic continuation of the regularized VEVs for several observables. In the conclusive Sections 6.6–6.9 we derive explicit expressions for the renormalized versions of these quantities in the physically interesting case of spatial dimension $d = 3$.

The literature on the configuration with two parallel planes is immense, both regarding local and global aspects; let us only mention some major contributions. In his seminal paper [34], using an exponential cut-off regularization along with Abel–Plana resummation, Casimir was the first to compute the total energy and the boundary forces for the case of two parallel planes. Concerning local aspects, the foremost derivation of the full stress-energy tensor VEV was given by Brown and Maclay [27], using a point-splitting technique.[a] Computation of both global and local quantities for this

---

[a] Actually, both Casimir and Brown–Maclay considered the case of an electromagnetic field; yet the methods employed by these authors can be trivially adapted to the case of a scalar field.

model was later reproposed by several authors, using various regularization techniques: see, e.g. the monographs by Milton [105], Elizalde *et al.* [53, 54], Bordag *et al.* [22] (with the works cited therein) and the papers by Zimerman *et al.* [129], by Esposito *et al.* [55] and by Romeo and Saharian [128]. For $d = 3$ and Dirichlet boundary conditions (a case analysed in Section 6.6), the above configuration is the one most typically considered when dealing with the (scalar) Casimir effect. In particular, an analysis based on zeta regularization of this model was already given in our previous work [58] (see also the references cited therein for different approaches); in that work, the required analytic continuation was performed using ad hoc, known results on the special functions related to this specific configuration (namely, the Riemann zeta and the polylogarithm). The case with Dirichlet boundary conditions on one plane and Neumann conditions on the other (discussed in Section 6.7) was originally considered in the electromagnetic case by Boyer [23], who derived the total energy; later on, computations of the total energy and boundary forces for a (massless or massive) scalar field at both zero and non-zero temperature were performed by Pinto *et al.* [42, 119] and Santos *et al.* [132] (see also [5, 22]). The above mentioned settings with Dirichlet and/or Neumann boundary conditions are also particular (yet, not elementary) limiting cases of the general configuration considered by Romeo and Saharian in [128], where Robin boundary conditions are considered. Finally, let us also mention the monograph by Fulling [67] where the stress-energy VEV for the model with periodic boundary conditions is given; see also [53, 54] and, again, [22, 105] for the derivation of the total energy in the same configuration.

Working in the framework of zeta regularization, in this chapter we present an "automatical" derivation of the required analytic continuations based on the schemes of Part 1; this allows to determine explicit expressions for the renormalized VEVs of several local and global observables.

6.1 Introducing the problem for arbitrary boundary conditions

As mentioned in Chapter 5, the segment configuration can be considered as the $d = 1$ case of a general, d-dimensional configuration with two parallel hyperplanes; this is the subject we are now going to analyze (with no external potential). So, we assume

$$\Omega := (0, a) \times \mathbf{R}^{d-1} \ , \quad a > 0 \ , \quad V = 0 \ ; \qquad (6.1)$$

these choices correspond to a massless scalar field confined between the two parallel hyperplanes[b]

$$\pi_0 = \{\mathbf{x} \in \mathbf{R}^d \mid x^1 = 0\} \ , \quad \pi_a = \{\mathbf{x} \in \mathbf{R}^d \mid x^1 = a\} \ . \qquad (6.2)$$

As we did in Chapter 5 for the segment configuration, we are going to consider separately the cases where the field fulfills Dirichlet, Neumann or periodic boundary conditions on the hyperplanes π_0, π_a. Throughout this chapter we assume

$$d \text{ odd} \ , \quad d \geqslant 3 \ ; \qquad (6.3)$$

this hypothesis is purely technical and will be motivated later (see the comments after Eq. (6.6)).

For any one of the boundary conditions mentioned above, keeping in mind the considerations of Section 4.1, we proceed in the manner explained in the following section. Before moving on, let us remark that, due to the results on slab configurations reported in Section 2.10, we just have to study the reduced one-dimensional problem where

$$\Omega_1 := (0, a) \subset \mathbf{R} \ , \quad \mathcal{A}_1 := -\partial_{x^1 x^1} \ , \qquad (6.4)$$

keeping into account the boundary conditions in $x^1 = 0$ and $x^1 = a$; in consequence of this, we can resort to the results of Chapter 5.

---

[b]Comments analogous to the ones in footnote b on page 81 of this book can be made; namely, in place of the standard Cartesian coordinates $\mathbf{x} \equiv (x^i)_{i=1,\dots,d}$, we could have used the set of rescaled coordinates (best fitting the features of the present configuration)

$$\mathbf{x}_\star \equiv (x^i_\star)_{i=1,\dots,d} \ , \quad \text{with} \quad x^1_\star := x^1/a \ , \quad x^i_\star := x^i \ \text{for} \ i \in \{2, \dots, d\} \ .$$

Also in this case, we choose not to employ the above rescaled coordinate system in order to render comparison with known results easier.

In Sections 6.2–6.5 we will present some general results on the configuration under analysis, holding for all the types of boundary conditions mentioned before. In Sections 6.6–6.9 we will consider specific boundary conditions, with a special attention for the case $d = 3$.

6.2   The reduced Dirichlet and cylinder kernels

According to Eqs. (2.134)–(2.137), the basic ingredients for the analysis of the d-dimensional problem are the Dirichlet kernel $D_s^{(1)}$ for the reduced one-dimensional problem at $s = (u - d)/2$, and its derivatives at $s = (u - d + 2)/2$. On the other hand, these functions can be expressed in terms of the one-dimensional cylinder kernel $T^{(1)}(\mathfrak{t}; x^1, y^1)$, which has been determined in Chapter 5 (where it was indicated simply with $T(\mathfrak{t}; x^1, y^1)$). Let us recall that this kernel and all its derivatives, when evaluated on the diagonal $y^1 = x^1$, are meromorphic functions of $\mathfrak{t}$ in a neighborhood of the positive real semiaxis with a unique pole singularity in $\mathfrak{t} = 0$; besides, they vanish exponentially for $\Re\mathfrak{t} \to +\infty$. For the reduced Dirichlet kernel and for the derivatives in which we are interested, Eq. (2.114) yields the following expressions:

$$D_{\frac{u-d}{2}}^{(1)}(x^1, y^1) = \frac{e^{-i\pi(u-d)}}{2\pi i} \frac{\Gamma(d+1-u)}{} \int_{\mathfrak{H}} d\mathfrak{t}\, \mathfrak{t}^{u-d-1}\, T^{(1)}(\mathfrak{t}; x^1, y^1) \;;$$

$$(6.5)$$

$$\partial_{zw} D_{\frac{u-d+2}{2}}^{(1)}(x^1, y^1)$$

$$= \frac{e^{-i\pi(u-d)}}{2\pi i} \frac{\Gamma(d-1-u)}{} \int_{\mathfrak{H}} d\mathfrak{t}\, \mathfrak{t}^{u-d+1}\, \partial_{zw} T^{(1)}(\mathfrak{t}; x^1, y^1)$$

$$\text{for } z, w \in \{x^1, y^1\} \;. \quad (6.6)$$

Consider the above relations along with Eqs. (2.134)–(2.137), relating the one-dimensional functions to the d-dimensional Dirichlet kernel $D_{\frac{u-1}{2}}(\mathbf{x}, \mathbf{y})$ or to the derivatives of $D_{\frac{u+1}{2}}(\mathbf{x}, \mathbf{y})$; these equations yield the sought-for meromorphic continuations in u to the whole complex plane. By direct inspection of the expressions thus obtained it appears that, with the assumption (6.3) on d, these meromorphic continuations are analytic for u in a neighborhood of the origin; so, the zeta strategy for renormalization is implemented by simply setting $u = 0$. In conclusion, we have the following integral representations for the renormalized Dirichlet kernel and for its renormalized derivatives:

$$D_{-\frac{1}{2}}(\mathbf{x}, \mathbf{y})\Big|_{\mathbf{y}=\mathbf{x}} = -\frac{C_d}{2\pi i} \int_{\mathfrak{H}} \frac{dt}{t^{d+1}}\, T^{(1)}(t; x^1, y^1)\Big|_{y^1=x^1} ; \qquad (6.7)$$

$$\partial_{zw} D_{\frac{1}{2}}(\mathbf{x}, \mathbf{y})\Big|_{\mathbf{y}=\mathbf{x}}$$

$$= -\frac{C_d}{(d-1)2\pi i} \int_{\mathfrak{H}} \frac{dt}{t^{d-1}}\, \partial_{zw} T^{(1)}(t; x^1, y^1)\Big|_{y^1=x^1}$$

$$\text{for } z, w \in \{x^1, y^1\}; \qquad (6.8)$$

$$\partial_{x_2^i y_2^j} D_{\frac{1}{2}}(\mathbf{x}, \mathbf{y})\Big|_{\mathbf{y}=\mathbf{x}} = -\partial_{x_2^i x_2^j} D_{\frac{1}{2}}(\mathbf{x}, \mathbf{y})\Big|_{\mathbf{y}=\mathbf{x}}$$

$$= -\partial_{y_2^i y_2^j} D_{\frac{1}{2}}(\mathbf{x}, \mathbf{y})\Big|_{\mathbf{y}=\mathbf{x}}$$

$$= \delta_{ij} \frac{C_d}{2\pi i} \int_{\mathfrak{H}} \frac{dt}{t^{d+1}} T^{(1)}(t; x^1, y^1)\Big|_{y^1=x^1}$$

$$\text{for } i, j \in \{1, ..., d-1\}; \qquad (6.9)$$

for the sake of brevity, in the above we have set[c]

$$C_d := (-\pi)^{-\frac{d-1}{2}} \Gamma\left(\frac{d+1}{2}\right) . \qquad (6.10)$$

Equations (6.7)–(6.9) are completed with Eq. (2.135), stating the vanishing of certain mixed derivatives.

Finally, recall that Eq. (3.19) (here employed with $d_2 = d - 1$) for the regularized, reduced bulk energy (see the subsequent Section 6.4) requires the evaluation of the trace $\operatorname{Tr} \mathcal{A}_1^{\frac{d-u}{2}}$. To this purpose, we first consider the one-dimensional cylinder trace $T^{(1)}(t)$, which has also been determined in Chapter 5 (where it was indicated simply with

---

[c]To obtain the expression (6.10) for C_d, some identities regarding the gamma function must be used.

$T(\mathfrak{t})$); we recall that, for all the boundary conditions considered in the following applications, the map $\mathfrak{t} \mapsto T^{(1)}(\mathfrak{t})$ admits a meromorphic extension to a neighborhood of $[0, +\infty)$ which only has a pole at $\mathfrak{t} = 0$ and vanishes exponentially for $\Re\mathfrak{t} \to +\infty$.

A discussion similar to the one carried over above for the Dirichlet kernel allows us to derive an explicit expression for the analytic continuation of $\operatorname{Tr} \mathcal{A}_1^{\frac{d-u}{2}}$ at $u = 0$; more precisely, using for the reduced problem a relation analogous to one in Eq. (2.118), we conclude

$$\operatorname{Tr} \mathcal{A}_1^{d/2} = \frac{(-1)^d\, \Gamma(d+1)}{2\pi i} \int_{\mathfrak{H}} d\mathfrak{t}\, \mathfrak{t}^{-(d+1)}\, T^{(1)}(\mathfrak{t}) . \qquad (6.11)$$

6.3   The stress-energy tensor

Substituting the relations (2.135) and (6.7)–(6.9) into Eqs. (2.28)–(2.30), we straightforwardly deduce the contour integral representations for the non-vanishing components of the renormalized stress-energy VEV; moreover, due to the meromorphic nature of the cylinder kernel (and of its derivatives), the resulting integrals along the Hankel contour can be explicitly evaluated via the residue theorem. The final expressions for the (non-zero) components of the renormalized stress-energy VEV are the following ones:

$$\langle 0|\widehat{T}_{00}(\mathbf{x})|0\rangle_{\mathrm{ren}} = -C_d \operatorname{Res}\left(\mathfrak{t}^{-(d+1)}\left[\left(\xi - \frac{d-2}{4d}\right) d\, T^{(1)}(\mathfrak{t}; x^1, y^1)\right.\right.$$

$$\left.\left. + \frac{\mathfrak{t}^2}{d-1}\left(\frac{1}{4} - \xi\right) \partial_{x^1 y^1} T^{(1)}(\mathfrak{t}; x^1, y^1)\right]_{y^1 = x^1} ; 0\right) ;$$

$$(6.12)$$

$$\langle 0|\widehat{T}_{11}(\mathbf{x})|0\rangle_{\mathrm{ren}}$$

$$= -C_d \operatorname{Res}\left(\mathfrak{t}^{-(d+1)}\left[\left(\frac{1}{4} - \xi\right) d\, T^{(1)}(\mathfrak{t}; x^1, y^1)\right.\right.$$

$$\left.\left. + \frac{\mathfrak{t}^2}{d-1}\left(\frac{1}{4} \partial_{x^1 y^1} - \xi \partial_{x^1 x^1}\right) T^{(1)}(\mathfrak{t}; x^1, y^1)\right]_{y^1 = x^1} ; 0\right) ; \quad (6.13)$$

$$\langle 0|\widehat{T}_{ij}(\mathbf{x})|0\rangle_{\text{ren}}$$

$$= \delta_{ij}\, C_d\, \text{Res}\left(\mathfrak{t}^{-(d+1)}\left[\left(\xi - \frac{d-2}{4d}\right) d\, T^{(1)}(\mathfrak{t};x^1,y^1)\right.\right.$$

$$\left.\left. + \frac{\mathfrak{t}^2}{d-1}\left(\frac{1}{4}-\xi\right)\partial_{x^1 y^1} T^{(1)}(\mathfrak{t};x^1,y^1)\right]_{y^1=x^1} ; 0\right)$$

$$\text{for } i,j \in \{2,...,d\}. \qquad (6.14)$$

We repeat that, in the above, $d \geqslant 3$ is an odd dimension. Starting from Section 6.6, for each one of the previously mentioned boundary conditions we will report the explicit expressions for the the stress-energy components arising from Eqs. (6.12)–(6.14) in the case of spatial dimension $d = 3$; again, we will give the final results in the form described in Subsection 1.4.2 (see, in particular, Eq. (1.32)), noting that (1.32) gives

$$\xi_3 = \frac{1}{6}. \qquad (6.15)$$

Some additional details related to these computations will be given, as examples, in the case of Dirichlet and periodic boundary conditions.

6.4 The reduced energy

Let us recall again that this is the energy per unit volume in the free dimensions and that it can be expressed in terms of the reduced bulk and boundary energies (see Subsection 3.1.1).

The general identity (3.19) allows us to represent the reduced bulk energy in terms of the renormalized trace $\text{Tr}\,\mathcal{A}_1^{d/2}$, corresponding to the reduced operator $\mathcal{A}_1$; on the other hand, Eq. (6.11) gives an explicit expression for the latter trace in terms of the reduced cylinder trace $T^{(1)}(\mathfrak{t})$. Evaluating the integral along the Hankel contour in the cited equation via the residue theorem (compare with Eq. (2.118)), we conclude

$$E_1^{\text{ren}} = \frac{(-1)^{d+1}\,\Gamma(d+1)\,\Gamma(-\frac{d}{2})}{2\,(4\pi)^{d/2}}\, \text{Res}\left(\mathfrak{t}^{-(d+1)}\, T^{(1)}(\mathfrak{t});0\right). \qquad (6.16)$$

Let us also mention that, for any one of the boundary conditions to be considered in the following, the (regularized and renormalized) reduced boundary energy always vanishes identically.

Finally we point out that, at least in spatial dimension $d = 3$, the results obtained via Eq. (6.16) for the renormalized, reduced bulk energy coincide with the integral over the interval $(0, a)$ of the conformal part of the corresponding renormalized energy density $\langle 0|\widehat{T}_{00}|0\rangle_{\mathrm{ren}}$; on the contrary, the non-conformal part of the latter appears to diverge in a non-integrable manner near the planes π_0, π_a. These facts closely resemble the anomalies pointed out for the total energy of the segment configuration in Chapter 5; as in that case, they will be checked by direct computation in Sections 6.6–6.9.

6.5   The boundary forces

The situation we meet in the present situation is of the kind described in general in Section 3.2, and already faced for the segment configuration in Chapter 5: in principle, we can define the renormalized pressure on the plates π_0, π_a in two alternative ways.

As usual, let $\mathbf{n}(\mathbf{x})$ denote the outer unit normal at points on the boundary, so that $\mathbf{n}(\mathbf{x}) = (-1, 0, ..., 0)$ on π_0 and $\mathbf{n}(\mathbf{x}) = (1, 0, ..., 0)$ on π_a. Then, on the one hand we can put

$$p_i^{\mathrm{ren}}(\mathbf{x}) := \langle 0|\widehat{T}_{ij}^u(\mathbf{x})|0\rangle\Big|_{u=0} \, n^j(\mathbf{x}) = \delta_{i1} \langle 0|\widehat{T}_{11}^u(\mathbf{x})|0\rangle\Big|_{u=0} \, ; \qquad (6.17)$$

on the other hand, we have the alternative definition

$$p_i^{\mathrm{ren}}(\mathbf{x}) := \left(\lim_{\mathbf{x}' \in \Omega, \mathbf{x}' \to \mathbf{x}} \langle 0|\widehat{T}_{ij}(\mathbf{x}')|0\rangle_{\mathrm{ren}} \right) n^j(\mathbf{x})$$

$$= \delta_{i1} \left(\lim_{\mathbf{x}' \in \Omega, \mathbf{x}' \to \mathbf{x}} \langle 0|\widehat{T}_{11}(\mathbf{x}')|0\rangle_{\mathrm{ren}} \right) . \qquad (6.18)$$

As a matter of fact, *definitions (6.17) and (6.18) will be found by explicit computations to yield the same result* for any one of the boundary conditions to be considered in the next sections.

6.6 Dirichlet boundary conditions

Let us first consider the case where the field fulfills Dirichlet boundary conditions on both the hyperplanes π_0, π_a, meaning that

$$\widehat{\phi}(t, \mathbf{x}) = 0 \qquad \text{for any } t \in \mathbf{R} \text{ and } \mathbf{x} \in \pi_0 \text{ or } \mathbf{x} \in \pi_a . \qquad (6.19)$$

Recall that in this case the cylinder kernel associated to the reduced problem is given by (see Eq. (5.20) on page 86)

$$T^{(1)}(t; x^1, y^1) = \frac{1}{2a} \left[\frac{\cos(\frac{\pi}{a}(x^1 - y^1)) - e^{-\frac{\pi}{a}t}}{\cosh(\frac{\pi}{a}t) - \cos(\frac{\pi}{a}(x^1 - y^1))} \right.$$

$$\left. - \frac{\cos(\frac{\pi}{a}(x^1 + y^1)) - e^{-\frac{\pi}{a}t}}{\cosh(\frac{\pi}{a}t) - \cos(\frac{\pi}{a}(x^1 + y^1))} \right] . \qquad (6.20)$$

To obtain the renormalized stress-energy VEV in any spatial dimension d, it suffices to substitute Eq. (6.20) into Eqs. (6.12)–(6.14) and to compute the residues therein. Let us make the final results explicit for $d = 3$; in this case the residues in (6.12)–(6.14), involving $T^{(1)}(t; x^1, y^1)|_{y^1 = x^1}$, can be derived from the $t \to 0$ expansion

$$T^{(1)}(t; x^1, y^1)\Big|_{y^1 = x^1} = \frac{1}{\pi t} + \frac{\pi(3 - \sin^2(\frac{\pi}{a} x^1))}{12 a^2 \sin^2(\frac{\pi}{a} x^1)} t$$

$$+ \frac{\pi^3(15(2 + \cos(\frac{2\pi}{a} x^1)) - \sin^4(\frac{\pi}{a} x^1))}{720 a^4 \sin^4(\frac{\pi}{a} x^1)} t^3$$

$$+ O(t^5) . \qquad (6.21)$$

Proceeding in a similar manner where the spatial derivatives of $T^{(1)}$ appear, one obtains the final result for the $d = 3$ renormalized VEV of the stress-energy tensor reported hereafter:

$$\langle 0|\widehat{T}_{\mu\nu}(\mathbf{x})|0\rangle_{\text{ren}}\Big|_{\mu,\nu=0,1,2,3} = A \begin{pmatrix} -1 & 0 & 0 & 0 \\ 0 & -3 & 0 & 0 \\ 0 & 0 & 1 & 0 \\ 0 & 0 & 0 & 1 \end{pmatrix} - \left(\xi - \frac{1}{6}\right) B(x^1) \begin{pmatrix} -1 & 0 & 0 & 0 \\ 0 & 0 & 0 & 0 \\ 0 & 0 & 1 & 0 \\ 0 & 0 & 0 & 1 \end{pmatrix} ,$$

$$A = \frac{\pi^2}{1440 a^4} , \qquad B(x^1) = \frac{\pi^2}{8 a^4} \frac{3 - 2\sin^2(\frac{\pi}{a} x^1)}{\sin^4(\frac{\pi}{a} x^1)} \qquad \text{for } x^1 \in (0, a) . \qquad (6.22)$$

This is a classical result, whose earlier derivations used point-splitting methods rather than zeta regularization. We refer, in particular, to the paper by Esposito *et al.* [55] and to the already cited monographs [18,67,105] (indeed, some of these references give expressions for the renormalized stress-energy VEV involving special functions, which we proved in [58] to be equivalent to (6.22)). Let us mention that the same result can also be obtained starting from the expressions derived by Romeo and Saharian [128] for the renormalized stress-energy VEV in the case of Robin boundary conditions; though, to this purpose one has to consider a proper limiting case,[d] proceeding similarly to what was done in Section 5.10 of this book. In our previous work [58] the expression (6.22) was obtained via zeta regularization, but the required analytic continuations were derived by *ad hoc* considerations, strictly related to the peculiar configuration under analysis. On the contrary, here the analytic continuation arises automatically from the general schemes that we have developed for an arbitrary geometry (in connection with the present approach, see also [59]).

Next, let us pass to the reduced bulk energy. To this purpose, we recall that the cylinder trace associated to the reduced problem under analysis is (see Eq. (5.25))

$$T^{(1)}(\mathfrak{t}) = \frac{1}{e^{\frac{\pi}{a}\mathfrak{t}} - 1} \; ; \tag{6.23}$$

now, the general relation (6.16) allows us to infer, for $d = 3$,

$$E_1^{\text{ren}} = -\frac{\pi^2}{1440a^3} \; . \tag{6.24}$$

Finally, using either definition (6.17) or (6.18), we obtain for the non-vanishing component of the pressure on the planes π_0 and π_a the following expressions, respectively:

$$p_1^{\text{ren}}(\mathbf{x})\Big|_{\pi_0} = \frac{\pi^2}{480a^4} \; , \qquad p_1^{\text{ren}}(\mathbf{x})\Big|_{\pi_a} = -\frac{\pi^2}{480a^4} \; . \tag{6.25}$$

---

[d]In the notation of [128], the present configuration with Dirichlet conditions on both planes formally corresponds to the limit $\beta_1 = \beta_2 \to 0$.

This means that in the present case, the forces on the boundary planes produced by the field in the interior region are attractive.[e]

6.7 Dirichlet–Neumann boundary conditions

Consider now the parallel hyperplanes configuration where the field fulfills Dirichlet and Neumann boundary conditions, respectively, on the hyperplanes π_0 and π_a; explicitly,

$$\widehat{\phi}(t, \mathbf{x}) = 0 \text{ for } (t, \mathbf{x}) \in \mathbf{R} \times \pi_0 , \quad \partial_{x^1}\widehat{\phi}(t, \mathbf{x}) = 0 \text{ for } (t, \mathbf{x}) \in \mathbf{R} \times \pi_a .$$
$$(6.26)$$

The cylinder kernel associated to the reduced operator $\mathcal{A}_1$ is (see Eq. (5.30) on page 88)

$$T^{(1)}(\mathfrak{t}; x^1, y^1) = \frac{1}{a} \left[\frac{\sinh(\frac{\pi}{2a}\,\mathfrak{t})\cos(\frac{\pi}{2a}(x^1 - y^1))}{\cosh(\frac{\pi}{a}\,\mathfrak{t}) - \cos(\frac{\pi}{a}(x^1 - y^1))} \right.$$
$$\left. - \frac{\sinh(\frac{\pi}{2a}\,\mathfrak{t})\cos(\frac{\pi}{2a}(x^1 + y^1))}{\cosh(\frac{\pi}{a}\,\mathfrak{t}) - \cos(\frac{\pi}{a}(x^1 + y^1))} \right] . \qquad (6.27)$$

Employing this kernel along with relations (6.12)–(6.14), we can determine the renormalized VEV of the stress-energy tensor; in particular, for $d = 3$ we obtain

$$\langle 0|\widehat{T}_{\mu\nu}(\mathbf{x})|0\rangle_{\text{ren}}\Big|_{\mu,\nu=0,1,2,3} = A \begin{pmatrix} 1 & 0 & 0 & 0 \\ 0 & 3 & 0 & 0 \\ 0 & 0 & -1 & 0 \\ 0 & 0 & 0 & -1 \end{pmatrix} - \left(\xi - \frac{1}{6}\right) B(x^1) \begin{pmatrix} -1 & 0 & 0 & 0 \\ 0 & 0 & 0 & 0 \\ 0 & 0 & 1 & 0 \\ 0 & 0 & 0 & 1 \end{pmatrix},$$

$$A = \frac{7\pi^2}{11520a^4} , \quad B(x^1) = \frac{\pi^2}{64\,a^4}\,\frac{23\cos(\frac{\pi}{a}\,x^1) + \cos(\frac{3\pi}{a}\,x^1)}{\sin^4(\frac{\pi}{a}\,x^1)} \quad \text{for } x^1 \in (0, a) .$$
$$(6.28)$$

To the best of our knowledge, an explicit expression in terms of elementary functions of the full stress-energy VEV for the present configuration has never been given in the previous literature, which

---

[e]Clearly, we are not taking into account any effect related to the outer region (see the comments at the end of Section 3.2 of Chapter 3). As a matter of fact, the forces produced by a massless scalar field in this region, fulfilling either Dirichlet or Neumann boundary conditions on the planes, vanish identically for $d = 3$; this result can be derived using the methods that will be developed in the subsequent Chapter 7.

typically focused on global computations (see the citations below). However, let us mention the notable paper [128] by Romeo and Saharian which derives integral formulas for the stress-energy tensor in the more general case of Robin conditions on the planes; with some effort, it can be proved by arguments analogous to those reported in Section 5.10 of this book that the results in Eq. (6.28) do, in fact, agree with those of [128] in the corresponding subcase with boundary conditions of Dirichlet type on one plane and of Neumann type on the other.[f]

Let us recall that the reduced cylinder trace in this case is (see Eq. (5.33))

$$T^{(1)}(\mathfrak{t}) = \frac{e^{\frac{\pi}{2a}\mathfrak{t}}}{e^{\frac{\pi}{a}\mathfrak{t}} - 1} \ . \tag{6.29}$$

The above expression, along with prescription (6.16), allows us to determine the renormalized, reduced bulk energy; for example, for $d = 3$ we obtain

$$E_1^{\mathrm{ren}} = \frac{7\pi^2}{11520a^3} \ . \tag{6.30}$$

Concerning the non-vanishing component of the boundary pressure, both prescriptions (6.17), (6.18) give

$$p_1^{\mathrm{ren}}(\mathbf{x})\Big|_{\pi_0} = -\frac{7\pi^2}{3840a^4} \ , \qquad p_1^{\mathrm{ren}}(\mathbf{x})\Big|_{\pi_a} = \frac{7\pi^2}{3840a^4} \ ; \tag{6.31}$$

let us stress that, similarly to the results found for the segment configuration in Chapter 5 (see Eqs. (5.27), (5.35) therein), the expressions in Eq. (6.31) have the opposite sign with respect to the analogous quantities corresponding to the case of Dirichlet boundary conditions on both the planes π_0, π_a (see Eq. (6.25)). This means that in the present Dirichlet–Neumann case, the forces on the boundary planes are repulsive.[g]

The results in Eqs. (6.30), (6.31) agree with those obtained, e.g. by Cougo-Pinto *et al.* [42,119] and Santos *et al.* [132] (see also [5,22]).

---

[f] In the notation of [128], this setting formally corresponds to the the limit where $\beta_1 \to 0$, $\beta_2 \to +\infty$.
[g] Recall the considerations of the previous footnote *e*.

6.8   Neumann boundary conditions

Assume that

$$\partial_{x^1}\widehat{\phi}(t,\mathbf{x})=0 \qquad \text{for any } t \in \mathbf{R} \text{ and } \mathbf{x} \in \pi_0 \text{ or } \mathbf{x} \in \pi_a . \qquad (6.32)$$

Recall that in this case, according to the considerations of Section 4.1, the reduced operator $\mathcal{A}_1$ must be viewed as acting in the Hilbert space $L_0^2(0,a)$ of square integrable functions on $(0,a)$ with mean zero (see Eq. (4.2)). The cylinder kernel of $\mathcal{A}_1$ is (see Eq. (5.38))

$$T^{(1)}(t;x^1,y^1)=\frac{1}{2a}\left[\frac{\cos(\frac{\pi}{a}(\mathbf{x}-\mathbf{y}))-e^{-t}}{\cosh t-\cos(\frac{\pi}{a}(\mathbf{x}-\mathbf{y}))}+\frac{\cos(\frac{\pi}{a}(\mathbf{x}+\mathbf{y}))-e^{-t}}{\cosh t-\cos(\frac{\pi}{a}(\mathbf{x}+\mathbf{y}))}\right].$$
$$(6.33)$$

Using this kernel along with relations (6.12-6.14), we obtain the renormalized stress-energy VEV; in particular, for $d=3$ the residue evaluation yields

$$\langle 0|\widehat{T}_{\mu\nu}(\mathbf{x})|0\rangle_{\text{ren}}\Big|_{\mu,\nu=0,1,2,3}$$

$$= A\begin{pmatrix} -1 & 0 & 0 & 0 \\ 0 & -3 & 0 & 0 \\ 0 & 0 & 1 & 0 \\ 0 & 0 & 0 & 1 \end{pmatrix} + \left(\xi-\frac{1}{6}\right)B(x^1)\begin{pmatrix} -1 & 0 & 0 & 0 \\ 0 & 0 & 0 & 0 \\ 0 & 0 & 1 & 0 \\ 0 & 0 & 0 & 1 \end{pmatrix}, \qquad (6.34)$$

where A and $B(x^1)$ are defined as in Eq. (6.22). Before proceeding, let us mention that the same result can be obtained starting from the general integral expressions of Romeo and Saharian [128] for the stress-energy VEV in the case of Robin boundary conditions on the planes; also in this case, one has to use methods similar to those described in Section 5.10 of this book.[h]

Let us pass to the reduced bulk energy; in Section 5.2 of Chapter 5 we noticed that in this case the cylinder trace $T^{(1)}(t)$ coincides with the one of Eq. (6.23), corresponding to the case of Dirichlet boundary conditions. In consequence of this the renormalized, reduced bulk

---

[h]The present configuration with Neumann boundary conditions on both planes corresponds, in the notation of [128], to the limit where $\beta_1 = \beta_2 \to +\infty$.

energy for $d = 3$ is the same as in the case of Dirichlet boundary conditions (see Eq. (6.24)):

$$E_1^{\mathrm{ren}} = -\frac{\pi^2}{1440a^3} \ .$$

Furthermore, concerning the pressure on the boundary, both definitions (6.17) and (6.18) also give the same result one obtains in the case of Dirichlet boundary conditions (see Eq. (6.25))

$$p_1^{\mathrm{ren}}(\mathbf{x})\Big|_{\pi_0} = \frac{\pi^2}{480a^4} \ , \qquad p_1^{\mathrm{ren}}(\mathbf{x})\Big|_{\pi_a} = -\frac{\pi^2}{480a^4} \ .$$

6.9   Periodic boundary conditions

The last case we consider is the one where

$$\widehat{\phi}(t, 0, x^2, ..., x^d) = \widehat{\phi}(t, a, x^2, ..., x^d) \ ,$$

$$\partial_{x^1}\widehat{\phi}(t, 0, x^2, ..., x^d) = \partial_{x^1}\widehat{\phi}(t, a, x^2, ..., x^d) \tag{6.35}$$

$$\text{for } t, x^2, ..., x^d \in \mathbf{R} \ .$$

Similarly to what was said for the segment with periodic boundary conditions, in agreement with the general analysis of Section 4.3, we remark that this configuration would be more properly formulated in terms of a free scalar field on the flat manifold $\Omega := \mathbf{T}_a^1 \times \mathbf{R}^{d-1}$, where the first factor is the torus $\mathbf{T}_a^1 := \mathbf{R}/(a\mathbf{Z})$.

The basic Hilbert space for the reduced problem is $L_0^2(\mathbf{T}_a^1)$ (see Eq. (4.2) of Section 4.1). We know that the cylinder kernel associated to $\mathcal{A}_1$ in this case is (see Eq. (5.43))

$$T^{(1)}(\mathfrak{t}; x^1, y^1) = \frac{\cos(\frac{2\pi}{a}(x^1 - y^1)) - e^{-\frac{2\pi}{a}\mathfrak{t}}}{a\left[\cosh(\frac{2\pi}{a}\mathfrak{t}) - \cos(\frac{2\pi}{a}(x^1 - y^1))\right]} \ . \tag{6.36}$$

Again, we can employ Eqs. (6.12)–(6.14) to evaluate the renormalized stress-energy VEV. Differently from the previous subcases, this time the expressions appearing in intermediate steps of the required calculations are simple enough to be reported; as an example, let us focus on the evaluation of the component $\langle 0|\widehat{T}_{00}(\mathbf{x})|0\rangle_{\text{ren}}$. First of all, notice that

$$T^{(1)}(t;x^1,x^1) = \frac{1}{a}\left[\coth\left(\frac{\pi}{a}t\right) - 1\right] ; \tag{6.37}$$

$$\partial_{x^1 y^1} T^{(1)}(t;x^1,x^1) = \frac{2\pi^2}{a^3}\coth\left(\frac{\pi}{a}t\right)\operatorname{csch}^2\left(\frac{\pi}{a}t\right) . \tag{6.38}$$

So, after some simple algebraic manipulations, Eq. (6.12) yields

$$\langle 0|\widehat{T}_{00}(\mathbf{x})|0\rangle_{\text{ren}} = -\frac{C_d}{4a}\operatorname{Res}\left(\frac{2 - (1 - 4\xi)d}{t^{d+1}}\left[\coth\left(\frac{\pi}{a}t\right) - 1\right]\right.$$
$$\left. + \frac{2(1-4\xi)}{(d-1)\,t^{d+1}}\left(\frac{\pi}{a}t\right)^2\coth\left(\frac{\pi}{a}t\right)\operatorname{csch}^2\left(\frac{\pi}{a}t\right);0\right) . \tag{6.39}$$

The function in the above expression, whose residue in $t = 0$ is required, is easily seen to be meromorphic with a pole of order $d+2$ in $t = 0$; more precisely, its Laurent expansion is

$$-\frac{d(d-3)+4(d^2-d-2)\xi}{4\pi(d-1)}\frac{1}{t^{d+2}} + \frac{(2-d)+4d\xi}{4a}\frac{1}{t^{d+1}}$$

$$+ \frac{\pi((2-d)+4d\xi)}{12a^2}\frac{1}{t^d} - \frac{\pi^3(-(d^2-3d-4)+4(d^2-d-6)\xi)}{180(d-1)\,a^4}\frac{1}{t^{d-2}}$$

$$+ O(t^{4-d}) . \tag{6.40}$$

For example, for $d = 3$ Eq. (6.39) yields ($C_3 = -1/\pi$, see Eq. (6.10))

$$\langle 0|\widehat{T}_{00}(\mathbf{x})|0\rangle_{\text{ren}} = -\frac{\pi^2}{90\,a^4} ; \tag{6.41}$$

proceeding similarly for the other components of the renormalized stress-energy VEV (for $d = 3$), we obtain

$$\langle 0|\widehat{T}_{\mu\nu}(\mathbf{x})|0\rangle_{\text{ren}}\Big|_{\mu,\nu=0,1,2,3} = \frac{\pi^2}{90a^4}\begin{pmatrix} -1 & 0 & 0 & 0 \\ 0 & -3 & 0 & 0 \\ 0 & 0 & 1 & 0 \\ 0 & 0 & 0 & 1 \end{pmatrix} . \tag{6.42}$$

Notice that the renormalized stress-energy tensor (6.42) does not depend explicitly on the periodic coordinate x^1. This comes as no surprise; indeed, it reflects the invariance of the theory under translations $x^1 \mapsto x^1 + \alpha$ (for arbitrary $\alpha \in \mathbf{R}$).

To conclude, let us consider the total bulk energy. First, recall that the reduced cylinder trace in this case is (see Eq. (5.46))

$$T(t) = \frac{2}{e^{\frac{2\pi}{a}t} - 1} \; ; \qquad (6.43)$$

then, using once more the prescription (6.16), we obtain (for $d = 3$)

$$E^{\mathrm{ren}} = -\frac{\pi^2}{90a^3} \; . \qquad (6.44)$$

In conclusion, let us stress that the above expression for the total energy agrees with the results reported in [22, 53, 54, 105].

Chapter 7

A massive field constrained by perpendicular hyperplanes

In this chapter we consider a massive field confined by an arbitrary number of perpendicular (hyper-)planes, fulfilling either Dirichlet or Neumann boundary conditions on each one of them. Also in this case, after giving a formal description of this setting (see Section 7.1), we show how to use the general methods of Part 1 to compute the analytic continuation of the regularized stress-energy VEV and pressure (see Sections 7.2–7.6). In Sections 7.7–7.9 we specialize the previous general results, determining the renormalized observables in two cases: a single plane and two perpendicular planes in spatial dimension $d = 3$. In the conclusive Section 7.10 we consider the zero mass limit in these two particular settings.

To the best of our knowledge, this type of configuration was only considered by Actor [3] and by Actor and Bender [5]. Both these papers deal with a scalar field fulfilling Dirichlet boundary conditions on at most three perpendicular planes (along with other similar models in three spatial dimensions, with boundaries consisting of flat, perpendicular parts). More precisely, in [3] the author fixes $d = 3$ and evaluates the renormalized effective Lagrangian, along with the VEV $\langle 0|\widehat{\phi}^2(x)|0\rangle_{\mathrm{ren}}$ (plus their analogues for non-zero temperature); in [5] the spatial dimension d is arbitrary and each component of the stress-energy VEV is computed. In both works cited above, a "zeta-like" approach based on the use of the heat kernel is employed. However, contrary to the regularization scheme considered here, that approach also involves the subtraction of terms (corresponding essentially to Minkowski spacetime contributions) which diverge for any value of the regulating parameter. In the sequel, whenever necessary we describe explicitly the connections with these works, making direct comparison.

7.1 Introducing the problem for arbitrary boundary conditions

We consider a scalar field of nonzero mass m fulfilling either Dirichlet or Neumann boundary conditions on d_1 orthogonal hyperplanes in spatial dimension $d = d_1 + d_2$. More precisely, we are interested in the case where

$$\Omega = (0, +\infty)^{d_1} \times \mathbf{R}^{d_2} , \qquad V(\mathbf{x}) = m^2 \quad (m > 0) . \qquad (7.1)$$

The domain Ω is bounded by the hyperplanes $\{x^1 = 0\}$, ..., $\{x^d = 0\}$ and its boundary is the union of the faces

$$\pi_n := \{\mathbf{x} \in \partial\Omega \mid x^n = 0\} \qquad (n \in \{1, ..., d_1\}) ; \qquad (7.2)$$

for each one of these faces, either Dirichlet or Neumann boundary conditions are prescribed.

Before proceeding, let us stress that also in this case we can use the results of Section 2.10 on slab configurations. Because of this, we will consider the reduced problem based on

$$\Omega_1 = (0, +\infty)^{d_1} , \qquad \mathcal{A}_1 := -\Delta_1 + m^2 , \qquad (7.3)$$

with the appropriate boundary conditions descending from the original d-dimensional problem.

7.2 The reduced heat kernel

In the approach we are going to considered, a basic step for the analysis of the reduced problem (7.3) is the computation of the heat kernel $K^{(1)}$ associated to $\mathcal{A}_1$. The final result for this object is[a]

$$K^{(1)}(t; \mathbf{x}_1, \mathbf{y}_1) = \frac{e^{-m^2 t}}{(4\pi t)^{d_1/2}} \prod_{n=1}^{d_1} \left(e^{-\frac{(x_1^n - y_1^n)^2}{4t}} + \alpha_n e^{-\frac{(x_1^n + y_1^n)^2}{4t}} \right) ,$$

$$(7.4)$$

---

[a] Here is one way to derive Eq. (7.4). First, notice that a complete orthonormal system of (improper) eigenfunctions of $\mathcal{A}_1 = -\Delta_1 + m^2$ on $\Omega_1 = (0, +\infty)^{d_1}$ fulfilling the prescribed

where, for any $n \in \{1, ..., d_1\}$, $\alpha_n \in \mathbf{R}$ is a parameter distinguishing between Dirichlet and Neumann boundary conditions on the face π_n; more precisely,

$$\alpha_n := \begin{cases} -1 \text{ for Dirichlet B.C. on } \pi_n \\ +1 \text{ for Neumann B.C. on } \pi_n \end{cases} \qquad (n \in \{1, ..., d_1\}) . \qquad (7.5)$$

Let us remark that Eq. (7.4) can be re-written as[b]

$$K^{(1)}(\mathfrak{t}; \mathbf{x}_1, \mathbf{y}_1)$$

$$= \frac{e^{-m^2 \mathfrak{t}}}{(4\pi \mathfrak{t})^{d_1/2}} \sum_{n=0}^{d_1} \frac{1}{(d_1-n)! \, n!}$$

$$\times \sum_{\sigma \in S_{d_1}} \alpha_{n,\sigma} \, e^{-\frac{1}{4\mathfrak{t}} \left(\sum_{i=1}^{n} \left(x_1^{\sigma(i)} - y_1^{\sigma(i)} \right)^2 + \sum_{j=n+1}^{d_1} \left(x_1^{\sigma(j)} + y_1^{\sigma(j)} \right)^2 \right)} . \qquad (7.6)$$

Here and in the following S_{d_1} denotes the symmetric group with d_1 elements and, by convention, the sums over i and j in (7.6) vanish for $n = 0$ and $n = d_1$, respectively; moreover, for any $\sigma \in S_{d_1}$, we

---

(either Dirichlet or Neumann) boundary conditions on $\partial \Omega_1$ is given by

$$F_{\mathbf{k}}(\mathbf{x}_1) := \frac{1}{(2\pi)^{d_1/2}} \prod_{n=1}^{d_1} \left(e^{ik_n x_1^n} + \alpha_n \, e^{-ik_n x_1^n} \right),$$

$$\omega_{\mathbf{k}} := \sqrt{|\mathbf{k}|^2 + m^2} \quad \text{for } \mathbf{k} \in \mathcal{K} \equiv (0, +\infty)^{d_1} ;$$

here, for any $n \in \{1, ..., d_1\}$, α_n is defined according to Eq. (7.5). Using the eigenfunction expansion (2.44) for the heat kernel, we obtain

$$K^{(1)}(\mathfrak{t}; \mathbf{x}_1, \mathbf{y}_1) = \frac{e^{-m^2 \mathfrak{t}}}{(2\pi)^{d_1}} \prod_{n=1}^{d_1} \int_0^{+\infty} dk \, e^{-\mathfrak{t} k^2} \left(e^{ik x_1^n} + \alpha_n \, e^{-ik x_1^n} \right) \left(e^{-ik y_1^n} + \alpha_n \, e^{ik y_1^n} \right)$$

$$= \frac{e^{-m^2 \mathfrak{t}}}{(2\pi)^{d_1}} \prod_{n=1}^{d_1} \int_{-\infty}^{+\infty} dk \, e^{-\mathfrak{t} k^2} \left(e^{ik(x_1^n - y_1^n)} + \alpha_n \, e^{ik(x_1^n + y_1^n)} \right) .$$

Then, Eq. (7.4) follows by evaluating explicitly every single Gaussian integral appearing in the last line of the above equation.

[b]This result follows from the identity

$$\prod_{n=1}^{d} (a_n + b_n) = \sum_{n=0}^{d} \frac{1}{(d-n)! \, n!} \sum_{\sigma \in S_d} \left(\prod_{i=1}^{n} a_{\sigma(i)} \right) \left(\prod_{j=n+1}^{d} b_{\sigma(j)} \right),$$

holding for any $d \in \{1, 2, 3, ...\}$ and $a_n, b_n \in \mathbf{R}$ ($n \in \{1, 2, ..., d\}$), where by convention we intend $\prod_{i=1}^{0} a_{\sigma(i)} := \prod_{j=d+1}^{d} b_{\sigma(j)} := 1$.

put

$$\alpha_{n,\sigma} := \prod_{l=n+1}^{d_1} \alpha_{\sigma(l)} \quad \text{for } n \in \{0, ..., d_1-1\}, \qquad \alpha_{d_1,\sigma} := 1. \quad (7.7)$$

7.3   The reduced Dirichlet kernel

In the following we will use the first identity in Eq. (2.85) to express the Dirichlet kernel $D_s^{(1)}$ associated to $\mathcal{A}_1$, along with its analytic continuation, in terms of $K^{(1)}$. Substituting the expression (7.6) for the heat kernel into Eq. (2.85), we obtain

$$D_s^{(1)}(\mathbf{x}_1, \mathbf{y}_1) = \frac{1}{(4\pi)^{d_1/2}\,\Gamma(s)} \sum_{n=0}^{d_1} \frac{1}{(d_1-n)!n!} \sum_{\sigma \in S_{d_1}} \alpha_{n,\sigma}\, \mathfrak{I}_{s,n,\sigma}(\mathbf{x}_1, \mathbf{y}_1),$$

$$(7.8)$$

where, for each $n \in \{0, ..., d_1\}$ and each $\sigma \in S_{d_1}$,

$$\mathfrak{I}_{s,n,\sigma}(\mathbf{x}_1, \mathbf{y}_1) := \int_0^{+\infty} dt\ t^{s-\frac{d_1}{2}-1}\, e^{-m^2 t - \frac{1}{4t}\mathfrak{N}_{n,\sigma}^2(\mathbf{x}_1,\mathbf{y}_1)}, \quad (7.9)$$

$$\mathfrak{N}_{n,\sigma}(\mathbf{x}_1, \mathbf{y}_1) := \left(\sum_{i=1}^{n} (x_1^{\sigma(i)} - y_1^{\sigma(i)})^2 + \sum_{j=n+1}^{d_1} (x_1^{\sigma(j)} + y_1^{\sigma(j)})^2 \right)^{1/2}.$$

$$(7.10)$$

Clearly $\mathfrak{N}_{n,\sigma}(\mathbf{x}_1, \mathbf{y}_1) \geqslant 0$, so that convergence conditions for the integral in Eq. (7.9) can be readily inferred; more precisely, if $\mathfrak{N}_{n,\sigma}(\mathbf{x}_1, \mathbf{y}_1) > 0$ the integral converges for any $s \in \mathbf{C}$ while if $\mathfrak{N}_{n,\sigma}(\mathbf{x}_1, \mathbf{y}_1) = 0$ (which, in the interior of Ω, happens if and only if $n = d_1$ and $\mathbf{y}_1 = \mathbf{x}_1$) it only converges for

$$\Re s > \frac{d_1}{2}. \quad (7.11)$$

Under these assumptions for convergence, Eq. (7.9) is strictly related to a known integral representation of the modified Bessel function of the second kind K_ν (see, e.g. [117], page 253, Eq. (10.32.10)); using this representation we obtain, for any $\mathbf{x}_1, \mathbf{y}_1 \in (0, +\infty)^{d_1}$,

$$\mathfrak{I}_{s,n,\sigma}(\mathbf{x}_1, \mathbf{y}_1) = 2^{\frac{d_1}{2}+1-s}\, m^{d_1-2s}\, \mathfrak{G}_{s-\frac{d_1}{2}}(m^2\, \mathfrak{N}_{n,\sigma}^2(\mathbf{x}_1, \mathbf{y}_1)), \quad (7.12)$$

where, for the sake of brevity, we have put[c]

$$\mathfrak{G}_\nu : [0, +\infty) \to \mathbf{C} , \qquad z \mapsto \mathfrak{G}_\nu(z) := z^{\nu/2} K_\nu(\sqrt{z}) . \qquad (7.13)$$

(Intend this as a limit for $z = 0$.) In conclusion,

$$D_s^{(1)}(\mathbf{x}_1, \mathbf{y}_1) = \frac{2^{1-s} m^{d_1-2s}}{(2\pi)^{d_1/2} \Gamma(s)} \sum_{n=0}^{d_1} \frac{1}{(d_1-n)!n!}$$

$$\times \sum_{\sigma \in S_{d_1}} \alpha_{n,\sigma} \, \mathfrak{G}_{s-\frac{d_1}{2}} (m^2 \mathfrak{N}_{n,\sigma}^2(\mathbf{x}_1, \mathbf{y}_1)) . \qquad (7.14)$$

The derivatives of $D_s^{(1)}$ of any order can be computed using the identity

$$\frac{d^n \mathfrak{G}_\nu}{dz^n}(z) = \left(-\frac{1}{2}\right)^n \mathfrak{G}_{\nu-n}(z) \qquad \text{for } n \in \{1, 2, 3, ...\} ; \qquad (7.15)$$

moreover, in the cases where $\mathfrak{N}_{n,\sigma}(\mathbf{x}_1, \mathbf{y}_1) = 0$, we can resort to the relation

$$\mathfrak{G}_\nu(0) = 2^{\nu-1} \Gamma(\nu) \qquad \text{for } \nu \in \mathbf{C}, \Re\nu > 0 . \qquad (7.16)$$

7.4 The *d*-dimensional Dirichlet kernel

Using the previous results and Eqs. (2.134)–(2.137), we obtain the following expressions for the Dirichlet kernel of the *d*-dimensional problem (7.1) and for its derivatives, along the diagonal $\mathbf{y} = \mathbf{x}$:

$$D_{\frac{u\pm1}{2}}(\mathbf{x}, \mathbf{y})\Big|_{\mathbf{y}=\mathbf{x}} = \frac{2^{\frac{2\mp1-u}{2}} m^{d\mp1-u}}{(2\pi)^{d/2} \Gamma(\frac{u\pm1}{2})} \sum_{n=0}^{d_1} \frac{1}{(d_1-n)!n!}$$

$$\times \sum_{\sigma \in S_{d_1}} \alpha_{n,\sigma} \, \mathfrak{G}_{\frac{u-d\pm1}{2}} (m^2 \mathfrak{N}_{n,\sigma}^2(\mathbf{x}_1, \mathbf{y}_1))\Big|_{\mathbf{y}_1=\mathbf{x}_1} ;$$

$$(7.17)$$

---

[c]See Appendix D for more details on the map $\mathfrak{G}_\nu$ of Eq. (7.13); therein it is also shown how to derive the forthcoming relations (7.15), (7.16).

$$\left. \partial_{z_1^i w_1^j} D_{\frac{u+1}{2}}(\mathbf{x}, \mathbf{y}) \right|_{\mathbf{y}=\mathbf{x}}$$

$$= \frac{2^{\frac{1-u}{2}} m^{d-1-u}}{(2\pi)^{d/2} \Gamma(\frac{u+1}{2})} \sum_{n=0}^{d_1} \frac{1}{(d_1 - n)! n!}$$

$$\times \left. \sum_{\sigma \in S_{d_1}} \alpha_{n,\sigma} \, \partial_{z_1^i w_1^j} \mathfrak{G}_{\frac{u-d+1}{2}}(m^2 \mathfrak{N}_{n,\sigma}^2(\mathbf{x}_1, \mathbf{y}_1)) \right|_{\mathbf{y}_1=\mathbf{x}_1}$$

$$\text{for } z, w \in \{x, y\} \text{ and } i, j \in \{1, ..., d_1\}; \qquad (7.18)$$

$$\left. \partial_{x_2^i y_2^j} D_{\frac{u+1}{2}}(\mathbf{x}, \mathbf{y}) \right|_{\mathbf{y}=\mathbf{x}}$$

$$= -\left. \partial_{x_2^i x_2^j} D_{\frac{u+1}{2}}(\mathbf{x}, \mathbf{y}) \right|_{\mathbf{y}=\mathbf{x}} = -\left. \partial_{y_2^i y_2^j} D_{\frac{u+1}{2}}(\mathbf{x}, \mathbf{y}) \right|_{\mathbf{y}=\mathbf{x}}$$

$$= \delta_{ij} \frac{2^{\frac{1-u}{2}} m^{d+1-u}}{(2\pi)^{d/2} \Gamma(\frac{u+1}{2})} \sum_{n=0}^{d_1} \frac{1}{(d_1 - n)! n!}$$

$$\times \left. \sum_{\sigma \in S_{d_1}} \alpha_{n,\sigma} \, \mathfrak{G}_{\frac{u-d-1}{2}}(m^2 \mathfrak{N}_{n,\sigma}^2(\mathbf{x}_1, \mathbf{y}_1)) \right|_{\mathbf{y}_1=\mathbf{x}_1}$$

$$\text{for } i, j \in \{1, ..., d_2\}. \qquad (7.19)$$

Note that, in each one of the sums over n appearing in the above expressions, the terms corresponding to $n \in \{0, 1, ..., d_1 - 1\}$ are analytic functions of u on the whole complex plane since $\mathfrak{N}_{n,\sigma}^2(\mathbf{x}_1, \mathbf{x}_1) > 0$ for any $\mathbf{x}_1 \in (0, +\infty)^{d_1}$. On the contrary, the terms corresponding to $n = d_1$ deserve particular attention; indeed, $\mathfrak{N}_{d_1,\sigma}^2(\mathbf{x}_1, \mathbf{x}_1) = 0$ for any $\mathbf{x}_1 \in (0, +\infty)^{d_1}$ so that, in order to evaluate these contributions, we have to resort to Eq. (7.16) (also recalling Eq. (7.15)). In this way we obtain the following identities:

$$\left. \mathfrak{G}_{\frac{u-d\pm1}{2}}(m^2 \mathfrak{N}_{d_1,\sigma}^2(\mathbf{x}_1, \mathbf{y}_1)) \right|_{\mathbf{y}_1=\mathbf{x}_1} = 2^{\frac{u-d\pm1}{2}-1} \Gamma\left(\frac{u-d\pm1}{2}\right); \quad (7.20)$$

$$\partial_{z_1^i w_1^j} \mathfrak{G}_{\frac{u-d+1}{2}} \left(m^2 \mathfrak{N}^2_{d_1,\sigma}(\mathbf{x}_1, \mathbf{y}_1) \right) \Big|_{\mathbf{y}_1=\mathbf{x}_1}$$

$$= -\delta_{ij} \left(2\delta_{zw} - 1 \right) m^2 \, 2^{\frac{u-d-3}{2}} \, \Gamma \left(\frac{u-d-1}{2} \right)$$

$$\text{for } z, w \in \{x, y\} \text{ and } i, j \in \{1, ..., d_1\}. \tag{7.21}$$

In principle, Eqs. (7.20) and (7.21) hold with the limitations on u arising from Eq. (7.16); more precisely, Eq. (7.20) holds for $\Re u > d\mp 1$ and Eq. (7.21) for $\Re u > d+1$. However, the right-hand sides of these equations are well defined and analytic on the whole complex plane with the exception of simple poles placed at

$$u \in \{d+1, d-1, d-3, ...\} ; \tag{7.22}$$

this remark gives the meromorphic continuation in u of the functions in Eqs. (7.20), (7.21) and, consequently, of the terms with $n = d_1$ in Eqs. (7.17)–(7.19).

7.5   The stress-energy tensor

Using Eqs. (2.28)–(2.30) along with Eqs. (7.17)–(7.19) (and Eqs. (7.20)–(7.21) for the terms with $n = d_1$), we obtain the analytic continuation to a meromorphic function of each component of the regularized stress-energy VEV, required in order to implement the local zeta approach. In particular, we have the following expressions for the non-vanishing components:

$$\langle 0 | \widehat{T}^u_{00}(\mathbf{x}) | 0 \rangle = \frac{2^{\frac{1-u}{2}} m^{d+1}}{(2\pi)^{d/2} \Gamma\left(\frac{u+1}{2}\right)} \left(\frac{m}{\kappa} \right)^{-u} \sum_{n=0}^{d_1} \frac{1}{(d_1-n)! n!} \sum_{\sigma \in S_{d_1}} \alpha_{n,\sigma}$$

$$\times \Bigg[\left(\frac{d-d_1-1+u}{4} - (d-d_1+1-u)\xi \right) \mathfrak{G}_{\frac{u-d-1}{2}} \left(m^2 \mathfrak{N}^2_{n,\sigma}(\mathbf{x}_1, \mathbf{y}_1) \right)$$

$$+ \left(\frac{1}{4} - \xi \right) (1 + m^{-2} \partial^{x_1^\ell}_{y_1^\ell}) \, \mathfrak{G}_{\frac{u-d+1}{2}} \left(m^2 \mathfrak{N}^2_{n,\sigma}(\mathbf{x}_1, \mathbf{y}_1) \right) \Bigg]_{\mathbf{y}_1=\mathbf{x}_1} ;$$

$$\tag{7.23}$$

$$\langle 0|\widehat{T}_{ij}^{u}(\mathbf{x})|0\rangle = \langle 0|\widehat{T}_{ji}^{u}(\mathbf{x})|0\rangle$$

$$= \frac{2^{\frac{1-u}{2}}\,m^{d+1}}{(2\pi)^{d/2}\,\Gamma(\frac{u+1}{2})}\left(\frac{m}{\kappa}\right)^{-u}\sum_{n=0}^{d_1}\frac{1}{(d_1-n)!\,n!}\sum_{\sigma\in S_{d_1}}\alpha_{n,\sigma}$$

$$\times\left[-\left(\frac{1}{4}-\xi\right)\delta_{ij}\left((d-d_1+1-u)\,\mathfrak{G}_{\frac{u-d-1}{2}}(m^2\mathfrak{N}_{n,\sigma}^2(\mathbf{x}_1,\mathbf{y}_1))\right)\right.$$

$$+\,\mathfrak{G}_{\frac{u-d+1}{2}}(m^2\mathfrak{N}_{n,\sigma}^2(\mathbf{x}_1,\mathbf{y}_1))\Big)+m^{-2}\left(-\left(\frac{1}{4}-\xi\right)\delta_{ij}\partial^{x_1^{\ell}}\partial_{y_1^{\ell}}\right.$$

$$\left.+\left(\frac{1}{2}-\xi\right)\partial_{x_1^{i}y_1^{j}}-\xi\,\partial_{x_1^{i}x_1^{j}}\right)\mathfrak{G}_{\frac{u-d+1}{2}}(m^2\mathfrak{N}_{n,\sigma}^2(\mathbf{x}_1,\mathbf{y}_1))\Big]_{\mathbf{y}_1=\mathbf{x}_1}$$

$$\text{for } i,j\in\{1,...,d_1\}\,;$$

$$(7.24)$$

$$\langle 0|\widehat{T}_{ij}^{u}(\mathbf{x})|0\rangle = \langle 0|\widehat{T}_{ji}^{u}(\mathbf{x})|0\rangle$$

$$= -\delta_{ij}\frac{2^{\frac{1-u}{2}}\,m^{d+1}}{(2\pi)^{d/2}\,\Gamma(\frac{u+1}{2})}\left(\frac{m}{\kappa}\right)^{-u}\sum_{n=0}^{d_1}\frac{1}{(d_1-n)!\,n!}\sum_{\sigma\in S_{d_1}}\alpha_{n,\sigma}$$

$$\times\left[\left(\frac{d-d_1-1-u}{4}-(d-d_1+1-u)\xi\right)\mathfrak{G}_{\frac{u-d-1}{2}}(m^2\mathfrak{N}_{n,\sigma}^2(\mathbf{x}_1,\mathbf{y}_1))\right.$$

$$\left.+\left(\frac{1}{4}-\xi\right)(1+m^{-2}\partial^{x_1^{\ell}}\partial_{y_1^{\ell}})\mathfrak{G}_{\frac{u-d+1}{2}}(m^2\mathfrak{N}_{n,\sigma}^2(\mathbf{x}_1,\mathbf{y}_1))\right]_{\mathbf{y}_1=\mathbf{x}_1}$$

$$\text{for } i,j\in\{d_1+1,...,d_1+d_2\equiv d\}\,.$$

$$(7.25)$$

The renormalized VEV of the stress-energy tensor is obtained sending u to zero in the above expressions; the only singularities appear in the terms corresponding to $n=d_1$, which must be treated resorting to Eqs. (7.20)–(7.22).

The conclusions are the following:

(i) For d *even* each component of the regularized stress-energy VEV is an analytic function of u near $u=0$; thus its renormalized version is obtained via the restricted zeta approach, i.e. by simply evaluating Eqs. (7.23)–(7.25) at $u=0$.

(ii) For d odd the regularized stress-energy VEV has a simple pole in $u = 0$, so that we have to resort to the extended zeta approach and consider the regular part at $u = 0$.

The manipulations indicated in (i) are trivial; the ones indicated in (ii) could be performed in principle for an arbitrary odd dimension, but the final expressions are too lengthy to be reported here. For this reason, we prefer to exemplify (ii) in two special cases with $d = 3$ (see the subsequent Sections 7.8 and 7.9).

7.6 The boundary forces

As in the previous chapter, following the general framework of Section 3.2, we give two alternative definitions for the pressure acting on the boundary of the spatial domain Ω.

Let us consider a point $\mathbf{x} \in \partial\Omega$; if $d_1 > 1$ we exclude $\mathbf{x}$ to be on a corner, where the outer normal is ill-defined. To fix our ideas, we assume that $\mathbf{x}$ is an inner point of the face

$$\pi_1 := \{x^1 = 0\} \cap \partial\Omega . \tag{7.26}$$

Let $\mathbf{n}(\mathbf{x})$ denote the outer unit normal at $\mathbf{x}$, so that $\mathbf{n}(\mathbf{x}) = (-1, 0, ..., 0)$.

On the one hand, we can define

$$p_i^{\text{ren}}(\mathbf{x}) := RP\Big|_{u=0} \langle 0|\widehat{T}_{ij}^u(\mathbf{x})|0\rangle\, n^j(\mathbf{x}) = -RP\Big|_{u=0} \langle 0|\widehat{T}_{i1}^u(\mathbf{x})|0\rangle . \tag{7.27}$$

On the other hand, we can consider the alternative prescription

$$p_i^{\text{ren}}(\mathbf{x}) := \left(\lim_{\mathbf{x}'\in\Omega, \mathbf{x}'\to\mathbf{x}} \langle 0|\widehat{T}_{ij}(\mathbf{x}')|0\rangle_{\text{ren}} \right) n^j(\mathbf{x})$$

$$= -\left(\lim_{\mathbf{x}'\in\Omega, \mathbf{x}'\to\mathbf{x}} \langle 0|\widehat{T}_{i1}(\mathbf{x}')|0\rangle_{\text{ren}} \right) . \tag{7.28}$$

As a matter of fact, *the alternatives (7.27), (7.28) give the same result for the renormalized pressure*; the rest of the present section is mainly devoted to the justification of this statement, which requires a nontrivial analysis.

Consider the expression (7.24) of $\langle 0|\widehat{T}_{i1}^u|0\rangle$. Due to the considerations in the previous sections, it appears that the terms in (7.24) deserving special attention when comparing the definitions (7.27), (7.28) for the pressure at $\mathbf{x} = (\mathbf{x}_1, \mathbf{x}_2) \in \pi_1$ are those with n, σ such that

$$\mathfrak{N}^2_{n,\sigma}(\mathbf{x}_1, \mathbf{x}_1) = 0 \quad \text{and} \quad \mathfrak{N}^2_{n,\sigma}(\mathbf{x}'_1, \mathbf{x}'_1) \neq 0 \quad \text{for} \quad \mathbf{x}' \equiv (\mathbf{x}'_1, \mathbf{x}'_2) \in \Omega \ ; \tag{7.29}$$

these are easily seen to correspond to the choices

$$n = d_1 - 1 \text{ and } \sigma \in S_{d_1} \text{ such that } \sigma(d_1) = 1 \ . \tag{7.30}$$

Indeed, all the terms corresponding to values of n, σ different from the above ones are straightforwardly seen to yield the same results according to both the prescriptions (7.27) and (7.28).

Now, let us focus on the potentially troublesome terms appearing in Eq. (7.24), which correspond to a choice of the form (7.30); in the sequel we will show that these terms *do not contribute to* $p_i^{\text{ren}}(\mathbf{x})$ for both the alternatives (7.27), (7.28). In order to prove this claim, let us denote with $\mathbf{x}' = (\mathbf{x}'_1, \mathbf{x}'_2)$ either the previously mentioned boundary point $\mathbf{x} \in \pi_1$ or a point of Ω. Let us pick any problematic term in the expression (7.24) for $\langle 0 | \widehat{T}^u_{i1}(\mathbf{x}') | 0 \rangle$ with n, σ as in Eq. (7.30); this term reads

$$\left[-\left(\frac{1}{4} - \xi\right) \delta_{i1} \left((d - d_1 + 1 - u) \, \mathfrak{G}_{\frac{u-d-1}{2}}(m^2 \mathfrak{N}^2_{d_1-1,\sigma}) + \mathfrak{G}_{\frac{u-d+1}{2}}(m^2 \mathfrak{N}^2_{d_1-1,\sigma}) \right) \right.$$

$$+ m^{-2} \left(-\left(\frac{1}{4} - \xi\right) \delta_{i1} \partial^{x'^\ell}_1 \partial_{y'^\ell_1} + \left(\frac{1}{2} - \xi\right) \partial_{x'^1_1 y'^1_1} \right.$$

$$\left. \left. - \xi \, \partial_{x'^i_1 x'^1_1} \right) \mathfrak{G}_{\frac{u-d+1}{2}}(m^2 \mathfrak{N}^2_{d_1-1,\sigma}) \right]_{\mathbf{y}'_1 = \mathbf{x}'_1} \tag{7.31}$$

(some of the arguments have been suppressed for the sake of brevity). With some effort, noting that $\mathfrak{N}^2_{d_1-1,\sigma}(\mathbf{x}'_1, \mathbf{y}'_1) = (x'^1_1 + y'^1_1)^2 + \sum_{i=2}^{d_2}(x'^i_1 - y'^i_1)^2$ (due to $\sigma(d_1) = 1$) and using Eq. (7.15), we can re-write expression (7.31) as

$$-\delta_{i1} \left(\frac{1}{4} - \xi\right) \left[(d+1-u) \mathfrak{G}_{\frac{u-d-1}{2}}(z^2) + \mathfrak{G}_{\frac{u-d+1}{2}}(z^2) - z^2 \mathfrak{G}_{\frac{u-d-3}{2}}(z^2) \right]_{z=2mx'^1_1}$$

$$\equiv f(u, x'^1_1) \ . \tag{7.32}$$

We now claim that

$$f(u,0)\Big|_{u=0} = 0 \tag{7.33}$$

and

$$\lim_{x'^1_1 \to 0} \left(f(u, x'^1_1)\Big|_{u=0} \right) = 0 . \tag{7.34}$$

Let us remark that in both the above equations the prescription of taking the regular part is superfluous, and $|_{u=0}$ indicates the analytic continuation at $u = 0$. Eqs. (7.33), (7.34) state that the problematic terms do not contribute to $p_i^{\mathrm{ren}}(\mathbf{x})$ as defined by Eqs. (7.27) and (7.28), respectively.

In order to prove Eq. (7.33) we note that

$$f(u,0) = -\delta_{i1}\left(\frac{1}{4} - \xi\right) 2^{\frac{u-d-1}{2}} \left[\frac{d+1-u}{2} \, \Gamma\left(\frac{u-d-1}{2}\right) + \Gamma\left(\frac{u-d+1}{2}\right)\right]$$

$$= 0 \tag{7.35}$$

where in the first passage we used Eq. (7.16), while equality to zero follows from the well-known relation $\Gamma(z+1) = z\,\Gamma(z)$.

To prove Eq. (7.34), recalling the definition (7.13) we infer (for all $x'^1_1 > 0$)

$$f(u, x'^1_1)\Big|_{u=0}$$

$$= -\delta_{i1}\left(\frac{1}{4} - \xi\right)\left[z^{-\frac{d+1}{2}}\left((d+1)K_{\frac{d+1}{2}}(z) + z\,K_{\frac{d-1}{2}}(z) - z\,K_{\frac{d+3}{2}}(z)\right)\right]_{z=2mx'^1_1}$$

$$= 0 ; \tag{7.36}$$

in this case equality to zero follows from the identity below, holding for Bessel functions K_ν of any order (see [117], page 251, Eq. (10.29.1)):

$$z\,K_{\nu+1}(z) - z\,K_{\nu-1}(z) = 2\nu\,K_\nu(z) . \tag{7.37}$$

In the above we assumed $\mathbf{x}$ to belong to the face with $x^1 = 0$ but, of course, similar considerations also hold for all the other boundary points not on the corners.

Now, let us spend a few words about points on the corners of $\partial\Omega$, which appear if $d_1 > 1$; we already noticed that the outer normal is ill-defined at these points, so that the notion of pressure is itself problematic. The natural strategies that could be guessed to overcome the problem make apparent some pathologies that we prefer to describe in an example, rather than in general: see Section 7.9.

In passing, let us mention that an analysis similar to present one is given in Chapter 10 (see, in particular, Section 10.5 therein), for the case of a massless field confined within a d-dimensional box and fulfilling Dirichlet conditions on the boundary. As in the present setting, the alternative definitions (3.22), (3.23) are found to agree at all boundary points except those on the corners, where pathologies appear.

7.7 Introducing two examples

The framework developed in the previous sections will be illustrated hereafter, for $d = 3$, in the cases where $\Omega := (0, +\infty) \times \mathbf{R}^2$, representing a half-space, and $\Omega := (0, +\infty)^2 \times \mathbf{R}$, representing a wedge bounded by orthogonal half-planes. In both cases, we consider Dirichlet and/or Neumann boundary conditions.

7.8 A half-space in spatial dimension $d = 3$

Let

$$\Omega := (0, +\infty) \times \mathbf{R}^2 \; ; \tag{7.38}$$

this is the subcase of the general setting (7.1) corresponding to $d = 3$ and

$$d_1 = 1 \, , \qquad d_2 = 2 \, . \tag{7.39}$$

With the above choices, the symmetric group appearing in the general framework of Section 7.5 consists of the sole idendity ($S_{d_1} = S_1 = \{id\}$). We have $\mathbf{x}_1 = (x_1^1) \equiv x^1$ and analogous relations for $\mathbf{y}_1$; besides, $\mathfrak{N}_{0,id}(\mathbf{x}_1, \mathbf{y}_1) = |x^1 + y^1|$ and $\mathfrak{N}_{1,id}(\mathbf{x}_1, \mathbf{y}_1) = |x^1 - y^1|$. Using the relations (7.17)–(7.19), (2.28)–(2.30) and Eqs. (7.15), (7.16) and (7.37), with some simple algebraic manipulations we obtain the following expressions for the non-vanishing components of the regularized stress-energy VEV (where $\mathbf{x} = (x^1, x^2, x^3)$):

$$\langle 0|\widehat{T}_{00}^u(\mathbf{x})|0\rangle = -\frac{m^4}{32\pi^{3/2}\,\Gamma(\frac{u+1}{2})}\left(\frac{m}{\kappa}\right)^{-u}\left[(1-u)\,\Gamma\left(\frac{u-4}{2}\right)\right.$$

$$-\,2^{5-\frac{u}{2}}\alpha_1\left(\left(\frac{1}{4}-\xi\right)\mathfrak{G}_{\frac{u-2}{2}}(z^2)\right.$$

$$\left.\left.+\left(\frac{1}{2}-(3-u)\xi\right)\mathfrak{G}_{\frac{u-4}{2}}(z^2)\right)_{z=2mx^1}\right]\,; \qquad (7.40)$$

$$\langle 0|\widehat{T}_{11}^u(\mathbf{x})|0\rangle = \frac{m^4}{32\pi^{3/2}\,\Gamma(\frac{u+1}{2})}\left(\frac{m}{\kappa}\right)^{-u}\Gamma\left(\frac{u-4}{2}\right)\,; \qquad (7.41)$$

$$\langle 0|\widehat{T}_{22}^u(\mathbf{x})|0\rangle = \langle 0|\widehat{T}_{33}^u(\mathbf{x})|0\rangle$$

$$= \frac{m^4}{32\pi^{3/2}\,\Gamma(\frac{u+1}{2})}\left(\frac{m}{\kappa}\right)^{-u}\left[\Gamma\left(\frac{u-4}{2}\right)\right.$$

$$-\,2^{5-\frac{u}{2}}\alpha_1\left(\left(\frac{1}{4}-\xi\right)\mathfrak{G}_{\frac{u-2}{2}}(z^2)\right.$$

$$\left.\left.+\left(\frac{2-u}{4}-(3-u)\xi\right)\mathfrak{G}_{\frac{u-4}{2}}(z^2)\right)_{z=2mx^1}\right]. \qquad (7.42)$$

The above expressions are easily seen to give the meromorphic continuation of the regularized stress-energy VEV to the whole complex plane, with poles determined by the terms where a gamma function appears. In particular, all the above components have a simple pole in $u = 0$; thus, we follow the extended version of the zeta approach and define the renormalized quantities to be the regular parts in $u = 0$. Recalling again that Eq. (1.32) gives

$$\xi_3 = \frac{1}{6}\,,$$

we write the final results in the form (1.32), obtaining

$$\langle 0|\widehat{T}_{00}(\mathbf{x})|0\rangle_{\mathrm{ren}} = \frac{m^4}{384\pi^2}\left(3\left(4\ln\left(\frac{m}{2\kappa}\right)+1\right)+32\,\alpha_1\,\frac{K_1(2mx^1)}{2mx^1}\right)$$

$$-\,\alpha_1\left(\xi-\frac{1}{6}\right)\frac{m^4}{\pi^2}$$

$$\times\left(\frac{(2mx^1)K_1(2mx^1)+3K_2(2mx^1)}{(2mx^1)^2}\right); \quad (7.43)$$

$$\langle 0|\widehat{T}_{11}(\mathbf{x})|0\rangle_{\mathrm{ren}} = -\frac{m^4}{128\pi^2}\left(4\ln\left(\frac{m}{2\kappa}\right)-3\right); \quad (7.44)$$

$$\langle 0|\widehat{T}_{22}(\mathbf{x})|0\rangle_{\mathrm{ren}} = \langle 0|\widehat{T}_{33}(\mathbf{x})|0\rangle_{\mathrm{ren}}$$

$$= -\frac{m^4}{384\pi^2}\left(3\left(4\ln\left(\frac{m}{2\kappa}\right)-3\right)+32\,\alpha_1\,\frac{K_1(2mx^1)}{2mx^1}\right)$$

$$-\,\alpha_1\left(\xi-\frac{1}{6}\right)\frac{m^4}{\pi^2}$$

$$\times\left(\frac{(2mx^1)K_1(2mx^1)+3K_2(2mx^1)}{(2mx^1)^2}\right). \quad (7.45)$$

Let us comment briefly on the above results. First of all, note that the renormalized VEV of the stress-energy tensor does not depend explicitly on the spatial coordinates x^2, x^3; this was to be expected due to the homogeneity with respect to these variables of the spatial configuration considered. Besides, in agreement with the general results of Section 7.6, the components $\langle 0|\widehat{T}^u_{11}|0\rangle$, $\langle 0|\widehat{T}_{11}|0\rangle_{\mathrm{ren}}$ are constant and the two alternative definitions for the pressure on the plane $\pi_1 = \{x_1 = 0\}$ (see Eqs. (7.27), (7.28)) give the same result; more explicitly, we obtain

$$p_i^{\mathrm{ren}} = -\,\delta_{i1}\langle 0|\widehat{T}_{11}|0\rangle_{\mathrm{ren}} \qquad (i \in \{1, 2, 3\})\,. \quad (7.46)$$

In conclusion, let us make a comparison with the results derived in [5] for the configuration with a single plane in arbitrary spatial dimension (to be considered here in the case $d = 3$); therein the attention is restricted to the "minimal" ($\xi = 0$) and "conformal" ($\xi = 1/6$) settings. In both cases the results derived here are found to agree with those reported in [5] (let us remark that the mass scale κ employed here does not coincide with the one considered therein; the latter is proportional, via a purely numerical coefficient, to m^2/κ).

7.9 The rectangular wedge

Let us pass to the case of a wedge in $\mathbf{R}^3$, bounded by two perpendicular half-planes; this is represented as

$$\Omega := (0, +\infty)^2 \times \mathbf{R} , \tag{7.47}$$

corresponding to the general framework (7.1) with $d = 3$ and

$$d_1 = 2 , \qquad d_2 = 1 . \tag{7.48}$$

In passing, let us mention that the rectangular wedge model is also considered by Actor and Bender in [5], for arbitrary spatial dimension; yet, these authors restrict the attention to a massless field, a case we discuss in the subsequent Section 7.10.

In the setting under analysis here, the symmetric group ($S_{d_1} = S_2$) of Section 7.5 consists of two elements, i.e. the identity id and the exchange $\mathfrak{p}$:

$$S_{d_1} \equiv S_2 = \{id, \mathfrak{p}\} , \quad id(1) = 1 , \;\; id(2) = 2 , \;\; \mathfrak{p}(1) = 2 , \;\; \mathfrak{p}(2) = 1 . \tag{7.49}$$

Moreover, setting $\mathbf{x}_1 = (x_1^1, x_1^2) \equiv (x^1, x^2)$ and using analogous notations for $\mathbf{y}_1$, we have

$$\begin{aligned}
\mathfrak{N}_{0,\sigma}(\mathbf{x}_1, \mathbf{y}_1) &= \left((x^1{+}y^1)^2 + (x^2{+}y^2)^2\right)^{1/2} \\
\mathfrak{N}_{2,\sigma}(\mathbf{x}_1, \mathbf{y}_1) &= \left((x^1{-}y^1)^2 + (x^2{-}y^2)^2\right)^{1/2}
\end{aligned} \qquad \text{for } \sigma \in S_2 = \{id, \mathfrak{p}\} ;$$

$$\tag{7.50}$$

$$\mathfrak{N}_{1,id}(\mathbf{x}_1, \mathbf{y}_1) = \left((x^1 - y^1)^2 + (x^2 + y^2)^2\right)^{1/2} ,$$

$$\mathfrak{N}_{1,\mathfrak{p}}(\mathbf{x}_1, \mathbf{y}_1) = \left((x^1 + y^1)^2 + (x^2 - y^2)^2\right)^{1/2} .$$

Also in this case, we can use the relations (2.28)–(2.30), (7.17)–(7.19) and the identities (7.15), (7.16) and (7.37) to deduce expressions for the non-vanishing components of the regularized stress-energy VEV. More precisely, we obtain the following (with $\mathbf{x} = (x^1, x^2, x^3)$)

$$\langle 0|\widehat{T}_{00}^u(\mathbf{x})|0\rangle = -\frac{m^4}{32\pi^{3/2}\,\Gamma(\frac{u+1}{2})}\left(\frac{m}{\kappa}\right)^{-u}\left[(1-u)\,\Gamma\left(\frac{u-4}{2}\right)\right.$$

$$- 2^{5-\frac{u}{2}}\sum_{i=1,2}\alpha_i\left(\left(\frac{1}{2} - (3-u)\xi\right)\mathfrak{G}_{\frac{u-4}{2}}(z^2)\right.$$

$$\left.+\left(\frac{1}{4} - \xi\right)\mathfrak{G}_{\frac{u-2}{2}}(z^2)\right)_{z=2mx^i}$$

$$- 2^{5-\frac{u}{2}}\,\alpha_1\alpha_2\left(\left(\frac{1}{4} - (2-u)\xi\right)\mathfrak{G}_{\frac{u-4}{2}}(z^2)\right.$$

$$\left.\left.+\left(\frac{1}{4} - \xi\right)\mathfrak{G}_{\frac{u-2}{2}}(z^2)\right)_{z=2m\sqrt{(x^1)^2+(x^2)^2}}\right]; \qquad (7.51)$$

$$\langle 0|\widehat{T}_{ij}^u(\mathbf{x})|0\rangle = \langle 0|\widehat{T}_{ji}^u(\mathbf{x})|0\rangle$$

$$= \frac{m^4}{32\pi^{3/2}\,\Gamma(\frac{u+1}{2})}\left(\frac{m}{\kappa}\right)^{-u}\left[\delta_{ij}\,\Gamma\left(\frac{u-4}{2}\right)\right.$$

$$- 2^{5-\frac{u}{2}}\,\alpha_{\mathsf{p}(j)}\,\delta_{ij}\left(\left(\frac{2-u}{4} - (3-u)\xi\right)\mathfrak{G}_{\frac{u-4}{2}}(z^2)\right.$$

$$\left.+\left(\frac{1}{4} - \xi\right)\mathfrak{G}_{\frac{u-2}{2}}(z^2)\right)_{z=2mx^{\mathsf{p}(j)}} - 2^{5-\frac{u}{2}}\,\alpha_1\alpha_2\left(\frac{1}{4} - \xi\right)$$

$$\times\left(\delta_{ij}\left((3-u)\mathfrak{G}_{\frac{u-4}{2}}(z^2) + \mathfrak{G}_{\frac{u-2}{2}}(z^2)\right)\right.$$

$$\left.\left.- 4\,m^2 x^i x^j\,\mathfrak{G}_{\frac{u-6}{2}}(z^2)\right)_{z=2m\sqrt{(x^1)^2+(x^2)^2}}\right]$$

$$\text{for } i, j \in \{1, 2\}; \qquad (7.52)$$

$$\langle 0|\widehat{T}^u_{33}(\mathbf{x})|0\rangle = \frac{m^4}{32\pi^{3/2}\,\Gamma(\frac{u+1}{2})} \left(\frac{m}{\kappa}\right)^{-u} \left[\Gamma\!\left(\frac{u-4}{2}\right)\right.$$

$$-\,2^{5-\frac{u}{2}} \sum_{i=1,2} \alpha_i\!\left(\!\left(\frac{2-u}{4} - (3-u)\xi\right)\mathfrak{G}_{\frac{u-4}{2}}(z^2)\right.$$

$$\left.+\left(\frac{1}{4}-\xi\right)\mathfrak{G}_{\frac{u-2}{2}}(z^2)\right)_{z=2mx^i}$$

$$-\,2^{5-\frac{u}{2}}\,\alpha_1\alpha_2\left(\!\left(\frac{1-u}{4} - (2-u)\xi\right)\mathfrak{G}_{\frac{u-4}{2}}(z^2)\right.$$

$$\left.\left.+\left(\frac{1}{4}-\xi\right)\mathfrak{G}_{\frac{u-2}{2}}(z^2)\right)_{z=2m\sqrt{(x^1)^2+(x^2)^2}}\right]. \quad (7.53)$$

The above expressions give the meromorphic continuation of the regularized stress-energy VEV to the whole complex plane, with poles determined by the terms involving a gamma function. In particular all components (7.51)–(7.53) happen to have a simple pole in $u = 0$, and we must resort to the extended zeta approach taking again the regular parts at $u = 0$. Recalling once more that Eq. (1.32) gives

$$\xi_3 = \frac{1}{6}\,,$$

we report separately the conformal and non-conformal parts of each component of the renormalized stress-energy VEV:

$$\langle 0|\widehat{T}^{\diamond}_{00}(\mathbf{x})|0\rangle_{\mathrm{ren}} = \frac{m^4}{384\pi^2}\left[3\left(4\ln\left(\frac{m}{2\kappa}\right)+1\right)+32\sum_{i=1,2}\alpha_i\left(\frac{K_1(z)}{z}\right)_{z=2mx^i}\right.$$

$$\left.+\,32\,\alpha_1\alpha_2\left(\frac{z\,K_1(z)-K_2(z)}{z^2}\right)_{z=2m\sqrt{(x^1)^2+(x^2)^2}}\right],$$

$$(7.54)$$

$$\langle 0|\widehat{T}_{00}^{\blacksquare}(\mathbf{x})|0\rangle_{\mathrm{ren}} = -\frac{m^4}{\pi^2}\left[\sum_{i=1,2}\alpha_i\left(\frac{z\,K_1(z)+3K_2(z)}{z^2}\right)_{z=2mx^i}\right.$$

$$\left. +\alpha_1\alpha_2\left(\frac{z\,K_1(z)+2K_2(z)}{z^2}\right)_{z=2m\sqrt{(x^1)^2+(x^2)^2}}\right];$$

$$(7.55)$$

$$\langle 0|\widehat{T}_{ij}^{\Diamond}(\mathbf{x})|0\rangle_{\mathrm{ren}}$$

$$= -\frac{m^4}{384\pi^2}\left[3\,\delta_{ij}\left(4\ln\left(\frac{m}{2\kappa}\right)-3\right)+32\,\alpha_{\mathfrak{p}(j)}\,\delta_{ij}\left(\frac{K_1(z)}{z}\right)_{z=2mx^{\mathfrak{p}(j)}}\right.$$

$$\left. +32\,\alpha_1\alpha_2\left(\delta_{ij}\,\frac{z\,K_1(z)+3K_2(z)}{z^2}-4m^2x^ix^j\,\frac{K_3(z)}{z^3}\right)_{z=2m\sqrt{(x^1)^2+(x^2)^2}}\right]$$

$$\text{for } i,j \in \{1,2\}\ ;$$

$$(7.56)$$

$$\langle 0|\widehat{T}_{ij}^{\blacksquare}(\mathbf{x})|0\rangle_{\mathrm{ren}}$$

$$= \frac{m^4}{\pi^2}\left[\alpha_{\mathfrak{p}(j)}\,\delta_{ij}\left(\frac{z\,K_1(z)+3K_2(z)}{z^2}\right)_{z=2mx^{\mathfrak{p}(j)}}\right.$$

$$\left. +\alpha_1\alpha_2\left(\delta_{ij}\,\frac{z\,K_1(z)+3K_2(z)}{z^2}-4m^2x^ix^j\,\frac{K_3(z)}{z^3}\right)_{z=2m\sqrt{(x^1)^2+(x^2)^2}}\right]$$

$$\text{for } i,j \in \{1,2\}\ ;$$

$$(7.57)$$

$$\langle 0|\widehat{T}_{33}^{\Diamond}(\mathbf{x})|0\rangle_{\mathrm{ren}} = -\frac{m^4}{384\pi^2}\left[3\left(4\ln\left(\frac{m}{2\kappa}\right)-3\right)+32\sum_{i=1,2}\alpha_i\left(\frac{K_1(z)}{z}\right)_{z=2mx^i}\right.$$

$$\left. +32\,\alpha_1\alpha_2\left(\frac{z\,K_1(z)-K_2(z)}{z^2}\right)_{z=2m\sqrt{(x^1)^2+(x^2)^2}}\right];$$

$$(7.58)$$

$$\langle 0|\widehat{T}_{33}^{\blacksquare}(\mathbf{x})|0\rangle_{\text{ren}} = \frac{m^4}{\pi^2}\left[\sum_{i=1,2}\alpha_i\left(\frac{z\,K_1(z)+3K_2(z)}{z^2}\right)_{z=2mx^i}\right.$$

$$\left.+\alpha_1\alpha_2\left(\frac{z\,K_1(z)+2K_2(z)}{z^2}\right)_{z=2m\sqrt{(x^1)^2+(x^2)^2}}\right].$$

(7.59)

As was to be expected, since the configuration (7.47) is invariant under translation along the x^3 direction, none of the expressions (7.54)–(7.59) depends on the spatial coordinate x^3.

Let us now discuss the pressure at points in the half-plane $\pi_1 = \{x^1 = 0,\, x^2 > 0\}$; note that, we are excluding points on the axis $\zeta := \{x^1 = x^2 = 0\}$. The two alternative definitions (7.27), (7.28) are easily seen to give the same result for the renormalized version of this quantity, in agreement with the general results stated in Section 7.6. Indeed, we can equivalently put $x^1 = 0$ in Eq. (7.52) for $\langle 0|\widehat{T}_{i1}^u(\mathbf{x})|0\rangle$ ($i \in \{1,2,3\}$) and then analytically continue up to $u = 0$, or directly evaluate the renormalized expressions (7.56), (7.57) for $\langle 0|\widehat{T}_{i1}(\mathbf{x})|0\rangle_{\text{ren}}$ in $x^1 = 0$; in both ways, we obtain

$$p_i^{\text{ren}}(\mathbf{x})\Big|_{\pi_1}$$

$$= \delta_{i1}\left[\frac{m^4}{384\pi^2}\left(3\left(4\ln\left(\frac{m}{2\kappa}\right)-3\right)+32\,\alpha_2\,\frac{(1+\alpha_1)z\,K_1(z)+3\alpha_1 K_2(z)}{z^2}\right)\right.$$

$$\left.-\left(\xi-\frac{1}{6}\right)\frac{m^4}{\pi^2}(1+\alpha_1)\alpha_2\,\frac{z\,K_1(z)+3K_2(z)}{z^2}\right]_{z=2mx^2}\quad (i\in\{1,2,3\}).$$

(7.60)

Let us stress that, differently from all the configurations considered so far within this book, in this case the renormalized pressure on the boundary depends in general on the parameter ξ; more precisely, this happens whenever Neumann boundary conditions are imposed on the half-plane π_1 (so that $\alpha_1 = +1$).

Now, let us consider the axis $\zeta = \{x^1 = x^2 = 0\}$; at any point of this axis the outer normal is ill defined, so that there is a basic obstruction to speaking of the pressure. However, we can discuss what happens if a point $\mathbf{x} = (0, x^2, x^3) \in \pi_1$ moves towards the axis ζ, i.e. if we consider the limit $x^2 \to 0^+$. In this limit $p_1^{\text{ren}}(\mathbf{x})$ is found to diverge; more precisely, Eq. (7.60) and the known asymptotic behavior of the Bessel function K_ν near zero (see [117], page 252, Eq. (10.30.2)) allow to infer the following:

$$p_1^{\text{ren}}(\mathbf{x}) = O\left(\frac{1}{(x^2)^4}\right) \qquad \text{for } \mathbf{x} \in \pi_1, \ x^2 \to 0^+ . \qquad (7.61)$$

7.10   The previous examples in the zero mass limit

Let us first consider the half-space configuration (7.38) bounded by the plane π_1, analyzed in Section 7.8. The expressions (7.43)–(7.45) for the components of the renormalized stress-energy VEV give, in the limit $m \to 0^+$,

$$\langle 0|\widehat{T}_{\mu\nu}(\mathbf{x})|0\rangle_{\text{ren}}\Big|_{\mu,\nu=0,1,2,3}$$

$$= \left(\xi - \frac{1}{6}\right) \frac{3\,\alpha_1}{8\pi^2 (x^1)^4} \begin{pmatrix} -1 & 0 & 0 & 0 \\ 0 & 0 & 0 & 0 \\ 0 & 0 & 1 & 0 \\ 0 & 0 & 0 & 1 \end{pmatrix} . \qquad (7.62)$$

As for the pressure on a point $\mathbf{x} \in \pi_1$, starting with Eq. (7.46) and taking the limit $m \to 0^+$, it is trivial to infer

$$p_i^{\text{ren}}(\mathbf{x}) = 0 \qquad (i \in \{1, 2, 3\}) . \qquad (7.63)$$

Let us pass to the case of a rectangular wedge, treated in Section 7.9. In the limit $m \to 0^+$, Eqs. (7.54)–(7.59) for the renormalized stress-energy VEV reduce to (recall that $\mathfrak{p}(1) = 2$, $\mathfrak{p}(2) = 1$)

$$\langle 0|\widehat{T}_{\mu\nu}(\mathbf{x})|0\rangle_{\text{ren}}\Big|_{\mu,\nu=0,1,2,3}$$

$$= \frac{\alpha_1\alpha_2}{96\pi^2\rho^4}\begin{pmatrix} -1 & 0 & 0 & 0 \\ 0 & A_1(\mathbf{x}) & B(\mathbf{x}) & 0 \\ 0 & B(\mathbf{x}) & A_2(\mathbf{x}) & 0 \\ 0 & 0 & 0 & 1 \end{pmatrix}$$

$$- \left(\xi - \frac{1}{6}\right)\left[\frac{\alpha_1\alpha_2}{8\pi^2\rho^4}\begin{pmatrix} -1 & 0 & 0 & 0 \\ 0 & A_1(\mathbf{x}) & B(\mathbf{x}) & 0 \\ 0 & B(\mathbf{x}) & A_2(\mathbf{x}) & 0 \\ 0 & 0 & 0 & 1 \end{pmatrix}\right.$$

$$\left. - \frac{3}{8\pi^2}\begin{pmatrix} -C_0(\mathbf{x}) & 0 & 0 & 0 \\ 0 & C_1(\mathbf{x}) & 0 & 0 \\ 0 & 0 & C_2(\mathbf{x}) & 0 \\ 0 & 0 & 0 & C_0(\mathbf{x}) \end{pmatrix}\right] ; \qquad (7.64)$$

$$A_i(\mathbf{x}) := 1 - \frac{4(x^{\mathfrak{p}(i)})^2}{\rho^2} , \quad C_i(\mathbf{x}) := \frac{\alpha_{\mathfrak{p}(i)}}{(x^{\mathfrak{p}(i)})^4} , \qquad \text{for } i = 1, 2 ;$$

$$B(\mathbf{x}) := \frac{4x^1x^2}{\rho^2} , \quad C_0(\mathbf{x}) := \frac{\alpha_1}{(x^1)^4} + \frac{\alpha_2}{(x^2)^4} + \frac{\alpha_1\alpha_2}{\rho^4} ,$$

$$\rho := \sqrt{(x^1)^2 + (x^2)^2} .$$

Next, consider the expression (7.60) for the pressure acting on a point $\mathbf{x}$ in the half-plane $\pi_1 = \{x^1 = 0, \, x^2 > 0\}$, for $m > 0$; using the asymptotic behaviour of the Bessel function K_ν near zero (see [117], page 252, Eq. (10.30.2)) we infer, in the limit $m \to 0^+$,

$$p_i^{\text{ren}}(\mathbf{x}) = \delta_{i1}\left[\frac{\alpha_1\alpha_2}{32\pi^2(x^2)^4} - \left(\xi - \frac{1}{6}\right)\frac{3(1+\alpha_1)\alpha_2}{8\pi^2(x^2)^4}\right] \qquad (i \in \{1, 2, 3\}) .$$

$$(7.65)$$

Notice that, as for the massive analogue (7.60), the above expression for the renormalized pressure depends explicitly on ξ if we assume Neumann boundary conditions on π_1 (so that $\alpha_1 = +1$).

In passing, let us remark that both results (7.63) and (7.65) for the pressure on π_1 could be determined equivalently via the prescription (3.23); according to the latter, we should have first considered the renormalized stress-energy VEV inside the corresponding spatial domain in the limit of zero mass (see Eqs. (7.62), (7.64)) and then move to the boundary (half-)plane π_1, i.e. take the limit $x^1 \to 0^{+}$.[d] Let us comment briefly on the construction described above, namely, that of taking the zero mass limit $(m \to 0^{+})$ of the renormalized results (7.43)–(7.45), (7.46) and (7.54)–(7.59), (7.60) for the massive field theory. As a matter of fact, this procedure corresponds to studying the case of a massless scalar field (in the same spatial configurations) with the technique of Section 4.2. Indeed, in the massless case $(m = 0)$ the spectrum of the fundamental operator $\mathcal{A} = -\Delta$ is $[0, +\infty)$, for both the settings (7.38) and (7.47); according to the framework of the cited section, we could treat these cases using the deformed operator $\mathcal{A}_\varepsilon := \mathcal{A} + \varepsilon^2$ (see Eq. (5.14)), and eventually taking the limit $\varepsilon \to 0^{+}$. On the other hand, if ε is identified with m, we recover the present constructions.[e]

Summing up: Eqs. (7.62)–(7.63) and (7.64)–(7.65) yield the renormalized VEVs of the stress-energy tensor and pressure for a $d = 3$ massless field, respectively confined within a half-space and a rectangular wedge, fulfilling either Dirichlet $(\alpha_i = -1)$ or Neumann $(\alpha_i = +1)$ boundary conditions.

In the Dirichlet case, the above results are found to agree with those derived by Actor and Bender [5] for $\xi = 0$ and $\xi = 1/6$, via a different version of the zeta approach (also involving, essentially, a subtraction of divergent contributions; see Subsections 3.1.1 and 4.1 of [5], setting $d = 3$ therein). Let us also mention that the massless half-space configuration in the case of Dirichlet boundary conditions was considered as well in our previous work [58]; therein the same

---

[d]Notice that not all of the components in Eqs. (7.62), (7.64) are finite for $x^1 \to 0^{+}$, but only those involved in the computation of the pressure on π_1 ; for example, in both cases we have

$$\lim_{x^1 \to 0^{+}} \langle 0|\widehat{T}_{22}(\mathbf{x})|0\rangle_{\text{ren}} = \infty \ .$$

[e]The same comments could be made in general for any spatial dimension d and for any number d_1 of faces.

results were obtained starting from the renormalized stress-energy VEV of a massless field between parallel planes, and taking the limit of infinite distance between the planes (see [58], page 430, Eq. (5·7); after an exchange $1 \leftrightarrow 3$ in the coordinate labels this becomes the present Eq. (7.62), with $\alpha_1 = -1$).

In the next chapter we show that the renormalized expressions (7.62)–(7.65), here deduced as the zero mass limit of a massive theory, can be obtained equivalently as particular cases of the theory of a massless scalar field confined within two half-planes forming an angle α of arbitrary width; more precisely, the present configurations with one single plane and two orthogonal planes correspond, respectively, to the limits $\alpha \to \pi$ and $\alpha \to \pi/2$.

Chapter 8

A massless field in a three-dimensional wedge

In the present chapter we consider a massless field confined within a wedge of arbitrary width in spatial dimension $d = 3$, for several types of boundary conditions. We also consider a variation of this configuration, which corresponds essentially to identify the sides of the wedge; this is the so-called case of the "cosmic string" (see Section 8.9). As usual, in the first Section 8.1 we give a formal description of the general configuration we are going to consider. In the subsequent Sections 8.2–8.5 we use the general techniques of Part 1 to determine the analytic continuation of the regularized observables VEVs, for several boundary conditions; the renormalized counterparts of these quantities are computed in Sections 8.6–8.9.

Some special cases of the wedge configuration have already been treated by Dowker *et al.* [47, 48], Deutsch and Candelas [45] (also discussing the electromagnetic case) and, more recently, by Saharian *et al.* [125, 131] and by Fulling *et al.* [73] (see also the references cited within these works and [25, 26, 114]); nearly all of these authors use the point-splitting approach, or some variant of it. More in detail, in [47] and [45] the attention is restricted to the conformal part of the stress-energy VEV for either Dirichlet or Neumann boundary conditions, while in [125, 131] the non-conformal part is considered as well, but in the Dirichlet case only; in [73], the authors present the graphs of the energy density and of the pressure components (for which no explicit expression is given), derived via a point-splitting approach for several configurations and various choices of the parameters describing the theory. The approach via zeta regularization employed in the present book allows to deal with several types of boundary conditions, both in the conformal and in the non-conformal case; moreover, in the subcases already analyzed in the literature cited above, it gives the same results.

8.1 Introducing the problem for arbitrary boundary conditions

We consider the case of a scalar field (with no external forces) confined within a three-dimensional wedge, meaning that the spatial domain is

$$\Omega := \left\{ \begin{array}{c} \text{the portion of } \mathbf{R}^3 \text{ enclosed by} \\ \text{two half-planes } \pi_0, \pi_\alpha \text{ forming an angle } \alpha \in (0, 2\pi] \end{array} \right\} ;$$

(8.1)

with no loss of generality, we assume

$$\pi_0 = \{x^2 = 0, \ x^1 \geqslant 0\} . \tag{8.2}$$

Suitable boundary conditions will be specified in the following. We choose to confine our attention to the massless case ($V = 0$) since, in this case, we are able to perform the explicit computations with a moderate effort for arbitrary values of the angle $\alpha \in (0, 2\pi]$; in the massive case ($V = m^2$) we could give an exhaustive analysis for rational values of α/π but this would require a big computational effort and produce cumbersome expressions for the final results.

In passing let us notice that, when either Dirichlet or Neumann conditions are prescribed on the boundary, the spatial domain under analysis corresponds for $\alpha = \pi$ and $\alpha = \pi/2$, respectively, to the configurations with a boundary made of a single plane and of two orthogonal half-planes. Besides, let us also stress that for $\alpha = 2\pi$ the two half-planes π_0, π_α overlap; because of this, for Dirichlet or Neumann boundary conditions, the boundary consists of a single half-plane, while, in the case of periodic boundary conditions, the spatial domain Ω can be identified with $\mathbf{R}^3$. So, in the latter case one is actually considering a massless scalar field on the whole Minkowski spacetime. We will comment further on each specific case in the next sections. In particular, we will show that in the cases with $\alpha = \pi$ and $\alpha = \pi/2$ one recovers the results of Section 7.10.

In order to deal with the present configuration, it is advisable to pass to a system of cylindrical coordinates

$$\mathbf{x} \mapsto \mathbf{q}(\mathbf{x}) \equiv (\rho(\mathbf{x}), \theta(\mathbf{x}), z(\mathbf{x})) \in (0, +\infty) \times (0, 2\pi) \times \mathbf{R} ; \qquad (8.3)$$

the inverse map will be written $\mathbf{q} \mapsto \mathbf{x}(\mathbf{q})$. These coordinates are chosen so that $z(\mathbf{x}) = x^3$ and the boundary $\partial\Omega$ corresponds to the limit values $\theta = 0$ and $\theta = \alpha$; the spatial line element reads

$$d\ell^2 = d\rho^2 + \rho^2 d\theta^2 + dz^2 . \qquad (8.4)$$

In order to avoid clumsy notations, given any function $\Omega \to Y$, $\mathbf{x} \mapsto f(\mathbf{x})$ (with Y any set), the composition $\mathbf{q} \in \mathbf{q}(\Omega) \mapsto f(\mathbf{x}(\mathbf{q}))$ will be written as $\mathbf{q} \mapsto f(\mathbf{q})$.

Since curvilinear coordinates are being employed, one should refer to the framework of Section 4.3. Writing ρ, θ, z for the coordinate labels, we see that the only non-vanishing Christoffel symbols associated to the line element (8.4) are $\gamma^\rho_{\theta\theta} = -\rho$ and $\gamma^\theta_{\rho\theta} = \gamma^\theta_{\theta\rho} = \frac{1}{\rho}$; so, the second order covariant derivatives of any scalar function f are given by

$$\nabla_{\rho\rho} f = \partial_{\rho\rho} f , \qquad \nabla_{\rho\theta} f = \partial_{\rho\theta} f - \frac{1}{\rho} \partial_\theta f , \qquad \nabla_{\rho z} f = \partial_{\rho z} f ,$$
$$\nabla_{\theta\theta} f = \partial_{\theta\theta} f + \rho \partial_\rho f , \qquad \nabla_{\theta z} f = \partial_{\theta z} f , \qquad \nabla_{zz} f = \partial_{zz} f . \qquad (8.5)$$

In conclusion, let us stress that the configuration under analysis could be dealt with as a slab configuration, where $\Omega = \Omega_1 \times \mathbf{R}$ and $\Omega_1 \subset \mathbf{R}^2$ corresponds to $(0, +\infty) \times (0, \alpha)$ in terms of the coordinates (ρ, θ); yet, this approach is not convenient. In fact, if one works directly on the three-dimensional spatial domain Ω (for any of the boundary conditions to be considered in the following), the modified cylinder kernel $\tilde{T}$ defined in Eq. (2.53) associated to the fundamental operator $\mathcal{A} = -\Delta$ can be expressed in terms of elementary functions and it is meromorphic in the variable $\mathfrak{t}$ on the whole complex plane. On the contrary, to treat the problem as a slab configuration we should use the analogous kernel $\tilde{T}$ for the reduced operator $\mathcal{A}_1$ on Ω_1, or any other integral kernel related to the latter; these kernels do not possess simple expressions (only integral representations are available), so that the whole analysis would become a lot more involved.

8.2   The Dirichlet kernel

For any of the boundary conditions to be considered in the following, the spectrum of $\mathcal{A}$ is $[0, +\infty)$. Since $\{0\}$ is a non-isolated point of the spectrum, we must resort to the methods discussed in Section 4.2 to determine the renormalized Dirichlet kernel (and its derivatives); we regularize the theory using the deformed operator $\mathcal{A}_\varepsilon = (\sqrt{\mathcal{A}} + \varepsilon)^2$ (see Eq. (4.17)), whose choice is found, *a posteriori*, to be more effective from the computational viewpoint. In the sequel (see Sections 8.6–8.9) we derive the explicit expression of the modified cylinder kernel $\tilde{T}$ of $\mathcal{A}$, for several types of boundary conditions; in any case $\tilde{T}$ is found to be a meromorphic function of t, decreasing faster than t^{-1} for $\Re t \to +\infty$; this allows us to proceed as explained in Section 4.2. In this way we obtain, for the Dirichlet kernels and its derivatives, the following renormalized expressions at $s = \pm 1/2$, respectively (see Eq. (4.21)):

$$D_{-\frac{1}{2}}(\mathbf{q}, \mathbf{p})\Big|_{\mathbf{p}=\mathbf{q}} = \text{Res}\left(2\,t^{-3}\,\tilde{T}(t;\mathbf{q},\mathbf{p})\Big|_{\mathbf{p}=\mathbf{q}};0\right); \qquad (8.6)$$

$$\nabla_{vw}D_{+\frac{1}{2}}(\mathbf{q}, \mathbf{p})\Big|_{\mathbf{p}=\mathbf{q}} = \text{Res}\left(t^{-1}\nabla_{vw}\tilde{T}(t;\mathbf{q},\mathbf{p})\Big|_{\mathbf{p}=\mathbf{q}};0\right),$$
$$\tag{8.7}$$

for v, w any two cylindrical coordinates .

Here and in the sequel, we are using the following notations:

$$\mathbf{q} \equiv (\rho, \theta, z) , \qquad \mathbf{p} \equiv (\rho', \theta', z') ;$$
$$f(\mathbf{x}(\mathbf{q}), \mathbf{x}(\mathbf{p})) \equiv f(\mathbf{q}, \mathbf{p}) \quad \text{for any } f : \Omega \times \Omega \to Y \ (Y \text{ a set}) .$$
$$\tag{8.8}$$

8.3   The stress-energy tensor

Relations (8.6), (8.7), along with Eqs. (2.28)–(2.30), can be used to obtain the following expressions for the renormalized VEV of the stress-energy tensor:

$$\langle 0|\widehat{T}_{00}(\mathbf{q})|0\rangle_{\text{ren}} = \text{Res}\left(t^{-3}\left[\left(\frac{1}{2} + 2\xi\right)\tilde{T}(t;\mathbf{q},\mathbf{p})\right.\right.$$
$$\left.\left. + \left(\frac{1}{4} - \xi\right)t^2\,\partial^{q^\ell}\partial_{p^\ell}\tilde{T}(t;\mathbf{q},\mathbf{p})\right]_{\mathbf{p}=\mathbf{q}};0\right); \qquad (8.9)$$

$$\langle 0|\widehat{T}_{ij}(\mathbf{q})|0\rangle_{\mathrm{ren}}$$

$$= \langle 0|\widehat{T}_{ji}(\mathbf{q})|0\rangle_{\mathrm{ren}}$$

$$= \mathrm{Res}\left(\mathfrak{t}^{-3} \left[\left(\frac{1}{2} - 2\xi \right) \delta_{ij}\, \tilde{T}(\mathfrak{t};\mathbf{q},\mathbf{p}) + \mathfrak{t}^2 \left(\left(\frac{1}{4} - \frac{\xi}{2} \right) \partial_{q^i p^j} \right. \right. \right.$$

$$\left. \left. \left. - \frac{\xi}{2} \nabla_{q^i q^j} - \left(\frac{1}{4} - \xi \right) \delta_{ij} \partial^{q^\ell} \partial_{p^\ell} \right) \tilde{T}(\mathfrak{t};\mathbf{q},\mathbf{p}) \right]_{\mathbf{p}=\mathbf{q}} ; 0 \right)$$

$$\text{for } i,j \in \{\rho,\theta,z\}\ . \tag{8.10}$$

In the following sections we will first compute $\tilde{T}$ and then explicitly evaluate the residues appearing in Eqs. (8.9), (8.10) for the cases where either Dirichlet, Neumann or periodic boundary conditions are prescribed; we will present the final results in the form (1.32), recalling once again that Eq. (1.32) implies

$$\xi_3 = \frac{1}{6}\ . \tag{8.11}$$

8.4 The boundary forces

In order to discuss this topic, since the spectrum of $\mathcal{A}$ contains $\{0\}$ as a non-isolated point, we must resort once more to the framework of Section 4.2; again, we choose to use the deformed operator $\mathcal{A}_\varepsilon := (\sqrt{\mathcal{A}} + \varepsilon)^2$. We can consider the two alternative definitions (4.25), (4.26) for the renormalized pressure on the boundary. Contrary to the configurations analyzed in the previous chapters, in this case the two mentioned alternatives *do not yield the same result*.

As an example, let us focus on the pressure acting on the half-plane π_α. We indicate with $\mathbf{n}(\mathbf{q})$ the outer unit normal at a point of this half-plane with coordinates $\mathbf{q} = (\rho,\alpha,z)$; this has components $(n^\rho, n^\theta, n^z)(\mathbf{q}) = (0, 1/\rho, 0)$. On the one hand, prescription (4.25) corresponds to put, for $i \in \{\rho,\theta,z\}$,

$$p_i^{\mathrm{ren}}(\mathbf{q}) := \lim_{\varepsilon \to 0+}\left(RP\Big|_{u=0} \langle 0|\widehat{T}_{ij}^{\varepsilon u}(\mathbf{q})|0\rangle\, n^j(\mathbf{q}) \right)$$

$$= \frac{1}{\rho} \lim_{\varepsilon \to 0+} RP\Big|_{u=0} \langle 0|\widehat{T}_{i\theta}^{\varepsilon u}(\mathbf{q})|0\rangle \tag{8.12}$$

(first consider the regularized stress-energy VEV on π_α, then analytically continue at $u = 0$ and finally take the limit $\varepsilon \to 0^+$). Similarly to Eq. (8.10), the prescription (8.12) yields

$$p_i^{\mathrm{ren}}(\mathbf{q}) = \frac{1}{\rho} \operatorname{Res}\left(\mathfrak{t}^{-3} \left[\left(\frac{1}{2} - 2\xi\right) \delta_{i\theta} \, \tilde{T}(\mathfrak{t}; \mathbf{q}, \mathbf{p}) \right.\right.$$

$$+ \mathfrak{t}^2 \left(\left(\frac{1}{4} - \frac{\xi}{2}\right) \partial_{q^i\theta'} - \frac{\xi}{2} \nabla_{q^i\theta} \right.$$

$$\left.\left. - \left(\frac{1}{4} - \xi\right) \delta_{i\theta} \partial^{q^\ell} \partial_{p^\ell} \right) \tilde{T}(\mathfrak{t}; \mathbf{q}, \mathbf{p}) \right]_{\mathbf{p} = \mathbf{q} \in \pi_\alpha} ; 0 \right) ; \quad (8.13)$$

let us stress that in the above expression, we first perform the evaluation on the boundary and then compute the residue.

On the other hand, the alternative prescription (4.26) yields, for $i \in \{\rho, \theta, z\}$,

$$p_i^{\mathrm{ren}}(\mathbf{q}) := \left(\lim_{\mathbf{q}' \to \mathbf{q}, \mathbf{x}(\mathbf{q}') \in \Omega} \langle 0 | \widehat{T}_{ij}(\mathbf{q}') | 0 \rangle_{\mathrm{ren}} \right) n^j(\mathbf{q})$$

$$= \frac{1}{\rho} \left(\lim_{\mathbf{q}' \to \mathbf{q}, \mathbf{x}(\mathbf{q}') \in \Omega} \langle 0 | \widehat{T}_{i\theta}(\mathbf{q}') | 0 \rangle_{\mathrm{ren}} \right). \quad (8.14)$$

In the cases to be considered in Sections 8.6–8.8, it will be apparent that the explicit expressions obtained for the renormalized stress-energy VEV inside the wedge diverge when approaching the boundary, in such a way to make divergent the renormalized pressure defined by (8.14). Because of this, we will always refer to Eq. (8.13) to deal with the pressure on the boundary.

8.5   Some remarks

For the computation of $\tilde{T}$ we will often refer to [73], where this kernel was already determined for the present configuration (but used in a different renormalization scheme, based on point-splitting as an alternative to zeta regularization; we note that the kernel $\overline{T}$ in [73] is the opposite of the kernel $\tilde{T}$ considered here).

At the end of each section dealing with Dirichlet and/or Neumann boundary conditions, we will comment briefly on the results obtained for the renormalized stress-energy VEV (8.9)–(8.10) and pressure (8.14) in the cases $\alpha = \pi$ and $\alpha = \pi/2$, respectively, describing a massless scalar field on a half-space and inside a wedge with orthogonal half-planes. Recall that these very same configurations were analyzed as the zero mass limit of a corresponding massive theory in Section 7.10; indeed, we will find that the same results obtained therein can be re-obtained from Eqs. (8.9)–(8.10) and (8.14), returning to the Cartesian coordinates $x^1 = \rho\sin\theta$, $x^2 = \rho\cos\theta$, $x^3 = z$ and considering the appropriate transformation laws.[a]

8.6 Dirichlet boundary conditions

Let us first consider the case where the field fulfills Dirichlet boundary conditions on the half-planes π_0, π_α. In this case a complete orthonormal system of (improper) eigenfunctions $(F_k)_{k\in\mathcal{K}}$ of $\mathcal{A} = -\Delta$, with eigenvalues $(\omega_k^2)_{k\in\mathcal{K}}$, is given by

$$F_k(\mathbf{q}) := \sqrt{\frac{\omega}{\pi\alpha}}\, J_{\lambda_n}(\omega\rho)\sin(\lambda_n\theta)\, e^{ihz}, \qquad \lambda_n := \frac{n\pi}{\alpha},$$

$$\omega_k^2 := \omega^2 + h^2 \qquad \text{for } k \equiv (n,\omega,h) \in \mathcal{K} \equiv \mathbf{N}\times(0,+\infty)\times\mathbf{R} \tag{8.15}$$

(here and elsewhere we are considering the Bessel functions, and the set of positive integers $\mathbf{N} := \{1,2,3,...\}$; $\mathcal{K}$ is endowed with the counting measure on $\mathbf{N}$ times the standard Lebesgue measure on $(0,+\infty)\times\mathbf{R}$, meaning that $\int_{\mathcal{K}}dk \equiv \sum_{n=1}^{+\infty}\int_0^{+\infty}d\omega\int_{-\infty}^{+\infty}dh$). The modified cylinder kernel $\tilde{T}$ can be evaluated starting from its eigenfunction expansion (2.53), which in the present setting reads

$$\tilde{T}(\mathsf{t}\,;\mathbf{q},\mathbf{p}) = \int_{\mathcal{K}}dk\,\frac{e^{-\mathsf{t}\omega_k}}{\omega_k}\, F_k(\mathbf{q})\overline{F_k}(\mathbf{p})$$

$$= \frac{1}{\pi\alpha}\sum_{n=1}^{+\infty}\sin(\lambda_n\theta)\sin(\lambda_n\theta')\int_0^{+\infty}d\omega\,\omega\,J_{\lambda_n}(\omega\rho)\,J_{\lambda_n}(\omega\rho')$$

$$\times\int_{-\infty}^{+\infty}dh\,\frac{e^{-\mathsf{t}\sqrt{\omega^2+h^2}}}{\sqrt{\omega^2+h^2}}\, e^{ih(z-z')} \,. \tag{8.16}$$

---

[a] Clearly, $\langle 0|\widehat{T}_{\mu\nu}(\mathbf{q})|0\rangle_{\text{ren}}$ and $p_i^{\text{ren}}(\mathbf{q})$ transform, respectively, as a rank-two tensor and a vector.

With some effort, the integrals over h and ω in the above expression can be explicitly evaluated to yield

$$\tilde{T}(t; \mathbf{q}, \mathbf{p}) = \frac{1}{\pi \alpha \, \rho \rho' \sinh v} \sum_{n=1}^{+\infty} \sin(\lambda_n \theta) \, \sin(\lambda_n \theta') \, e^{-\lambda_n v} \,,$$

$$v := -\ln\left(\frac{r_+ - r_-}{r_+ + r_-}\right), \qquad r_\pm := \sqrt{(\rho \pm \rho')^2 + (z - z')^2 + t^2} \,.$$

(8.17)

We refer to [73] for more details on the computations giving the above result; the notations $v, r_\pm$ are borrowed from this reference (see also [95]). The series in Eq. (8.17) can be re-expressed via four geometric series writing the trigonometric functions in terms of complex exponentials; in this way we obtain

$$\tilde{T}(t; \mathbf{q}, \mathbf{p}) = \frac{1}{4\pi \alpha \, \rho\rho' \sinh v} \left(\frac{\cos(\frac{\pi}{\alpha}(\theta - \theta')) - e^{-\frac{\pi}{\alpha}v}}{\cosh(\frac{\pi}{\alpha}v) - \cos(\frac{\pi}{\alpha}(\theta - \theta'))} \right.$$

$$\left. - \frac{\cos(\frac{\pi}{\alpha}(\theta + \theta')) - e^{-\frac{\pi}{\alpha}v}}{\cosh(\frac{\pi}{\alpha}v) - \cos(\frac{\pi}{\alpha}(\theta + \theta'))} \right).$$

(8.18)

Now, we resort to Eqs. (8.9)–(8.10) for the renormalized stress-energy VEV; evaluating the residues appearing therein, we obtain

$$\langle 0|\widehat{T}_{\mu\nu}(\mathbf{q})|0\rangle_{\mathrm{ren}}\Big|_{\mu,\nu=0,\rho,\theta,z}$$

$$= A(\mathbf{q}) \begin{pmatrix} -1 & 0 & 0 & 0 \\ 0 & 1 & 0 & 0 \\ 0 & 0 & -3\rho^2 & 0 \\ 0 & 0 & 0 & 1 \end{pmatrix} - \left(\xi - \frac{1}{6}\right)$$

$$\times \begin{pmatrix} -(B(\mathbf{q})+C(\mathbf{q})) & 0 & 0 & 0 \\ 0 & B(\mathbf{q}) & -\rho\,E(\mathbf{q}) & 0 \\ 0 & -\rho\,E(\mathbf{q}) & \rho^2\,C(\mathbf{q}) & 0 \\ 0 & 0 & 0 & B(\mathbf{q})+C(\mathbf{q}) \end{pmatrix},$$

$$A(\mathbf{q}) := \frac{\pi^4 - \alpha^4}{1440\pi^2 \alpha^4 \, \rho^4},$$

$$B(\mathbf{q}) := \frac{9\pi^4 - 3\pi^2(2\pi^2+\alpha^2)\sin^2(\frac{\pi\theta}{\alpha}) + \alpha^2(\pi^2-\alpha^2)\sin^4(\frac{\pi\theta}{\alpha})}{24\pi^2\alpha^4 \sin^4(\frac{\pi\theta}{\alpha})\,\rho^4},$$

$$C(\mathbf{q}) := \frac{3\pi^2 - (\pi^2-\alpha^2)\sin^4(\frac{\pi\theta}{\alpha})}{8\pi^2\alpha^2 \sin^2(\frac{\pi\theta}{\alpha})\,\rho^4}, \qquad E(\mathbf{q}) := \frac{3\pi\cos(\frac{\pi\theta}{\alpha})}{8\alpha^3 \sin^3(\frac{\pi\theta}{\alpha})\,\rho^4}.$$

$$(8.19)$$

It can be easily checked that the above result agrees with the one derived by Saharian and Tarloyan by means of point-splitting regularization in [131] (see, in particular, Section 3 therein); see also the former papers by Dowker *et al.* [47, 48] and by Deutsch and Candelas [45], where point-splitting regularization is used for the computation of the conformal stress-energy VEV alone.

Let us briefly comment on the explicit expression (8.19) obtained for the renormalized VEV $\langle 0|\widehat{T}_{\mu\nu}(\mathbf{q})|0\rangle_{\mathrm{ren}}$. First of all, notice that the function $A(\mathbf{q})$ multiplying the conformal part of the renormalized VEV is positive for $\alpha < \pi$, negative for $\alpha > \pi$ and vanishes for $\alpha = \pi$. Next, let us remark that both the conformal and non-conformal parts diverge quartically in ρ in the proximity of the axis $\{\rho = 0\}$.

The non-conformal part also diverges near the half-planes π_0, π_α, that is for $\theta \to 0^+$ or $\theta \to \alpha^-$. Because of this, the pressure on these half-planes evaluated according to Eq. (8.14) is infinite. On the other hand, the alternative definition (8.12)–(8.13) (first move to the boundary, and then take the analytic continuation) gives a finite pressure on π_α with components

$$p_i^{\mathrm{ren}}(\mathbf{q}) = -\,\delta_{i\theta}\,\frac{\pi^4 - \alpha^4}{480\pi^2\alpha^4\rho^3}. \qquad (8.20)$$

To conclude, we consider the special cases $\alpha = \pi$ and $\alpha = \pi/2$ (a space domain bounded by a plane or by two perpendicular half-planes), and compare the present results with the ones derived in Section 7.10. This can be done with the procedure outlined in Section 8.5 (i.e. returning to Cartesian coordinates via the appropriate transformation rules for tensor coefficients). Indeed, the expressions (8.19), (8.20) for the renormalized stress-energy VEV and pressure are easily seen to yield for $\alpha = \pi$ and $\alpha = \pi/2$, respectively, Eqs. (7.62), (7.63) with $\alpha_1 = -1$, and Eqs. (7.64), (7.65) with $\alpha_1 = \alpha_2 = -1$.

8.7   Dirichlet–Neumann boundary conditions

Let us now pass to the analysis of the wedge configuration where Dirichlet and Neumann boundary conditions are prescribed, respectively, on the half-planes π_0 and π_α. To the best of our knowledge, this case has never been considered before in the literature.

A complete orthonormal system of (improper) eigenfunctions $(F_k)_{k \in \mathcal{K}}$ of $\mathcal{A}$, with eigenvalues $(\omega_k^2)_{k \in \mathcal{K}}$, is given by

$$F_k(\mathbf{q}) := \sqrt{\frac{\omega}{\pi\alpha}}\, J_{\lambda_n}(\omega\rho)\sin(\lambda_n\theta)\, e^{ihz}\,, \quad \lambda_n := \left(n + \frac{1}{2}\right)\frac{\pi}{\alpha}\,, \tag{8.21}$$

$$\omega_k^2 := \omega^2 + h^2 \quad \text{for } k \equiv (n, \omega, h) \in \mathcal{K} \equiv \mathbf{N}_0 \times (0, +\infty) \times \mathbf{R}$$

($\mathbf{N}_0 := \{0, 1, 2, ...\}$ is the set of non-negative integers and the measure on the label space $\mathcal{K}$ is such that $\int_\mathcal{K} dk \equiv \sum_{n=0}^{+\infty} \int_0^{+\infty} d\omega \int_{-\infty}^{+\infty} dh$). Resorting again to Eq. (2.53) and proceeding similarly to the case with Dirichlet boundary conditions, we can express the modified cylinder kernel as

$$\tilde{T}(\mathfrak{t}; \mathbf{q}, \mathbf{p}) = \frac{1}{2\pi\alpha\, \rho\rho' \sinh v}\left(\frac{\sinh(\frac{\pi}{2\alpha}v)\cos(\frac{\pi}{2\alpha}(\theta - \theta'))}{\cosh(\frac{\pi}{\alpha}v) - \cos(\frac{\pi}{\alpha}(\theta - \theta'))}\right.$$

$$\left. - \frac{\sinh(\frac{\pi}{2\alpha}v)\cos(\frac{\pi}{2\alpha}(\theta + \theta'))}{\cosh(\frac{\pi}{\alpha}v) - \cos(\frac{\pi}{\alpha}(\theta + \theta'))}\right)\,, \tag{8.22}$$

where v is defined as in Eq. (8.17). Using Eqs. (8.9)–(8.10) we obtain the renormalized VEV of the stress-energy tensor:

$$\langle 0|\widehat{T}_{\mu\nu}(\mathbf{q})|0\rangle_{\text{ren}}\Big|_{\mu,\nu=0,\rho,\theta,z} = A(\mathbf{q})\begin{pmatrix} 1 & 0 & 0 & 0 \\ 0 & -1 & 0 & 0 \\ 0 & 0 & 3\rho^2 & 0 \\ 0 & 0 & 0 & -1 \end{pmatrix} - \left(\xi - \frac{1}{6}\right)$$

$$\times \begin{pmatrix} -(B(\mathbf{q}) + C(\mathbf{q})) & 0 & 0 & 0 \\ 0 & B(\mathbf{q}) & -\rho\, E(\mathbf{q}) & 0 \\ 0 & -\rho\, E(\mathbf{q}) & \rho^2 C(\mathbf{q}) & 0 \\ 0 & 0 & 0 & B(\mathbf{q}) + C(\mathbf{q}) \end{pmatrix}\,,$$

$$A(\mathbf{q}) := \frac{7\pi^4 + 8\alpha^4}{11520\pi^2\alpha^4\rho^4} \, ,$$

$$B(\mathbf{q}) := \frac{-3\pi^2 \cos(\frac{\pi\theta}{\alpha})(11\pi^2 - 2\alpha^2 + (\pi^2 + 2\alpha^2)\cos(\frac{2\pi\theta}{\alpha})) + 2\alpha^2(\pi^2 + 2\alpha^2)\sin^4(\frac{\pi\theta}{\alpha})}{96\pi^2\alpha^4\sin^4(\frac{\pi\theta}{\alpha})\rho^4} \, ,$$

$$C(\mathbf{q}) := \frac{6\pi^2 \cos(\frac{\pi\theta}{\alpha}) + (\pi^2 + 2\alpha^2)\sin^2(\frac{\pi\theta}{\alpha})}{16\pi^2\alpha^2\sin^2(\frac{\pi\theta}{\alpha})\rho^4} \, , \quad E(\mathbf{q}) := \frac{3\pi(3 + \cos(\frac{2\pi\theta}{\alpha}))}{32\alpha^3\sin^3(\frac{\pi\theta}{\alpha})\rho^4} \, .$$

$$(8.23)$$

Let us compare the above results with the ones of Eq. (8.19), holding for the case of Dirichlet boundary conditions on both the half-planes π_0, π_α. As in Eq. (8.19), both the conformal and non-conformal parts of the renormalized stress-energy VEV diverge for $\rho \to 0$; the latter also diverges for $\theta \to 0^+$ and $\theta \to \alpha^-$, so that Eq. (8.14) yields again a divergent pressure on the boundary. On the other hand, the present results differ from the ones derived in the previous section because of some crucial features; in particular, the conformal part has an overall minus sign and the function $A(\mathbf{q})$ in Eq. (8.23) is always strictly positive (whereas the one in Eq. (8.19) changes sign for $\alpha < \pi$ and $\alpha > \pi$).

As for the boundary forces on π_α, resorting to Eq. (8.13), in this case we obtain

$$p_i^{\mathrm{ren}} = \delta_{i\theta} \frac{1}{8\pi^2\rho^3} \left[\frac{7\pi^4 + 8\alpha^4}{480\alpha^4} - \left(\xi - \frac{1}{6}\right) \frac{\pi^2 + 2\alpha^2}{\alpha^2} \right]. \qquad (8.24)$$

We notice that also in this case the parameter ξ appears in the final expression for the renormalized pressure; because of this the resulting boundary forces can be either attractive or repulsive, depending on the value of ξ.

Again, we conclude comparing the results obtained for the renormalized stress-energy VEV and pressure for $\alpha = \pi/2$ with the analogous ones deduced in Section 7.10. Also this time, Eqs. (8.23), (8.24) (with $\alpha = \pi/2$) are found to give, respectively, Eqs. (7.64), (7.65) with $\alpha_1 = 1$, $\alpha_2 = -1$.

8.8 Neumann boundary conditions

We now analyze the case where the field fulfills Neumann boundary conditions on both the half-planes π_0, π_α. A complete orthonormal system of (improper) eigenfunctions $(F_k)_{k\in\mathcal{K}}$ in $L^2(\Omega)$ of $\mathcal{A}$, with related eigenvalues $(\omega_k^2)_{k\in\mathcal{K}}$, is

$$F_k(\mathbf{q}) = \sqrt{\frac{\omega}{\pi\alpha}}\, J_{\lambda_n}(\omega\rho)\cos(\lambda_n\theta)\, e^{ihz}\,, \qquad \lambda_n := \frac{n\pi}{\alpha}\,,$$

$$\omega_k^2 = \omega^2 + h^2 \qquad \text{for } k \equiv (n,\omega,h) \in \mathcal{K} \equiv \mathbf{N}_0\times(0,+\infty)\times\mathbf{R} \tag{8.25}$$

(recall that $\mathbf{N}_0 := \{0,1,2,...\}$; again, we assume the measure on the label space $\mathcal{K}$ is such that $\int_{\mathcal{K}} dk \equiv \sum_{n=0}^{+\infty}\int_0^{+\infty} d\omega \int_{-\infty}^{+\infty} dh$). Also in this case, the modified cylinder kernel $\tilde{T}$ can be evaluated according to Eq. (2.53). More precisely, proceeding as we did in Section 8.6 for the case of Dirichlet boundary conditions (see, in particular, the derivation of Eq. (8.18)), we obtain

$$\tilde{T}(\mathfrak{t};\mathbf{q},\mathbf{p}) = \frac{1}{4\pi\alpha\,\rho\rho'\sinh v}\left(\frac{e^{\frac{\pi}{\alpha}v}-\cos(\frac{\pi}{\alpha}(\theta-\theta'))}{\cosh(\frac{\pi}{\alpha}v)-\cos(\frac{\pi}{\alpha}(\theta-\theta'))}\right.$$

$$\left.+\frac{e^{\frac{\pi}{\alpha}v}-\cos(\frac{\pi}{\alpha}(\theta+\theta'))}{\cosh(\frac{\pi}{\alpha}v)-\cos(\frac{\pi}{\alpha}(\theta+\theta'))}\right); \tag{8.26}$$

again, v si defined as in Eq. (8.17). Now, we can resort once more to Eqs. (8.9)–(8.10) to evaluate the renormalized VEV of the stress-energy tensor; the final result is

$$\langle 0|\widehat{T}_{\mu\nu}(\mathbf{q})|0\rangle_{\text{ren}}\Big|_{\mu,\nu=0,\rho,\theta,z} = A(\mathbf{q})\begin{pmatrix}-1 & 0 & 0 & 0\\ 0 & 1 & 0 & 0\\ 0 & 0 & -3\rho^2 & 0\\ 0 & 0 & 0 & 1\end{pmatrix} + \left(\xi-\frac{1}{6}\right)$$

$$\times\left[\begin{pmatrix}-(B(\mathbf{q})+C(\mathbf{q})) & 0 & 0 & 0\\ 0 & B(\mathbf{q}) & -\rho\,E(\mathbf{q}) & 0\\ 0 & -\rho\,E(\mathbf{q}) & \rho^2\,C(\mathbf{q}) & 0\\ 0 & 0 & 0 & B(\mathbf{q})+C(\mathbf{q})\end{pmatrix}\right.$$

$$\left.+\,G(\mathbf{q})\begin{pmatrix}-2 & 0 & 0 & 0\\ 0 & -1 & 0 & 0\\ 0 & 0 & 3\rho^2 & 0\\ 0 & 0 & 0 & 2\end{pmatrix}\right], \tag{8.27}$$

where the functions A, B, C, E are defined as in Eq. (8.19) and we set

$$G(\mathbf{q}) := \frac{\pi^2 - \alpha^2}{12\pi^2 \alpha^2 \rho^4} \ . \tag{8.28}$$

We notice that, in accordance with the existing literature (see, e.g. [45]), the conformal part of the renormalized stress-energy VEV coincides with the analogous contribution derived for Dirichlet boundary conditions in Section 8.6; besides, comments analogous to those made at the end of the cited section also hold in this case. Let us only remark that the VEV $\langle 0|\widehat{T}_{\mu\nu}(\mathbf{q})|0\rangle_{\mathrm{ren}}$ has an additional term proportional to the function $G(\mathbf{q})$; this function changes sign for either $\alpha < \pi$ or $\alpha > \pi$ and diverges for $\rho \to 0$.

Concerning the pressure on the boundary, also in this case Eq. (8.14) clearly yields a divergent result; on the other hand, using Eq. (8.13), we obtain

$$p_i^{\mathrm{ren}} = -\,\delta_{i\theta}\,\frac{1}{4\pi^2\rho^3}\left[\frac{\pi^4 - \alpha^4}{120\alpha^4} - \left(\xi - \frac{1}{6}\right)\frac{\pi^2 - \alpha^2}{\alpha^2}\right] . \tag{8.29}$$

As in the previous sections, we find that the renormalized pressure depends explicitly on the parameter ξ.

Proceeding as explained in Section 8.5, the renormalized stress-energy VEV (8.27) and pressure (8.29) are easily seen to give for $\alpha = \pi$ and $\alpha = \pi/2$, respectively, Eqs. (7.62), (7.63) with $\alpha_1 = 1$, and Eqs. (7.64), (7.65) with $\alpha_1 = \alpha_2 = 1$.

8.9 Periodic boundary conditions (the cosmic string)

Finally, let us consider the case where the field fulfills periodic boundary conditions on the half-planes π_0, π_α, meaning that

$$\widehat{\phi}(t, \rho, 0, z) = \widehat{\phi}(t, \rho, \alpha, z) \ , \quad \partial_\theta \widehat{\phi}(t, \rho, 0, z) = \partial_\theta \widehat{\phi}(t, \rho, \alpha, z)$$

$$\text{for } t, z \in \mathbf{R}, \ \rho \in (0, +\infty) \ . \tag{8.30}$$

In passing, let us mention that the same framework was also analyzed by Dowker [48] and by Fulling *et al.* [73], both employing a point-splitting approach; more precisely, in [48] the conformal part of the energy density alone is computed, while in [73] the authors only report the graphs of the energy density and pressure for $\xi = 1/4$.

Similarly to the cases of the segment and parallel hyperplanes configurations with periodic boundary conditions (considered, respectively, in Section 5.9 of Chapter 5 and in Section 6.9 of Chapter 6), the spatial domain Ω for the present setting is more properly addressed as a flat Riemannian manifold. The manifold Ω admits a global coordinate system $\mathbf{q} = (\rho, \theta, z) : \Omega \to (0, +\infty) \times \mathbf{T}_\alpha^1 \times \mathbf{R}$, $\mathbf{x} \mapsto \mathbf{q}(\mathbf{x})$ where the second factor is the one-dimensional torus $\mathbf{T}_\alpha^1 := \mathbf{R}/(\alpha\mathbf{Z})$; the line element in these coordinates has the form (8.4).[b] The corresponding spacetime $\mathbf{R} \times \Omega$ (with the line element $ds^2 = -dt^2 + d\ell^2$) is usually described in terms of a "cosmic string" due to the presence of a one-dimensional topological defect coinciding with the axis $\{\rho = 0\}$.

A complete orthonormal system of (improper) eigenfunctions $(F_k)_{k \in \mathcal{K}}$ of $\mathcal{A}$ in $L^2(\Omega)$, with the related eigenvalues $(\omega_k^2)_{k \in \mathcal{K}}$, is given by

$$F_k(\mathbf{q}) := \sqrt{\frac{\omega}{2\pi\alpha}}\, J_{|\lambda_n|}(\omega\rho)\, e^{i\lambda_n\theta}\, e^{ihz}\,, \qquad \lambda_n := \frac{2n\pi}{\alpha}\,, \tag{8.31}$$

$$\omega_k^2 := \omega^2 + h^2 \qquad \text{for } k \equiv (n, \omega, h) \in \mathcal{K} \equiv \mathbf{Z} \times (0, +\infty) \times \mathbf{R}$$

(similarly to the previous sections, we are assuming $\mathcal{K}$ to be a measure space such that $\int_{\mathcal{K}} dk \equiv \sum_{n=-\infty}^{+\infty} \int_0^{+\infty} d\omega \int_{-\infty}^{+\infty} dh$). The modified cylinder kernel $\tilde{T}$ can then be evaluated starting from its eigenfunction expansion (2.53):

$$\tilde{T}(\mathbf{t}; \mathbf{q}, \mathbf{p}) = \int_{\mathcal{K}} dk\, \frac{e^{-t\omega_k}}{\omega_k}\, F_k(\mathbf{x})\overline{F_k}(\mathbf{y})$$

$$= \frac{1}{2\pi\alpha} \sum_{n=-\infty}^{+\infty} e^{i\lambda_n(\theta-\theta')} \int_0^{+\infty} d\omega\, \omega\, J_{|\lambda_n|}(\omega\rho)\, J_{|\lambda_n|}(\omega\rho')$$

$$\times \int_{-\infty}^{+\infty} dh\, \frac{e^{-t\sqrt{\omega^2+h^2}}}{\sqrt{\omega^2+h^2}}\, e^{ih(z-z')}\,. \tag{8.32}$$

---

[b]In other words, Ω is a quotient space of the Dowker manifold Ω_∞; this is an infinite-sheeted Riemannian surface that can be described in terms of a global coordinate system

$$\mathbf{q} : \Omega_\infty \to (0, +\infty) \times \mathbf{R} \times \mathbf{R}\,, \qquad \mathbf{x} \mapsto \mathbf{q}(\mathbf{x}) = (\rho(\mathbf{x}), \theta(\mathbf{x}), z(\mathbf{x}))$$

(and of the line element (8.4)).

Evaluating the integrals in h and ω as in [73] and considering separately the terms with positive and negative values of n, we obtain the expression

$$\tilde{T}(\mathsf{t};\mathbf{q},\mathbf{p}) = \frac{1}{2\pi\alpha\rho\rho'\sinh v}\left(1 + \sum_{n=1}^{+\infty} e^{-\lambda_n v}\, e^{i\lambda_n(\theta-\theta')}\right.$$

$$\left. + \sum_{n=1}^{+\infty} e^{-\lambda_n v}\, e^{-i\lambda_n(\theta-\theta')}\right), \qquad (8.33)$$

which in turn, summing the geometric series, yields (again, v is defined as in Eq. (8.17))

$$\tilde{T}(\mathsf{t};\mathbf{q},\mathbf{p}) = \frac{1}{2\pi\alpha\,\rho\rho'\sinh v}\,\frac{\sinh(\frac{2\pi}{\alpha}v)}{\cosh(\frac{2\pi}{\alpha}v)-\cos(\frac{2\pi}{\alpha}(\theta-\theta'))}\,. \qquad (8.34)$$

Using Eqs. (8.9)–(8.10) once more, we find the following expression for the renormalized stress-energy VEV:

$$\langle 0|\widehat{T}_{\mu\nu}(\mathbf{q})|0\rangle_{\mathrm{ren}}\Big|_{\mu,\nu=0,\rho,\theta,z}$$

$$= A(\mathbf{q})\begin{pmatrix} -1 & 0 & 0 & 0 \\ 0 & 1 & 0 & 0 \\ 0 & 0 & -3\rho^2 & 0 \\ 0 & 0 & 0 & 1 \end{pmatrix} + \left(\xi-\frac{1}{6}\right)G(\mathbf{q})\begin{pmatrix} -2 & 0 & 0 & 0 \\ 0 & -1 & 0 & 0 \\ 0 & 0 & 3\rho^2 & 0 \\ 0 & 0 & 0 & 2 \end{pmatrix}, \qquad (8.35)$$

$$A(\mathbf{q}) = \frac{(2\pi)^4 - \alpha^4}{1440\pi^2\alpha^4\rho^4}\,, \qquad G(\mathbf{q}) = \frac{(2\pi)^2 - \alpha^2}{24\pi^2\alpha^2\rho^4}\,.$$

Let us observe that the above result does not depend explicitly on the angular variable θ; this was to be expected due to the homogeneity of the considered configuration with respect to this coordinate. Besides, as for the cases of Dirichlet and Neumann boundary conditions, both the conformal and non-conformal part of the renormalized VEV of the stress-energy tensor diverge near the axis $\{\rho = 0\}$, that is in the proximity of the cosmic string.

In conclusion, we notice that for $\alpha = 2\pi$, in which case the considered configuration is equivalent to that of a scalar field on the whole Minkowski spacetime, the expressions in Eq. (8.35) reduce to $A(\mathbf{q}) = G(\mathbf{q}) = 0$. So, we have this result with its own interest: when zeta regularization is applied to a massless scalar field on the whole $(3 + 1)$-dimensional Minwkowski spacetime, the renormalized stress-energy VEV vanishes identically.

Chapter 9

A scalar field with a harmonic background potential

Hereafter we proceed to describe a more engaging application of the general techniques developed in Part 1 of this book, which requires some numerical computations. The results reported here were first derived in a previous work of ours [60]; they concern the case of a massless scalar field living on the whole space $\mathbf{R}^d$, and confined by a background harmonic potential. For simplicity, the latter is assumed to be isotropic, i.e. to be proportional to the squared radius $|\mathbf{x}|^2$ ($\mathbf{x} \in \mathbf{R}^d$); nevertheless, the approach described here could be extended with little effort to anisotropic harmonic potentials and to the case of a massive field theory.

Adhering to the general scheme adopted in the previous chapters of Part 2, in the opening Section 9.1 we give a formal description of the model under analysis. In the subsequent Sections 9.2–9.5 we derive explicit integral representations for the renormalized stress-energy VEV, using the techniques of Part 1 and the fact that the heat kernel in the case of a harmonic potential is the well-known Mehler kernel [102]. In Sections 9.6 and 9.7 we describe general methods allowing to determine the asymptotics of the previously mentioned VEV for small and large values of the radial coordinate. The renormalized VEVs of the total energy is derived in Section 9.8. Finally, in Sections 9.9 and 9.10 we consider the cases of spatial dimensions $d = 2$ and $d = 3$, respectively; in particular, the above mentioned integral representations for the renormalized stress-energy VEV are evaluated numerically using `Mathematica`.

Before proceeding, let us mention that we refer to [60] for the proof of some technical results which are stated in the present chapter (this work also gives a detailed analysis of the case $d = 1$, here omitted for brevity).

The idea to replace sharp boundaries with suitable background potentials is well-known in the literature on the Casimir effect. Typically, delta-like potentials are introduced in order to mimic boundary conditions in a physically more realistic framework (see, e.g. [7, 8, 13, 20, 24, 31, 79, 100, 112, 113, 141, 144]); the ultimate purpose is to obtain renormalized quantities with less singular behaviors, avoiding, e.g. boundary divergences such as the ones pointed out in Section 3.5 of Part 1 and found in the preceding Chapters 5–8. The case of a scalar field interacting with an external harmonic potential was originally considered by Actor and Bender [1, 4]. These authors were able to determine the renormalized VEV of the total energy via a global zeta approach based on ad hoc results, giving the analytic continuation of some particular kind of "generalized" zeta functions befitting the configuration under analysis. On the contrary, following our previous work [60], in the present chapter we compute the renormalized VEV of the total energy applying our general procedures in a mechanical way. Our methods and those of [1, 4] could be proved with some effort to be equivalent in any spatial dimension d; we limit ourselves to check by direct comparison that, in the case $d = 3$,[a] the numerical results we obtain agree with the ones derived in [4] (see Section 9.10).

The computation of the renormalized stress-energy VEV is mentioned only in few lines of [4], where it is indicated as an interesting, though non simple problem.

9.1   Introducing the problem

We consider the case of a massless scalar field on $\mathbf{R}^d$ in presence of a classical, isotropic harmonic potential. More precisely, we assume

$$\Omega := \mathbf{R}^d, \qquad V(\mathbf{x}) := k^4 |\mathbf{x}|^2 , \qquad (9.1)$$

---

[a] As a matter of fact, this is the only model discussed both here and in [4].

where $k > 0$ and $|\mathbf{x}| := \sqrt{(x^1)^2 + ... + (x^d)^2}$; the constant k is, dimensionally, a mass (or an inverse length) like the parameter κ employed for the field regularization (see Eq. (1.17) and the corresponding remarks).

All the results reported in the present chapter could be generalized, with some computational effort, to the anisotropic case $V(\mathbf{x}) = \sum_{i=1}^{d} (k_i)^4 (x^i)^2$ where $k_i \geqslant 0$ for all $i \in \{1, ..., d\}$, also including slab cases where some of the k_i's vanish. A further variation would concern a massive field in one of the above mentioned configurations, so that, e.g. in place of (9.1) we would have $V(\mathbf{x}) := m^2 + k^4 |\mathbf{x}|^2$. None of these generalizations will be considered here, for the sake of computational simplicity.

9.2 The heat kernel

Even though the configuration (9.1) is patently spherically symmetric, we choose to postpone the use of spherical coordinates to the next section; here we work in standard Cartesian coordinates $(x^i)_{i=1,...,d}$, for reasons that will soon become apparent.

First of all, let us remark that (9.1) is a product type configuration: the Hilbert space of the system and the fundamental operator $\mathcal{A} := -\Delta + V$ have, respectively, the representations

$$L^2(\mathbf{R}^d) = \bigotimes_{i=1}^{d} L^2(\mathbf{R}) \, , \qquad (9.2)$$

$$\mathcal{A} = \mathcal{A}_1 \otimes \mathbf{1} \otimes ... \otimes \mathbf{1} + ... + \mathbf{1} \otimes ... \otimes \mathbf{1} \otimes \mathcal{A}_1 \, ,$$
$$\mathcal{A}_1 := -\frac{d^2}{dx^2} + k^4 x^2 \quad \text{on } L^2(\mathbf{R}) \, . \qquad (9.3)$$

In view of the general considerations of Section 2.9 (see, in particular, Eq. (2.127)), in this situation, the heat kernel of $\mathcal{A}$ is given by

$$K(\mathfrak{t}; \mathbf{x}, \mathbf{y}) = \prod_{i=1}^{d} K_1(\mathfrak{t}; x^i, y^i) \qquad (9.4)$$

where K_1 is the heat kernel corresponding to $\mathcal{A}_1$; this the familiar *Mehler kernel*, that is[b]

$$K_1(\mathsf{t};x,y) = \frac{k}{\sqrt{2\pi \sinh(2k^2\mathsf{t})}} \exp\left[-k^2\left(\frac{x^2+y^2}{2\tanh(2k^2\mathsf{t})} - \frac{x\,y}{\sinh(2k^2\mathsf{t})}\right)\right].$$
(9.5)

Writing $\mathbf{x}\cdot\mathbf{y} := \sum_{i=1}^{d} x^i\, y^i$, Eqs. (9.4) and (9.5) imply

$$K(\mathsf{t};\mathbf{x},\mathbf{y}) = \left(\frac{k}{\sqrt{2\pi\sinh(2k^2\mathsf{t})}}\right)^d$$

$$\times \exp\left[-k^2\left(\frac{|\mathbf{x}|^2+|\mathbf{y}|^2}{2\tanh(2k^2\mathsf{t})} - \frac{\mathbf{x}\cdot\mathbf{y}}{\sinh(2k^2\mathsf{t})}\right)\right].$$
(9.6)

A similar analysis can be performed for the heat trace $K(\mathsf{t})$. We refer to [60] for more details on the related computation; here we only report the final result, giving the explicit expression

$$K(\mathsf{t}) = \left(\frac{1}{2\sinh(k^2\mathsf{t})}\right)^d.$$
(9.7)

9.3   Rescaled spherical coordinates

As anticipated in the previous section, we now pass to a set of curvilinear coordinates which best fit the symmetries of the problem (see Section 4.3); more precisely, we introduce the spherical "k-rescaled" coordinates

$$\mathbf{x} \mapsto \mathbf{q}(\mathbf{x}) \equiv (r(\mathbf{x}), \theta_1(\mathbf{x}), ..., \theta_{d-2}(\mathbf{x}), \theta_{d-1}(\mathbf{x})) \in (0, +\infty)$$

$$\times (0,\pi) \times ... \times (0,\pi) \times (0,2\pi)$$
(9.8)

whose inverse map $\mathbf{q} \mapsto \mathbf{x}(\mathbf{q})$ is described by the equations

$$k\,x^1 = r\cos(\theta_1)\,,$$
$$k\,x^2 = r\sin(\theta_1)\cos(\theta_2)\,,$$
$$\vdots$$
(9.9)
$$k\,x^{d-1} = r\sin(\theta_1)...\sin(\theta_{d-2})\cos(\theta_{d-1})\,,$$
$$k\,x^d = r\sin(\theta_1)...\sin(\theta_{d-1})\,.$$

---

[b]We refer, in particular, to [32]; the equations reported therein are here employed making the replacement $(t,x,y) \to (k^2t, kx, ky)$. See also [17, 44, 80].

Needless to say, for $d = 1$ the system is invariant under the parity symmetry $x^1 \to -x^1$; in this case, we set

$$r := k\,|x^1| \ . \tag{9.10}$$

Let us stress that, for any spatial dimension d, we have

$$r = k\,|\mathbf{x}| \ ; \tag{9.11}$$

this shows that the coordinate r is in fact a dimensionless radius. In order to avoid cumbersome notations, given a function $\Omega \to Y$, $\mathbf{x} \mapsto f(\mathbf{x})$ (with Y any set), we indicate the composition $\mathbf{q} \mapsto f(\mathbf{x}(\mathbf{q}))$ as $\mathbf{q} \mapsto f(\mathbf{q})$; we will use similar conventions for functions on $\Omega \times \Omega$.

Let us now consider the Dirichlet kernel $D_s(\mathbf{x}, \mathbf{y})$ and the heat kernel $K(\mathfrak{t}; \mathbf{x}, \mathbf{y})$; let

$$\mathbf{q} = (r, \theta_1, ..., \theta_{d-1}) \equiv (r, \boldsymbol{\theta}) \ , \quad \mathbf{p} = (r', \theta'_1, ..., \theta'_{d-1}) \equiv (r, \boldsymbol{\theta}') \ , \tag{9.12}$$

$$\tau := k^2\,\mathfrak{t} \in (0, +\infty) \ . \tag{9.13}$$

In the sequel we write $D_s(\mathbf{q}, \mathbf{p})$ and $K(\tau; \mathbf{q}, \mathbf{p})$, respectively, for the Dirichlet and heat kernels at two points $\mathbf{x}, \mathbf{y}$ of (rescaled) spherical coordinates $\mathbf{q}, \mathbf{p}$, with τ related to $\mathfrak{t}$ via Eq. (9.13). Using these notations, Eq. (9.6) can be rephrased as

$$K(\tau; \mathbf{q}, \mathbf{p}) = \left(\frac{k}{\sqrt{2\pi \sinh(2\tau)}} \right)^d$$

$$\times \exp\left[-\left(\frac{r^2 + r'^{\,2}}{2 \tanh(2\tau)} - \frac{r\,r'\,S(\boldsymbol{\theta})S(\boldsymbol{\theta}')}{\sinh(2\tau)} \right) \right] \tag{9.14}$$

where $S(\boldsymbol{\theta})$ and $S(\boldsymbol{\theta}')$ are the products of sines and cosines of the angular coordinates $(\theta_1, ..., \theta_{d_1})$ and $(\theta'_1, ..., \theta'_{d_1})$ of Eq. (9.12), corresponding to the scalar product $\mathbf{x} \cdot \mathbf{y}$.

From Eq. (2.85) it follows that

$$D_s(\mathbf{q}, \mathbf{p}) = \frac{k^{-2s}}{\Gamma(s)} \int_0^{+\infty} d\tau\,\tau^{s-1}\,K(\tau; \mathbf{q}, \mathbf{p}) \ . \tag{9.15}$$

Substituting this relation into the analogues of Eqs. (2.28)–(2.30) for curvilinear coordinates (see Section 4.3), we get

$$\langle 0|\widehat{T}^u_{\mu\nu}(\mathbf{q})|0\rangle = \frac{k^{d+1}}{\Gamma(\frac{u+1}{2})}\left(\frac{\kappa}{k}\right)^u \int_0^{+\infty} d\tau\, \tau^{\frac{u-d-3}{2}}\, \mathcal{H}^{(u)}_{\mu\nu}(\tau;\mathbf{q})$$

$$(\mu,\nu \in \{0,r,\theta_1,...,\theta_{d-1}\}) \quad (9.16)$$

where the coefficients $\mathcal{H}^{(u)}_{\mu\nu}(\tau;\mathbf{q})$ are as follows (for $i,j,h,\ell \in \{\rho,\theta_1,...,\theta_{d-1}\}$; D indicates the spatial covariant derivative)

$$\mathcal{H}^{(u)}_{00}(\tau;\mathbf{q}) := \left(\frac{\tau}{k^2}\right)^{d/2}\left[\left(\frac{1}{4}+\xi\right)\left(\frac{u-1}{2}\right)\right.$$

$$\left.+\left(\frac{1}{4}-\xi\right)\tau\left(a^{h\ell}(\mathbf{q})D_{q^h p^\ell}+r^2\right)\right]\Bigg|_{\mathbf{p}=\mathbf{q}} K(\tau;\mathbf{q},\mathbf{p}), \quad (9.17)$$

$$\mathcal{H}^{(u)}_{0i}(\tau;\mathbf{q}) = \mathcal{H}^{(u)}_{i0}(\tau;\mathbf{q}) := 0, \quad (9.18)$$

$$\mathcal{H}^{(u)}_{ij}(\tau;\mathbf{q}) = \mathcal{H}^{(u)}_{ji}(\tau;\mathbf{q})$$

$$:= \left(\frac{\tau}{k^2}\right)^{d/2}\left[\left(\frac{1}{4}-\xi\right)a_{ij}(\mathbf{q})\left(\frac{u-1}{2}-\tau\left(a^{h\ell}(\mathbf{q})D_{q^h p^\ell}+r^2\right)\right)\right.$$

$$\left.+\left(\frac{\tau}{k^2}\right)\left(\left(\frac{1}{2}-\xi\right)D_{q^i p^j}-\xi\, D_{q^i q^j}\right)\right]\Bigg|_{\mathbf{p}=\mathbf{q}} K(\tau;\mathbf{q},\mathbf{p}).$$

$$(9.19)$$

Here and in the reminder of this chapter, we are implicitly understanding the dependence on the parameter ξ for simplicity of notation: so, $\mathcal{H}^{(u)}_{\mu\nu}(\tau;\mathbf{q})$ stands for $\mathcal{H}^{(u)}_{\mu\nu}(\tau;\mathbf{q};\xi)$.

9.4   Regularity properties of some coefficients

Hereafter we point out some notable properties of the coefficients $\mathcal{H}^{(u)}_{\mu\nu}(\tau;\mathbf{q})$, defined in Eqs. (9.17)–(9.19) of the previous section. These properties will be employed in the subsequent Section 9.5 to evaluate the renormalized stress-energy VEV.

Using Eq. (9.14) for the heat kernel expressed in rescaled, spherical coordinates, one can prove the following facts:

(i) For fixed $\tau, \mathbf{q}$ and any $\mu, \nu \in \{0, r, \theta_1, ..., \theta_{d-1}\}$, $\mathcal{H}_{\mu\nu}^{(u)}(\tau; \mathbf{q})$ is polynomial of degree 1 in u.

(ii) For fixed $\mathbf{q}$ and any $u \in \mathbf{C}$, $\mu, \nu \in \{0, r, \theta_1, ..., \theta_{d-1}\}$, the map $\tau \mapsto \mathcal{H}_{\mu\nu}^{(u)}(\tau; \mathbf{q})$ is smooth on $[0, +\infty)$ and exponentially vanishing for $\tau \to +\infty$.

(iii) The final expressions for the coefficients $\mathcal{H}_{\mu\nu}^{(u)}$ do not depend on the parameter k, even though this appears in the right-hand sides of Eqs. (9.17)–(9.19).

(iv) There holds

$$\mathcal{H}_{\mu\nu}^{(u)} = 0 \qquad \text{for } \mu \neq \nu \,,$$

$$\mathcal{H}_{\theta_{d-1}\theta_{d-1}}^{(u)} = \sin^2(\theta_{d-2}) \, \mathcal{H}_{\theta_{d-2}\theta_{d-2}}^{(u)} = \ ... \ = \sin^2(\theta_{d-2}) \, ... \, \sin^2(\theta_1) \, \mathcal{H}_{\theta_1\theta_1}^{(u)} \,. \tag{9.20}$$

(v) For $\mu = \nu \in \{0, r, \theta_1\}$, there holds

$$\mathcal{H}_{\mu\nu}^{(u)}(\tau; \mathbf{q}) = e^{-r^2 \tanh \tau} \mathcal{M}_{\mu\nu}^{(u)}(\tau; r) \tag{9.21}$$

where $\mathcal{M}_{\mu\nu}^{(u)}(\tau; r)$ is a polynomial in r, u of degree 1 in both these variables, with coefficients depending smoothly on τ.

Before moving on, let us make a couple of remarks.

(a) Consider the integral representation (9.16) for the regularized stress-energy VEV; due to item (iii), the latter VEV only depends on the parameter k through the multiplicative coefficient $k^{d+1}(\kappa/k)^u$ in front of the integral in the cited equation.

(b) Eq. (9.20) of item (iv) indicates that the VEV $\langle 0|\widehat{T}_{\mu\nu}^u(\mathbf{q})|0\rangle$ is diagonal and that the only independent components are those with $\mu = \nu \in \{0, r, \theta_1\}$; these components only depend on the rescaled radial coordinate r (and not on the angles $\{\theta_1, ..., \theta_{d-1}\}$), in agreement with the spherical symmetry of the configuration under analysis.

9.5 Analytic continuation of the regularized stress-energy tensor

Let us move on to determine the analytic continuation of the regularized VEV (9.16). To this purpose, we refer to the general framework of Subsection 2.8.2; using Eq. (2.100) with $\mathcal{H} = \mathcal{H}_{\mu\nu}^{(u)}(\ ; \mathbf{q})$ and

$\rho = (d + 3 - u)/2$, we obtain

$$\langle 0|\widehat{T}_{\mu\nu}^u(\mathbf{q})|0\rangle = \frac{k^{d+1}}{\Gamma(\frac{u+1}{2})} \frac{(-1)^n}{(\frac{u-d-3}{2}+1)...(\frac{u-d-3}{2}+n)} \left(\frac{\kappa}{k}\right)^u$$

$$\times \int_0^{+\infty} d\tau \; \tau^{\frac{u-d-3}{2}+n} \; \partial_\tau^n \mathcal{H}_{\mu\nu}^{(u)}(\tau; \mathbf{q}) \; . \qquad (9.22)$$

Notice that, due to the features of $\mathcal{H}_{\mu\nu}^{(u)}(\tau; \mathbf{q})$ pointed out in the previous section, the integral in the above expression converges for all $u \in \mathbf{C}$ with

$$\Re u > d + 1 - 2n \; ; \qquad (9.23)$$

therefore, Eq. (9.22) yields the required analytic continuation of $\langle 0|\widehat{T}_{\mu\nu}^u(\mathbf{q})|0\rangle$ to the very same region, with the possible exception of poles determined by the denominator in the right-hand side of Eq. (9.22). Since we are interested in evaluating (the regular part of) this analytic continuation at $u = 0$, we choose $n \in \mathbf{N}$ so that the condition in Eq. (9.23) holds for $u = 0$, i.e.

$$n > \frac{d+1}{2} \; . \qquad (9.24)$$

Following Eq. (1.24), in general we define

$$\langle 0|\widehat{T}_{\mu\nu}(\mathbf{q})|0\rangle_{\text{ren}} := RP\Big|_{u=0} \langle 0|\widehat{T}_{\mu\nu}^u(\mathbf{q})|0\rangle \; . \qquad (9.25)$$

To proceed, let us fix $n \in \mathbf{N}$ fulfilling Eq. (9.24) and consider the expression on the right-hand side of Eq. (9.22). For any even spatial dimension d this expression is regular at $u = 0$, so that we can apply the zeta regularization in its restricted version to obtain the renormalized stress-energy VEV. On the other hand, for odd d we must resort to the extended version of the zeta technique, since the function under analysis has a simple pole in $u = 0$.

Because of the pole singularity, in the case of odd d the procedure of evaluating the regular part of Eq. (9.22) in $u = 0$ implies the appearance of a logarithmic term in τ in the corresponding integrand function (see footnote e on page 12 of [60] for more details on this statement). Simple but rather lengthy computations give the following results, for d either odd or even and $\mu, \nu \in \{0, r, \theta_1\}$:

$$\langle 0|\widehat{T}_{\mu\nu}(\mathbf{q})|0\rangle_{\text{ren}} = k^{d+1}\left(T_{\mu\nu}^{(0)}(r) + M_{\kappa,k}\, T_{\mu\nu}^{(1)}(r)\right), \quad \text{where}$$

$$T_{\mu\nu}^{(0)}(r) := \int_0^{+\infty} d\tau\; \tau^{n-\frac{d+3}{2}}\; e^{-r^2\tanh\tau}\left[\mathcal{P}_{\mu\nu}^{(0)}(\tau\,;r) + \ln\tau\; \mathcal{P}_{\mu\nu}^{(1)}(\tau;r)\right],$$

$$T_{\mu\nu}^{(1)}(r) := \int_0^{+\infty} d\tau\; \tau^{n-\frac{d+3}{2}}\; e^{-r^2\tanh\tau}\; \mathcal{P}_{\mu\nu}^{(1)}(\tau;r),$$

$$M_{\kappa,k} := \gamma_{EM} + 2\ln\left(\frac{2\kappa}{k}\right)$$

(9.26)

($\gamma_{EM} \simeq 0.577$ is the Euler–Mascheroni constant). In the above $\mathcal{P}_{\mu\nu}^{(0)}(\tau\,;r)$ and $\mathcal{P}_{\mu\nu}^{(1)}(\tau\,;r)$ are suitable functions determined by $\mathcal{H}_{\mu\nu}^{(u)}(\tau;\mathbf{q})$; these functions are in fact polynomials in r^2 of the form

$$\mathcal{P}_{\mu\nu}^{(a)}(\tau\,;r) = \sum_{i=0}^{n+1} p_{i,\mu\nu}^{(a)}(\tau)\, r^{2i} \quad (a \in \{0,1\}), \quad (9.27)$$

for some smooth functions $p_{i,\mu\nu}^{(0)}, p_{i,\mu\nu}^{(1)} : [0, +\infty) \to \mathbf{R}$ ($i \in \{0, ..., n+1\}$). Let us stress that

$$\mathcal{P}_{\mu\nu}^{(1)}(\tau\,;r) = 0 \quad \text{and} \quad T_{\mu\nu}^{(1)}(r) = 0 \quad \text{for } d \text{ even}, \quad (9.28)$$

a fact corresponding to the previous comments on the logarithmic terms.

Next notice that, in consequence of the remarks made in Section 9.4, the renormalized VEV $\langle 0|\widehat{T}_{\mu\nu}(\mathbf{q})|0\rangle_{\text{ren}}$ only depends on the parameter k through the coefficients k^{d+1} and $M_{\kappa,k}$ in the first equation of (9.26). In particular, the functions $T_{\mu\nu}^{(a)}(r)$ ($a \in \{0,1\}$) introduced therein do not depend on k and they can be evaluated computing the integrals in Eq. (9.26) numerically, for any fixed value of the rescaled radial coordinate $r \in (0, +\infty)$.

To conclude, in agreement with the considerations of Subsection 1.4.2, we define the conformal and non-conformal parts of the functions $r \mapsto T_{\mu\nu}^{(0)}(r), T_{\mu\nu}^{(1)}(r)$ respectively as

$$T_{\mu\nu}^{(a,\Diamond)} := T_{\mu\nu}^{(a)}\Big|_{\xi=\xi_d}, \quad T_{\mu\nu}^{(a,\blacksquare)} := \frac{1}{\xi - \xi_d}\left(T_{\mu\nu}^{(a)} - T_{\mu\nu}^{(a,\Diamond)}\right) \quad (a \in \{0,1\})$$

(9.29)

where ξ_d is defined according to (1.32). In Sections 9.9 and 9.10 we write explicitly the functions $\mathcal{P}_{\mu\nu}^{(a)}$ ($a \in \{0,1\}$) and present the graphs (obtained via numerical integration) of the functions defined in Eq. (9.29) for $d = 2$ and $d = 3$, respectively.

9.6   Asymptotics for small values of the radial coordinate

Hereafter we briefly describe a result which allows to determine the asymptotic expansion for the renormalized stress-energy VEV in the limit $r = k|\mathbf{x}| \to 0^+$. More details on the derivation of the said result can be found in [60] (see, in particular, Section 3.6 and the related Appendix A therein).

Let us consider a function $F : (0, +\infty) \to \mathbf{R}$ defined by

$$F(r) := \int_0^{+\infty} d\tau \, e^{-r^2 h(\tau)} \, P(\tau \, ; r) \, , \qquad (9.30)$$

where $h(\tau)$ is a positive bounded function and $P(\tau \, ; r)$ is a polynomial of degree N in r^2 with integrable coefficients; more precisely, assume that

$$P(\tau \, ; r) = \sum_{i=0}^{N} p_i(\tau) \, r^{2i} \, , \qquad (9.31)$$

for some integrable functions $p_i : (0, +\infty) \to \mathbf{R}$ ($i \in \{0, ..., N\}$). Substituting the exponential in Eq. (9.30) with the corresponding Taylor series, one obtains an expansion of the form

$$F(r) = \sum_{i=0}^{N} a_i \, r^{2i} + O(r^{2(N+1)}) \qquad \text{for } r \to 0^+ \, , \qquad (9.32)$$

where the coefficients $a_i \in \mathbf{R}$ ($i \in \{0, ..., N\}$) are defined as follows:

$$a_i := \sum_{j=0}^{i} \frac{(-1)^{i-j}}{(i-j)!} \int_0^{+\infty} d\tau \, p_j(\tau) \, h(\tau)^{i-j} \, . \qquad (9.33)$$

Now, let us consider the expression (9.26) for the independent components of the renormalized stress-energy VEV; recalling that $\mathcal{P}_{\mu\nu}^{(0)}$ and $\mathcal{P}_{\mu\nu}^{(1)}$ are both polynomials in r^2 of degree $n+1$ (see

Eq. (9.27)), we see that each component (for given μ, ν) has the form (9.30), (9.31) with

$$h(\tau) = \tanh(\tau) \,,$$

$$P(\tau; r) = \tau^{n - \frac{d+3}{2}} \left(\mathcal{P}_{\mu\nu}^{(0)}(\tau; r) + (M_{\kappa, k} + \ln \tau) \, \mathcal{P}_{\mu\nu}^{(1)}(\tau; r) \right) \,, \quad N = n + 1 \,.$$

$$(9.34)$$

Thus, using the above approach, we can infer the expansion for the renormalized stress-energy VEV in the limit $r = k|\mathbf{x}| \to 0^+$; we defer the explicit results for $d = 2$ and $d = 3$ to Sections 9.9 and 9.10, where the integrals in Eq. (9.33) are evaluated numerically for the corresponding choices of h and P.

9.7 Asymptotics for large values of the radial coordinate

In the present section we discuss some general techniques allowing to determine the asymptotic behavior of renormalized stress-energy VEV in the limit $r = k|\mathbf{x}| \to +\infty$. A detailed analysis of this topic is given in [60] (see, in particular, Section 3.7 and the related Appendix A therein).

Let $\lambda > -1$ and consider the function $F_\lambda : (0, +\infty) \to \mathbf{R}$ defined by

$$F_\lambda(r) := \int_0^1 dv \, e^{-r^2 v} \, v^\lambda \, Q(v; r) \,, \tag{9.35}$$

where $Q(v; r)$ is a polynomial in r^2 of the form

$$Q(v; r) = Q^{(0)}(v; r) + Q^{(1)}(v; r) \ln v \,,$$

$$Q^{(a)}(v; r) = \sum_{i=0}^{N} q_i^{(a)}(v) \, r^{2i} \quad (a \in \{0, 1\}) \,, \tag{9.36}$$

for some smooth integrable functions $q_i^{(0)}, q_i^{(1)} : [0, 1) \to \mathbf{R}$ ($i \in \{0, ..., N\}$).

The asymptotic expansion of $F(r)$ for large values of r can be obtained using the standard theory of Laplace integrals [41, 63, 116,

142]. With some lengthy computations (which can be found in [60]), for any $K \in \{1, 2, 3, ...\}$, one infers

$$F_\lambda(r) = \sum_{i=0}^{N} \sum_{m=0}^{K+i-1} \left[\alpha_{i,m} + \beta_{i,m} \ln r^2 \right] r^{2(i-m-\lambda-1)}$$

$$+ O(r^{-2(K+\lambda+1)} \ln r^2) \qquad \text{for } r \to +\infty \; ; \quad (9.37)$$

here, the real coefficients $\alpha_{i,m}, \beta_{i,m}$ (for $i \in \{0, ..., N\}$, $m \in \{0, 1, 2, ...\}$) are given by

$$\alpha_{i,m} := \Gamma(m + \lambda + 1) \left[q_{i,m}^{(0)} + \psi(m + \lambda + 1) \, q_{i,m}^{(1)} \right] ,$$

$$\beta_{i,m} := -\Gamma(m + \lambda + 1) \, q_{i,m}^{(1)} , \qquad (9.38)$$

where $\psi(s) := \partial_s \ln \Gamma(s)$ denotes the digamma function, and we have put

$$q_{i,m}^{(a)} := \left. \frac{d^m}{dv^m} \right|_{v=0} q_i^{(a)}(v) \qquad (a \in \{0, 1\}) . \qquad (9.39)$$

Let us now consider the expression (9.26) for the renormalized VEV of the stress-energy tensor and perform in the integrals appearing therein the change of variable

$$\tau = \operatorname{arcth}(v) , \quad v \in (0, 1) \qquad (9.40)$$

("arcth" denotes the inverse hyperbolic tangent function); in this way we obtain

$$T_{\mu\nu}^{(0)}(r) = \int_0^1 dv \, e^{-r^2 v} \frac{(\operatorname{arcth} v)^{n - \frac{d+3}{2}}}{1 - v^2}$$

$$\times \left[\mathcal{P}_{\mu\nu}^{(0)}(\operatorname{arcth} v \, ; r) + \ln(\operatorname{arcth} v) \, \mathcal{P}_{\mu\nu}^{(1)}(\operatorname{arcth} v \, ; r) \right] ,$$

$$T_{\mu\nu}^{(1)}(r) = \int_0^1 dv \, e^{-r^2 v} \frac{(\operatorname{arcth} v)^{n - \frac{d+3}{2}}}{1 - v^2} \, \mathcal{P}_{\mu\nu}^{(1)}(\operatorname{arcth} v \, ; r) .$$

$$(9.41)$$

Since $\mathcal{P}_{\mu\nu}^{(0)}$ and $\mathcal{P}_{\mu\nu}^{(1)}$ are polynomials in r^2 of degree $n + 1$ (see Eq. (9.27)), we can re-express each component of the renormalized stress-energy VEV (for given μ, ν) in the form (9.35), where $Q(v;r)$ is as in Eq. (9.36) with

$$\lambda := n - \frac{d+3}{2}, \qquad N = n+1 ; \tag{9.42}$$

$$Q^{(0)}(v;r) := \frac{1}{1-v^2} \left(\frac{\mathrm{arcth}\, v}{v}\right)^{n-\frac{d+3}{2}} \left[\mathcal{P}^{(0)}_{\mu\nu}(\mathrm{arcth}\, v\,;r)\right.$$

$$\left. + \left(\ln\left(\frac{\mathrm{arcth}\, v}{v}\right) + M_{\kappa,k}\right)\mathcal{P}^{(1)}_{\mu\nu}(\mathrm{arcth}\, v\,;r)\right] ; \tag{9.43}$$

$$Q^{(1)}(v;r) := \frac{1}{1-v^2} \left(\frac{\mathrm{arcth}\, v}{v}\right)^{n-\frac{d+3}{2}} \mathcal{P}^{(1)}_{\mu\nu}(\mathrm{arcth}\, v\,;r) . \tag{9.44}$$

In this way we can derive asymptotic expansions analogous to the one in Eq. (9.37) for the renormalized stress-energy VEV in the limit $r = k|\mathbf{x}| \to +\infty$; we report in Sections 9.9 and 9.10 the explicit results of this computation for the cases with $d = 2$ and $d = 3$, respectively.

9.8 The total energy

Let us recall that the total energy can always be expressed as the sum of a bulk and a boundary contribution according to Eq. (3.1); in the following Subsections 9.8.1 and 9.8.2 we are going to discuss these two contributions separately.

9.8.1 *The bulk energy*

Let us first point out that, in view of the general identity (3.6) and of the second relation in Eq. (2.85), the regularized bulk energy E^u is completely determined by the heat trace $K(t) := \mathrm{Tr}\, e^{-t\mathcal{A}}$; according to Eq. (9.7), the latter has the form

$$K(t) = \frac{1}{t^d} H(t) \qquad \text{with} \qquad H(t) := \left(\frac{t}{2\sinh(k^2 t)}\right)^d . \tag{9.45}$$

It is patent that the map $t \mapsto H(t)$ is smooth on $[0, +\infty)$ and exponentially vanishing for $t \to +\infty$. Therefore, following the general prescriptions of Subsection 2.8.2, we obtain for the regularized bulk energy the expression

$$E^u = \frac{(-1)^n \, \kappa^u}{2\,\Gamma(\frac{u-1}{2})(\frac{u-1}{2}-d)...(\frac{u-1}{2}-d+n-1)} \int_0^{+\infty} dt \, t^{\frac{u-3}{2}-d+n} \frac{d^n}{dt^n} H(t) \, . \tag{9.46}$$

The above relation holds for any $n \in \{1, 2, 3, ...\}$ and the integral appearing therein converges for all $u \in \mathbf{C}$ with $\Re u > 2(d - n) + 1$. Thus, for any integer $n > d + 1/2$, Eq. (9.46) gives the analytic continuation of E^u in a neighborhood of $u = 0$; since no singularity appears at this point, we can obtain the renormalized bulk energy simply by setting $u = 0$ in Eq. (9.46). Making again the change of integration variable $\tau := k^2 t$, we infer

$$E^{\text{ren}} := E^u\big|_{u=0} = -\frac{k}{2^{d+2-n}\sqrt{\pi}} \left(\prod_{i=0}^{n-1} \frac{1}{2(d-i)+1} \right)$$

$$\times \int_0^{+\infty} d\tau \, \tau^{n-d-\frac{3}{2}} \frac{d^n}{d\tau^n} \mathcal{H}(\tau)$$

$$\text{for any } \; n > d + \frac{1}{2}, \quad \text{with} \quad \mathcal{H}(\tau) := \left(\frac{\tau}{\sinh \tau}\right)^d. \tag{9.47}$$

The above expression is fully explicit and holds for any spatial dimension d; in the subsequent Sections 9.9 and 9.10 we are going to evaluate numerically the integral appearing in Eq. (9.47) for $d = 2$ and $d = 3$, respectively, making the minimal choice $n = d + 1$.

Before proceeding, let us point out that in the case under analysis, where the potential is assumed to be isotropic, an alternative representation for the bulk energy E^{ren} could be derived in terms of the (analytically continued) Riemann zeta function; we refer to Appendix B of [60] for a detailed analysis of this topic. Here we prefer to focus the attention on the approach (9.47) for the computation of E^{ren} since it is more general; in fact, it could be adapted with very little effort to treat cases where the background potential is not isotropic.

9.8.2 The boundary energy

Contrary to all the cases treated in the previous Chapters 5–8, where the boundary energy was easily found to vanish identically in consequence of the boundary conditions, in the present setting a more careful analysis is required. Since the spatial domain Ω is $\mathbf{R}^d$, we must resort to the procedure pointed out in Section 3.1 and intend the integral over $\partial\Omega$ in the definition (3.7) as the limit of integrals over the boundary of suitable, bounded subdomains Ω_ℓ ($\ell \in \{0, 1, 2, ...\}$). Due to the spherical symmetry and to the characteristic scale of the problem under analysis, it is natural to choose for Ω_ℓ the balls with center in the origin and radius ℓ/k:

$$\Omega_\ell := B^{(d)}_{\ell/k}(\mathbf{0}) \equiv \{\mathbf{x} \in \mathbf{R}^d \mid |\mathbf{x}| \leqslant \ell/k\} \qquad \text{for } \ell \in \{0, 1, 2, ...\} .$$
(9.48)

Then, we proceed to rephrase Eq. (3.7) for the boundary energy as

$$B^u = \kappa^u \left(\frac{1}{4} - \xi\right) \lim_{\ell \to +\infty} \int_{\partial B^{(d)}_{\ell/k}(\mathbf{0})} da(\mathbf{x}) \; \partial_{r_\mathbf{y}} D_{\frac{u+1}{2}}(\mathbf{x}, \mathbf{y}) \Big|_{\mathbf{y}=\mathbf{x}}$$
(9.49)

where $\partial B^{(d)}_{\ell/k}(\mathbf{0})$ is a $(d-1)$-dimensional spherical hypersurface, while $\partial_{r_\mathbf{y}}$ denotes the derivative in the radial direction with respect to the variable $\mathbf{y}$.

Using the integral representation in Eq. (2.85) for the Dirichlet kernel in terms of the heat kernel (again, with the change of variable $\tau := k^2 t$) and noting that $\partial_{r_\mathbf{y}} K(t; \mathbf{x}, \mathbf{y})\big|_{\mathbf{y}=\mathbf{x}}$ does not depend on the angular variables $(\theta_1, ..., \theta_{d-1})$, one infers

$$B^u = -\frac{k}{2^{\frac{d}{2}-1}\Gamma(\frac{d}{2})\,\Gamma(\frac{u+1}{2})} \left(\frac{\kappa}{k}\right)^u \left(\frac{1}{4} - \xi\right)$$

$$\times \lim_{\ell \to +\infty} \int_0^{+\infty} d\tau \; \tau^{\frac{u-d+1}{2}} \left[e^{-\ell^2 \tanh \tau} \frac{\ell^d \, \tau^{d/2-1} \tanh \tau}{(\sinh(2\tau))^{d/2}}\right].$$
(9.50)

For any $d \in \{1, 2, 3, ...\}$, $\ell \geqslant 0$, the expression within the square brackets in the above equation is an analytic function of τ on $[0, +\infty)$ and vanishes exponentially for $\tau \to +\infty$. In consequence of this, we easily infer that the integral in Eq. (9.50) converges for any complex u with $\Re u > d - 3$; within this region, by the Lebesgue dominated

convergence theorem (see footnote (j) on page 19 of [60] for more details) we can take the limit under the integral sign in Eq. (9.50), thus
proving that B^u vanishes identically for $\Re u > d - 3$. Summing up, by
analytic continuation we conclude that the renormalized boundary
energy vanishes:

$$B^{\mathrm{ren}} := B^u \Big|_{u=0} = 0 \ . \tag{9.51}$$

9.8.3   *A remark on energy anomalies*

Before moving on, let us point out a fact that we anticipated with
much more generality in Section 3.5 of Part 1: the renormalized total
energy $\mathcal{E}^{\mathrm{ren}}$ (which, due to Eq. (9.51), in the present setting is equal
to E^{ren}) does not coincide with the integral $\int_{\mathbf{R}^d} d\mathbf{x} \, \langle 0|\widehat{T}_{00}(\mathbf{x})|0\rangle_{\mathrm{ren}}$.
This fact is patently exemplified by the results to be reported in
Sections 9.9 and 9.10, dealing with a scalar field in presence of
an isotropic harmonic potential in spatial dimension $d = 2, 3$. Indeed, on the one hand, the forthcoming Eqs. (9.68), (9.86) state
that the renormalized total energy $\mathcal{E}^{\mathrm{ren}} = E^{\mathrm{ren}}$ is always finite; on
the other hand, Eqs. (9.65), (9.83) (along with Eqs. (9.59), (9.76))
show that $\langle 0|\widehat{T}_{00}(\mathbf{x})|0\rangle_{\mathrm{ren}}$ diverges in a non-integrable way for $|\mathbf{x}|$
$(= r/k) \to +\infty$. Let us remark once more that the "energy anomaly"
$\mathcal{E}^{\mathrm{ren}} \neq \int d\mathbf{x}\langle 0|\widehat{T}_{00}(\mathbf{x})|0\rangle_{\mathrm{ren}}$ is not at all a specific fact of the present
configuration with an external harmonic potential; on the contrary,
this anomaly and other similar ones appear in several configurations (including cases with a bounded space domain: see, for example, the results of Chapter 5 about the case of a segment), even if
one uses alternative approaches (e.g. point-splitting) to renormalize
$\langle 0|\widehat{T}_{\mu\nu}(\mathbf{x})|0\rangle$.

9.9   **A harmonic potential in spatial dimension $d = 2$**

Consider the general framework described in Subsection 9.5; in this
case we use the rescaled polar coordinates $\mathbf{q} = (r, \theta_1) \in (0, +\infty) \times$

$(0, 2\pi)$, fulfilling (see Eq. (9.9))^c

$$k\,x^1 = r\cos\theta_1\,, \qquad k\,x^2 = r\sin\theta_1\,. \tag{9.52}$$

With some effort, it is possible to obtain the following integral representation for the zeta-regularized stress-energy VEV (compare with Eq. (9.16) and recall that dependence on ξ is understood):

$$\langle 0|\widehat{T}^u_{\mu\nu}(\mathbf{q})|0\rangle = \frac{k^3}{\Gamma(\frac{u+1}{2})}\left(\frac{\kappa}{k}\right)^u\int_0^{+\infty}d\tau\;\tau^{-\frac{5}{2}+\frac{u}{2}}\,\mathcal{H}^{(u)}_{\mu\nu}(\tau\,;\mathbf{q}) \quad (\mu,\nu\in\{0,r,\theta_1\})\,, \tag{9.53}$$

where the tensor $\mathcal{H}^{(u)}_{\mu\nu}$ is diagonal with diagonal components

$$\mathcal{H}^{(u)}_{00}(\tau\,;\mathbf{q})$$
$$:= A_2(\tau,r)\left[-(1-u)(1+4\xi) + (1-4\xi)\left(\frac{2\tau}{\sinh 2\tau}\right)\left(2 + r^2\,\frac{\sinh 4\tau}{2\cosh^2\tau}\right)\right]\,, \tag{9.54}$$

$$\mathcal{H}^{(u)}_{rr}(\tau\,;\mathbf{q}) := A_2(\tau,r)\left[8\xi\,\frac{\tau}{\tanh\tau} - (1-4\xi)\left(1 - u + r^2\,\frac{2\tau}{\cosh^2\tau}\right)\right]\,, \tag{9.55}$$

$$\mathcal{H}^{(u)}_{\theta_1\theta_1}(\tau\,;\mathbf{q})$$
$$:= \left(\frac{r}{k}\right)^2 A_2(\tau,r)\left[8\xi\,\frac{\tau}{\tanh\tau} - (1-4\xi)\left(1 - u + r^2\frac{2\tau\cosh 2\tau}{\cosh^2\tau}\right)\right]\,. \tag{9.56}$$

In the above, for simplicity of notation we have put

$$A_2(\tau,r) := \frac{1}{64\,\pi}\,e^{-r^2\tanh\tau}\left(\frac{2\tau}{\sinh 2\tau}\right)\,. \tag{9.57}$$

Again, the features indicated in Section 9.4 are all possessed by the expressions (9.54)–(9.57) for $\mathcal{H}^{(u)}_{\mu\nu}$ ($\mu,\nu\in\{0,r,\theta_1\}$). So, we can analytically continue in u the expression in Eq. (9.53) integrating by

---

^{c}Of course, the spatial line element in this coordinate system reads $d\ell^2 = k^{-2}(dr^2 + r^2 d\theta^2)$; this determines the Christoffel symbols in Eq. (4.32) for the derivatives D_{ij}.

parts n times, for any $n > 3/2$ (see Eqs. (9.22), (9.24)); we make the minimal choice $n = 2$, giving

$$\langle 0|\widehat{T}^u_{\mu\nu}(\mathbf{q})|0\rangle = \frac{k^3}{\Gamma(\frac{u+1}{2})} \frac{1}{(\frac{u}{2}-\frac{3}{2})(\frac{u}{2}-\frac{1}{2})} \left(\frac{\kappa}{k}\right)^u \int_0^{+\infty} d\tau\, \tau^{\frac{u}{2}-\frac{1}{2}}\, \partial^2_\tau \mathcal{H}^{(u)}_{\mu\nu}(\tau; \mathbf{q})\,.$$

(9.58)

It appears that each component of $\langle 0|\widehat{T}^u_{\mu\nu}|0\rangle$ is regular at $u = 0$; thus, we can obtain the renormalized version of the stress-energy VEV by simply putting $u = 0$ in Eq. (9.58) (we already noticed this property in general for the cases with even spatial dimension d; see the comments below Eq. (9.24)). In conclusion, we have

$$\langle 0|\widehat{T}_{\mu\nu}(\mathbf{q})|0\rangle_{\text{ren}} = k^3\, T^{(0)}_{\mu\nu}(r)\,,$$

$$T^{(0)}_{\mu\nu}(r) := \int_0^{+\infty} d\tau\, e^{-r^2 \tanh \tau}\, \mathcal{P}^{(0)}_{\mu\nu}(\tau; r)\,,$$

(9.59)

$$\mathcal{P}^{(0)}_{\mu\nu}(\tau; r) := \frac{4}{3\sqrt{\pi\,\tau}}\, e^{r^2 \tanh \tau}\, \partial^2_\tau \mathcal{H}^{(0)}_{\mu\nu}(\tau; \mathbf{q})$$

(compare with Eqs. (9.26)–(9.28)). It can be easily checked that $\mathcal{P}^{(0)}_{\mu\nu}$ is a polynomial of degree $N = 3$ in r^2.

Next, we proceed to evaluate numerically the integrals in Eq. (9.59) and separate the conformal and non-conformal parts $\Diamond, \blacksquare$ of each component (see Eq. (9.29)), keeping in mind that Eq. (1.32) gives

$$\xi_2 = \frac{1}{8}\,.$$

(9.60)

The forthcoming Figs. 9.1–9.3 (reprinted, as a courtesy of World Scientific, from [60]) show the graphs of the functions

$$r \mapsto T^{(0,\times)}_{\mu\nu}(r)\ (\mu=\nu\in\{0,r\})\,,\ (k/r)^2\, T^{(0,\times)}_{\theta_1\theta_1}(r) \qquad \text{for } \times \in \{\Diamond, \blacksquare\}\,.$$

(9.61)

Now, let us consider the small and large r asymptotics of the functions in Eq. (9.61). On the one hand, Eqs. (9.30)–(9.33) (with $N = 3$) give, for $r = k|\mathbf{x}| \to 0^+$,

$$T^{(0,\Diamond)}_{00}(r) = -0.0017 - 0.0134\, r^2 - 0.0154\, r^4 + 0.0027\, r^6 + O(r^8)\,,$$

$$T^{(0,\blacksquare)}_{00}(r) = 0.1649 - 0.1069\, r^2 + 0.0516\, r^4 - 0.0141\, r^6 + O(r^8)\,;$$

(9.62)

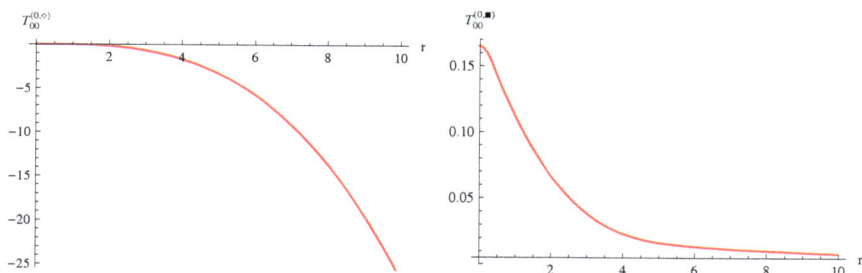

Fig. 9.1 $d = 2$: graphs of $T_{00}^{(0,\diamond)}$ and $T_{00}^{(0,\blacksquare)}$.

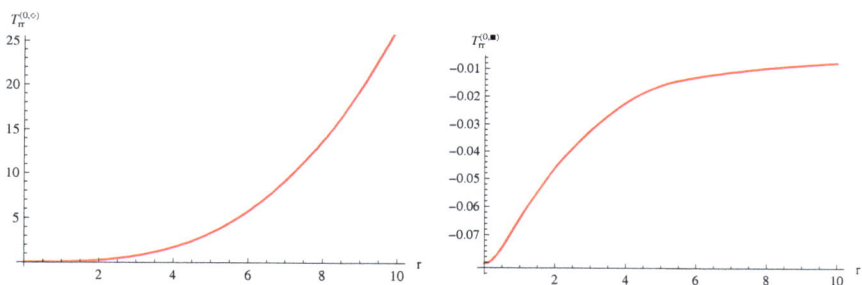

Fig. 9.2 $d = 2$: graphs of $T_{rr}^{(0,\diamond)}$ and $T_{rr}^{(0,\blacksquare)}$.

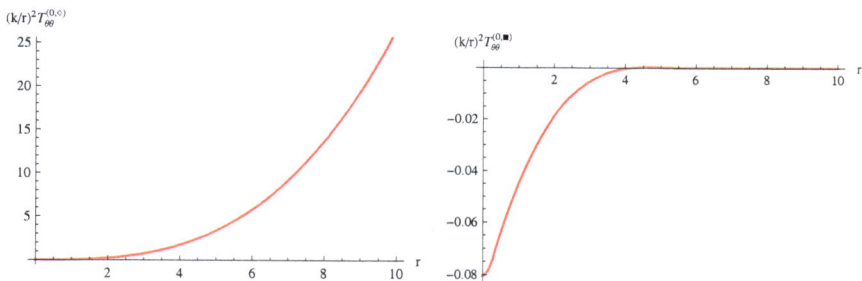

Fig. 9.3 $d = 2$: graphs of $(k/r)^2 T_{\theta_1\theta_1}^{(0,\diamond)}$ and $(k/r)^2 T_{\theta_1\theta_1}^{(0,\blacksquare)}$.

$$T_{rr}^{(0,\diamond)}(r) = -0.0010 + 0.0207\,r^2 + 0.0114\,r^4 - 0.0013\,r^6 + O(r^8)\ ,$$

$$T_{rr}^{(0,\blacksquare)}(r) = -0.0806 + 0.0267\,r^2 - 0.0133\,r^4 + 0.0018\,r^6 + O(r^8)\ ;$$

$$(9.63)$$

$$(k/r)^2 \, T^{(0,\Diamond)}_{\theta_1\theta_1}(r) = -0.0010 + 0.0140 \, r^2 + 0.0153 \, r^4 - 0.0027 \, r^6 + O(r^8) \, ,$$

$$(k/r)^2 \, T^{(0,\blacksquare)}_{\theta_1\theta_1}(r) = -0.0806 + 0.0802 \, r^2 - 0.0440 \, r^4 + 0.0124 \, r^6 + O(r^8) \, .$$

$$(9.64)$$

The (approximate) values of the coefficients appearing in the above expansions have been obtained evaluating numerically the integrals in Eq. (9.33).

On the other hand, using Eqs. (9.35), (9.37) and (9.38) (with $K = 5$, $\lambda = -1/2$; note that no logarithmic term $\ln r^2$ appears in this case), we obtain the following asymptotic expansions for $r = k|\mathbf{x}| \to +\infty$:

$$T^{(0,\Diamond)}_{00}(r) = -\frac{r^3}{12\pi} - \frac{19}{2560\pi \, r^5} + O(r^{-9}) \, ,$$

$$T^{(0,\blacksquare)}_{00}(r) = \frac{1}{4\pi \, r} + \frac{3}{32\pi \, r^5} + O(r^{-9}) \, ;$$

$$(9.65)$$

$$T^{(0,\Diamond)}_{rr}(r) = \frac{r^3}{12\pi} + \frac{1}{48\pi \, r} - \frac{17}{2560\pi \, r^5} + O(r^{-9}) \, ,$$

$$T^{(0,\blacksquare)}_{rr}(r) = -\frac{1}{4\pi \, r} + \frac{1}{32\pi \, r^5} + O(r^{-9}) \, ;$$

$$(9.66)$$

$$(k/r)^2 \, T^{(0,\Diamond)}_{\theta_1\theta_1}(r) = \frac{r^3}{12\pi} - \frac{1}{96\pi \, r} + + \frac{33}{2560\pi \, r^5} + O(r^{-9}) \, ,$$

$$(k/r)^2 \, T^{(0,\blacksquare)}_{\theta_1\theta_1}(r) = -\frac{1}{8\pi \, r^5} + O(r^{-9}) \, .$$

$$(9.67)$$

Let us pass to discuss the bulk energy; using the expression (9.47) with $n = 3$ and evaluating numerically the integral appearing therein, we obtain

$$E^{\text{ren}} = -(0.0180207591 \pm 10^{-10}) \, k \, .$$

$$(9.68)$$

9.10   A harmonic potential in spatial dimension $d = 3$

In this case we use the coordinates $\mathbf{q} = (r, \theta_1, \theta_2) \in (0, +\infty) \times (0, \pi) \times (0, 2\pi)$, which are related to the Cartesian coordinates

$\mathbf{x} \equiv (x_1, x_2, x_3)$ via (see Eq. (9.9))[d]

$$k\,x^1 = r \cos\theta_1 \,, \quad k\,x^2 = r \sin\theta_1 \cos\theta_2 \,, \quad k\,x^3 = r \sin\theta_1 \sin\theta_2 \,.$$
$$(9.69)$$

Similarly to what we did in the previous section, after lengthy computations we can express the zeta-regularized stress-energy VEV as (compare with Eq. (9.16) and recall that dependence on ξ is understood)

$$\langle 0|\widehat{T}^u_{\mu\nu}(\mathbf{q})|0\rangle$$
$$= \frac{k^4}{\Gamma(\frac{u+1}{2})} \left(\frac{\kappa}{k}\right)^u \int_0^{+\infty} d\tau\, \tau^{-3+\frac{u}{2}}\, \mathcal{H}^{(u)}_{\mu\nu}(\tau\,;\mathbf{q}) \quad (\mu,\nu \ln\{0, r, \theta_1, \theta_2\})\,,$$
$$(9.70)$$

where $\mathcal{H}^{(u)}_{\mu\nu}$ is diagonal and we only have to consider the independent components

$$\mathcal{H}^{(u)}_{00}(\tau\,;\mathbf{q})$$
$$:= A_3(\tau, r) \left[-(1-u)(1+4\xi) + (1-4\xi)\left(\frac{2\tau}{\sinh 2\tau}\right)\left(3 + r^2\, \frac{\sinh 3\tau - \sinh\tau}{\cosh\tau}\right) \right],$$
$$(9.71)$$

$$\mathcal{H}^{(u)}_{rr}(\tau\,;\mathbf{q})$$
$$:= A_3(\tau, r) \left[8\xi\, \frac{\tau}{\tanh\tau} - (1-4\xi)\left(1 - u + \left(\frac{2\tau}{\sinh 2\tau}\right)\left(1 + 2r^2 \tanh\tau\right)\right) \right],$$
$$(9.72)$$

$$\mathcal{H}^{(u)}_{\theta_1\theta_1}(\tau\,;\mathbf{q})$$
$$:= \left(\frac{r}{k}\right)^2 A_3(\tau, r) \left[8\xi\, \frac{\tau}{\tanh\tau} - (1-4\xi)\left(1 - u + \left(\frac{2\tau}{\sinh 2\tau}\right)\left(1 + \frac{r^2}{\cosh^2 2\tau}\right)\right) \right]$$
$$(9.73)$$

(for the remaining diagonal component, i.e. $\mathcal{H}^{(u)}_{\theta_2\theta_2}(\tau\,;\mathbf{q})$, see Eq. (9.20)). In the above, for simplicity of notation we have put

$$A_3(\tau, r) := \frac{1}{64\,\pi^{3/2}}\, e^{-r^2 \tanh\tau} \left(\frac{2\tau}{\sinh 2\tau}\right)^{3/2}. \qquad (9.74)$$

---

[d]Here the framework of Subsection 4.3 must be employed using the spatial line element $d\ell^2 = k^{-2}(dr^2 + r^2(d\theta_1^2 + \sin^2\theta_1\, d\theta_2^2))$ and the corresponding Christoffel symbols.

Again, the functions $\mathcal{H}_{\mu\nu}^{(u)}$ defined by Eqs. (9.71)–(9.74) are found to possess the properties indicated in Section 9.4; thus, we can obtain the analytic continuation in u of the expression in Eq. (9.70) integrating by parts n times, for any $n > 2$ (see Eq. (9.24)). With the minimal choice $n = 3$, Eq. (9.22) reads

$$\langle 0|\widehat{T}_{\mu\nu}^{u}(\mathbf{q})|0\rangle = -\frac{k^4}{\Gamma(\frac{u+1}{2})}\,\frac{1}{(\frac{u}{2}-2)(\frac{u}{2}-1)\frac{u}{2}}\left(\frac{\kappa}{k}\right)^u\int_0^{+\infty} d\tau\;\tau^{\frac{u}{2}}\,\partial_\tau^3\mathcal{H}_{\mu\nu}^{(u)}(\tau;\mathbf{q})\,.$$

(9.75)

As in all cases with odd spatial dimension, the analytic continuation of the regularized stress-energy VEV given in Eq. (9.75) has a simple pole in $u = 0$ (recall Subsection 9.5). In consequence of this, we have to adopt the extended version of the zeta approach to define the renormalized VEV $\langle 0|\widehat{T}_{\mu\nu}(\mathbf{q})|0\rangle_{\text{ren}}$, taking the regular part in $u = 0$ of Eq. (9.75) (see Eq. (9.25)); with some effort, we obtain (compare with Eq. (9.26))

$$\langle 0|\widehat{T}_{\mu\nu}(\mathbf{q})|0\rangle_{\text{ren}} = k^4\left(T_{\mu\nu}^{(0)}(r) + M_{\kappa,k}\,T_{\mu\nu}^{(1)}(r)\right)\,,$$

$$T_{\mu\nu}^{(0)}(r) := \int_0^{+\infty} d\tau\,e^{-r^2\tanh\tau}\left[\mathcal{P}_{\mu\nu}^{(0)}(\tau;r) + \ln\tau\,\mathcal{P}_{\mu\nu}^{(1)}(\tau;r)\right]\,,$$

$$T_{\mu\nu}^{(1)}(r) := \int_0^{+\infty} d\tau\,e^{-r^2\tanh\tau}\,\mathcal{P}_{\mu\nu}^{(1)}(\tau;r)\,,$$

(9.76)

$$M_{\kappa,k} := \gamma_{EM} + 2\ln\left(\frac{2\mu}{k}\right)\,,$$

where

$$\mathcal{P}_{\mu\nu}^{(0)}(\tau;r) := -\frac{1}{4\sqrt{\pi}}\,e^{r^2\tanh\tau}\left[3\,\partial_\tau^3\mathcal{H}_{\mu\nu}^{(0)}(\tau;\mathbf{q}) + 4\partial_u\Big|_{u=0}\partial_\tau^3\mathcal{H}_{\mu\nu}^{(u)}(\tau;\mathbf{q})\right]\,,$$

$$\mathcal{P}_{\mu\nu}^{(1)}(\tau;r) := -\frac{1}{2\sqrt{\pi}}\,e^{r^2\tanh\tau}\,\partial_\tau^3\mathcal{H}_{\mu\nu}^{(0)}(\tau;\mathbf{q})\,.$$

(9.77)

We readily infer that $\mathcal{P}_{\mu\nu}^{(0)}$, $\mathcal{P}_{\mu\nu}^{(1)}$ are polynomials of degree $N = 4$ in r^2.

To proceed, we evaluate numerically the integrals in Eq. (9.76) and distinguish between the conformal and non-conformal parts $\Diamond$,

■ of each component; once more we refer to Eq. (9.29), recalling that for $d = 3$ we have (see Eq. (1.32))

$$\xi_3 = \frac{1}{6} \ . \tag{9.78}$$

The forthcoming Figs. 9.4–9.9 (again, reprinted from [60]) show the graphs of the functions

$$r \mapsto T_{00}^{(a,\times)}(r), \ T_{rr}^{(a,\times)}(r), \ (k/r)^2 \, T_{\theta_1\theta_1}^{(a,\times)}(r) \quad \text{for } a \in \{0,1\}, \times \in \{\Diamond, \blacksquare\} \ . \tag{9.79}$$

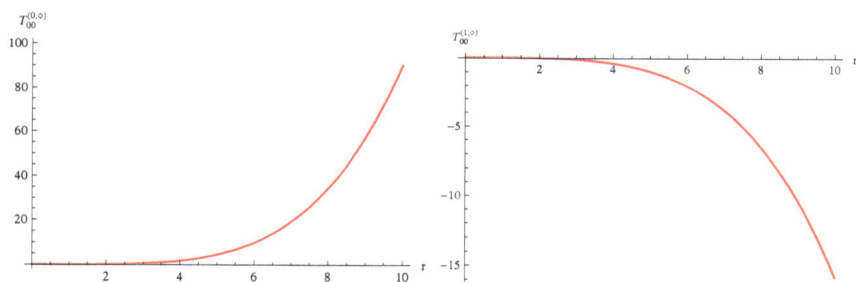

Fig. 9.4 $d = 3$: graphs of $T_{00}^{(0,\Diamond)}$ and $T_{00}^{(1,\Diamond)}$.

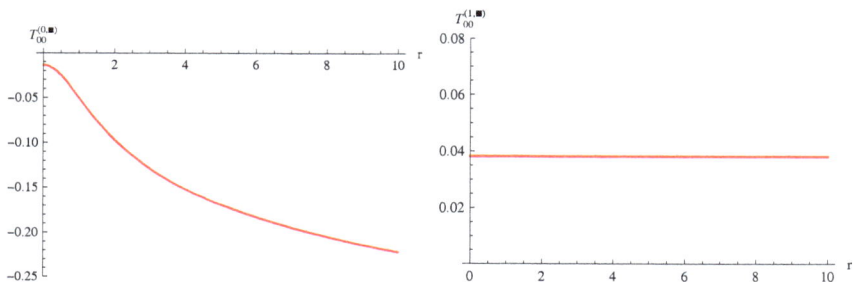

Fig. 9.5 $d = 3$: graphs of $T_{00}^{(0,\blacksquare)}$ and $T_{00}^{(1,\blacksquare)}$.

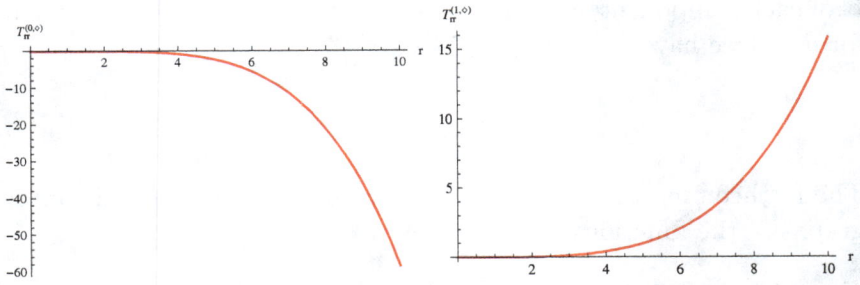

Fig. 9.6 $d = 3$: graphs of $T_{rr}^{(0,\Diamond)}$ and $T_{rr}^{(1,\Diamond)}$.

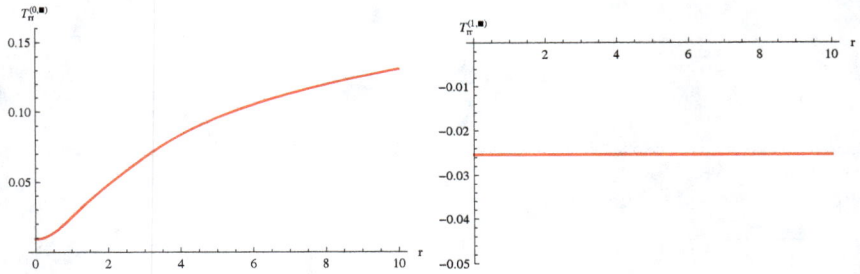

Fig. 9.7 $d = 3$: graphs of $T_{rr}^{(0,\blacksquare)}$ and $T_{rr}^{(1,\blacksquare)}$.

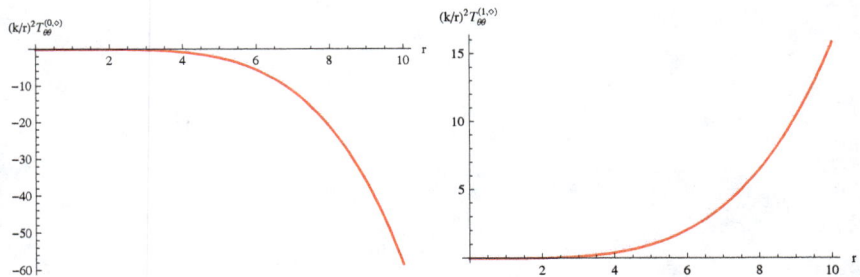

Fig. 9.8 $d = 3$: graphs of $(k/r)^2 T_{\theta_1\theta_1}^{(0,\Diamond)}$ and $(k/r)^2 T_{\theta_1\theta_1}^{(1,\Diamond)}$.

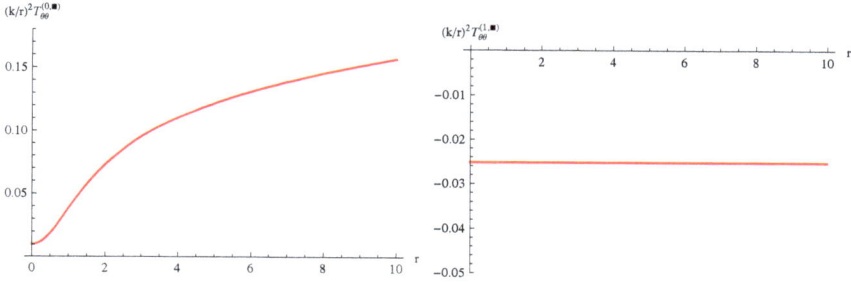

Fig. 9.9 $d = 3$: graphs of $(k/r)^2 T_{\theta_1 \theta_1}^{(0,\blacksquare)}$ and $(k/r)^2 T_{\theta_1 \theta_1}^{(1,\blacksquare)}$.

In conclusion, let us discuss the small and large r asymptotics of the functions in Eq. (9.79). On the one hand, Eqs. (9.30)–(9.33) (with $N = 4$) yield, for $r = k|\mathbf{x}| \to 0^+$,

$$
\begin{aligned}
T_{00}^{(0,\Diamond)}(r) &= -0.0047 - 0.0024\, r^2 + 0.0028\, r^4 + 0.0006\, r^6 \\
&\quad - 0.0001\, r^8 + O(r^{10})\,, \\
T_{00}^{(1,\Diamond)}(r) &= -0.0016\, r^4 + O(r^{10})\,, \\
T_{00}^{(0,\blacksquare)}(r) &= -0.0143 - 0.0468\, r^2 + 0.0134\, r^4 - 0.0033\, r^6 \\
&\quad + 0.0007\, r^8 + O(r^{10})\,, \\
T_{00}^{(1,\blacksquare)}(r) &= 0.0380 + O(r^{10})\,;
\end{aligned}
\tag{9.80}
$$

$$
\begin{aligned}
T_{rr}^{(0,\Diamond)}(r) &= -0.0016 + 0.0039\, r^2 - 0.0003\, r^4 \\
&\quad - 0.0005\, r^6 + O(r^{10})\,, \\
T_{rr}^{(1,\Diamond)}(r) &= 0.0016\, r^4 + O(r^{10})\,, \\
T_{rr}^{(0,\blacksquare)}(r) &= 0.0095 + 0.0188\, r^2 - 0.0038\, r^4 + 0.0007\, r^6 \\
&\quad - 0.0001\, r^8 + O(r^{10})\,, \\
T_{rr}^{(1,\blacksquare)}(r) &= -0.0253 + O(r^{10})\,;
\end{aligned}
\tag{9.81}
$$

$$(k/r)^2 \, T^{(0,\Diamond)}_{\theta_1\theta_1}(r) = -0.0016 + 0.0023\,r^2 + 0.0004\,r^4$$
$$- 0.0006\,r^6 + 0.0001\,r^8 + O(r^{10}),$$

$$(k/r)^2 \, T^{(1,\Diamond)}_{\theta_1\theta_1}(r) = 0.0016\,r^4 + O(r^{10}),$$

$$(k/r)^2 \, T^{(0,\blacksquare)}_{\theta_1\theta_1}(r) = 0.0095 + 0.0375\,r^2 - 0.0115\,r^4$$
$$+ 0.0030\,r^6 - 0.0006\,r^8 + O(r^{10}),$$

$$(k/r)^2 \, T^{(1,\blacksquare)}_{\theta_1\theta_1}(r) = -0.0253 + O(r^{10}) \tag{9.82}$$

(again, the values of the coefficients appearing above are approximate, and have been obtained via numerical computations of the integrals in Eq. (9.33)).

On the other hand, Eqs. (9.35), (9.37) and (9.38) (with $K = 4$, $\lambda = 0$) allow us to infer the following asymptotic expansions, for $r = k|\mathbf{x}| \to +\infty$:

$$T^{(0,\Diamond)}_{00}(r) = \frac{r^4}{64\pi^2}\left(\ln r^2 + \gamma_{EM} + \frac{1}{2}\right) - \frac{5}{96\pi^2}$$
$$- \frac{23}{2880\pi^2 r^4} + O(r^{-8}\ln r^2),$$

$$T^{(1,\Diamond)}_{00}(r) = -\frac{r^4}{64\pi^2} + O(r^{-8}\ln r^2),$$

$$T^{(0,\blacksquare)}_{00}(r) = -\frac{3}{8\pi^2}\left(\ln r^2 + \gamma_{EM} + \frac{2}{3}\right)$$
$$+ \frac{1}{12\pi^2 r^4} + O(r^{-8}\ln r^2),$$

$$T^{(1,\blacksquare)}_{00}(r) = \frac{3}{8\pi^2} + O(r^{-8}\ln r^2); \tag{9.83}$$

$$T^{(0,\Diamond)}_{rr}(r) = -\frac{r^4}{64\pi^2}\left(\ln r^2 + \gamma_{EM} - \frac{3}{2}\right) + \frac{1}{96\pi^2}$$
$$- \frac{49}{2880\pi^2 r^4} + O(r^{-8}\ln r^2),$$

$$T^{(1,\Diamond)}_{rr}(r) = \frac{r^4}{64\pi^2} + O(r^{-8}\ln r^2),$$

$$T_{rr}^{(0,\blacksquare)}(r) = \frac{1}{4\pi^2}\left(\ln r^2 + \gamma_{EM}\right) + \frac{1}{6\pi^2 r^4} + O(r^{-8}\ln r^2) \,,$$

$$T_{rr}^{(1,\blacksquare)}(r) = -\frac{1}{4\pi^2} + O(r^{-8}\ln r^2) \,;$$

$$(9.84)$$

$$(k/r)^2\, T_{\theta_1\theta_1}^{(0,\diamond)}(r) = -\frac{r^4}{64\pi^2}\left(\ln r^2 + \gamma_{EM} - \frac{3}{2}\right) - \frac{1}{96\pi^2}$$

$$+ \frac{31}{2880\pi^2 r^4} + O(r^{-8}\ln r^2) \,,$$

$$(k/r)^2\, T_{\theta_1\theta_1}^{(1,\diamond)}(r) = \frac{r^4}{64\pi^2} + O(r^{-8}\ln r^2) \,,$$

$$(9.85)$$

$$(k/r)^2\, T_{\theta_1\theta_1}^{(0,\blacksquare)}(r) = \frac{1}{4\pi^2}\left(\ln r^2 + \gamma_{EM} + 1\right)$$

$$- \frac{1}{6\pi^2 r^4} + O(r^{-8}\ln r^2) \,,$$

$$(k/r)^2\, T_{\theta_1\theta_1}^{(1,\blacksquare)}(r) = -\frac{1}{4\pi^2} + O(r^{-8}\ln r^2) \,.$$

Finally, Eq. (9.47) with $n = 4$ and numerical evaluation of the corresponding integral allow us to derive the following value for the renormalized bulk energy:

$$E^{\mathrm{ren}} = -(0.0078607119 \pm 10^{-10})\,k \,. \qquad (9.86)$$

This result is found to agree with the one obtained by Actor and Bender [4] using a different method.[e]

---

[e]To check this, one must compare the numerical value in the above Eq. (9.86) with the one reported in Eq. (4.4) on page 3586 of [4]. Let us stress that conventions different from ours are used therein. In fact, using our language, the bulk energy is formally defined in [4] as $E := \sum_k \omega_k$, while our general prescription (3.6) gives $E = \frac{1}{2}\sum_k \omega_k$; moreover the parameter α of [4] and our parameter k are related by $\alpha = \sqrt{2}\,k$. Summing up, the "total energy" derived in [4] has to be multiplied by $1/2$ in order to obtain our E^{ren}.

Chapter 10

A massless field inside a rectangular box

In this final chapter we consider the case of a massless field confined within a d-dimensional rectangular spatial domain. We summarize here the results obtained in [61]; following this reference, for the sake of simplicity we restrict attention to the case where the field fulfills Dirichlet conditions on the boundary. Nevertheless, the techniques we employ could be generalized with little effort to treat different settings including, e.g. the case of Neumann or periodic boundary conditions.

In the opening Section 10.1 we give a formal description of the model under analysis. In Section 10.2 we introduce two alternative series representations for the heat kernel, which capture the asymptotic behavior of this kernel in different regimes. These expansions are used in Section 10.3 to produce series expansions for the Dirichlet kernel and for its derivatives, ultimately yielding their analytic continuation to the whole complex plane. The results thus obtained are employed in Sections 10.4–10.7 to determine by pure analytic continuation the renormalized VEVs of several observables, namely: the stress-energy tensor, the pressure on the boundary, the total energy and the total force acting on any side of the box. All these objects are represented as sums of series converging with exponential speed, for which quantitative remainder estimates are reported in Section 10.8. Section 10.9 contains a few remarks about some useful scaling properties. All the results presented in Sections 10.1–10.9 hold for an arbitrary spatial dimension d; we finally specialize them to the case $d = 2$ in Section 10.10, where we also produce several graphs for the above mentioned renormalized VEVs and discuss some of their qualitative features with the aid of the software **Mathematica** for some related numerical computations.

In order to be concise, we have omitted here the proofs of a number of technical results; a detailed account on these proofs can be found in [61].

Let us make a comparison with the existing literature on the configuration discussed in the present chapter. The renormalized total energy for a field in a rectangular box has been discussed in a great number of works. The foremost computation was performed by Lukosz [94, 95] for the electromagnetic field, by means of exponential regularization and Abel–Plana formula; the same techniques were used by Mamaev and Trunov [97, 98] (see also [99, 100]) to discuss, among other models, the case of a conformal scalar field. Alternative derivations of the total energy for a scalar field, based on global zeta regularization, were given by Ruggerio, Villani and Zimerman [129, 130] for $d = 2$ and $d = 3$, and by Ambjørn and Wolfram [9] for arbitrary d and several boundary conditions (see also [10] for the electromagnetic case). The same configurations were later re-examined by X. Li, Cheng, J. Li and Zhai [92] by means of a zeta-type strategy, and by Edery [50, 51] using a so called "multidimensional cut-off technique". Let us also cite the papers [57, 72] by Fulling, Kirsten *et al.*, on which we return later. Finally, let us mention the monographs of Elizalde *et al.* [53, 54] and Bordag *et al.* [22], which can be taken as standard references for the study of global aspects. The series representations for the total energy and integrated force reported in Sections 10.6, 10.7 of the present chapter are different, but equivalent to the ones of [22] and converge exponentially, likewise.

Concerning local aspects, let us first mention the two seminal papers [2] and [3] by Actor; therein the framework is Euclidean and the author renormalizes, mostly by analytic continuation, the effective Lagrangian density and the vacuum polarization for a scalar field in spatial dimension $d = 3$. The renormalized stress-energy VEV is derived for the case of a three-dimensional, infinite rectangular waveguide in a work of Svaiter *et al.* [127], where the components of the said VEV are expressed as sums of series converging with polynomial speed. In [127] some of the divergences which appear are eliminated by analytic continuation techniques, using a series representation of the heat kernel equivalent to the one we write in Eq. (10.15). On the other hand, contrary to the results reported in the present chapter, *ad hoc* prescriptions are used in both [2, 3] and [127] to remove some additional divergences, which are interpreted as "empty space" contributions also appearing when there is no boundary. To go on, let us return to the already mentioned works by Fulling, Kirsten *et al.* [57, 72], where the case of a two-dimensional rectangular box is analyzed for both Dirichlet and Neumann boundary conditions. The authors derive by the method of images a regularized version of the stress-energy VEV, based on an exponential cutoff approach; the regularized total energy and regularized force on a side are obtained by integration of the said VEV. The position of principle adhered to in [57, 72] is that the theory with a cutoff yields a more realistic description of the physical system under investigation; nonetheless, the authors point out that the renormalization of the considered VEVs can be obtained retaining only the non-divergent terms of the expansions with respect to the cut-off parameter (an idea somehow related to what we call the "extended zeta approach"). Renormalization along these lines is carried out for global observables like the total energy, and hinted at for local observables (namely, for the energy density).

10.1 Introducing the problem

The model we analyze in the present chapter concerns a massless scalar field confined inside a d-dimensional box, with no external potential; more precisely, we assume

$$\Omega = \times_{i=1}^d (0, a_i) \quad \text{with } a_i > 0 \text{ for } i \in \{1, ..., d\} , \qquad V = 0 . \quad (10.1)$$

The boundary $\partial\Omega$ of the spatial domain is composed by the sides

$$\pi_{p,\lambda} := \{\mathbf{x} \in \mathbf{R}^d \mid x^p = \lambda\, a_p , \, x^i \in [0, a_i] \text{ for } i \neq p\} \\ \text{for } p \in \{1, ..., d\}, \lambda \in \{0, 1\} . \quad (10.2)$$

As mentioned previously, we restrict attention to the case where the field fulfills Dirichlet boundary conditions on each one of these sides, meaning that

$$\widehat{\phi}(t,\mathbf{x})=0 \quad \text{for any } t\in\mathbf{R} \text{ and all } \mathbf{x}\in\pi_{p,\lambda} \ (p\in\{1,...,d\}, \ \lambda\in\{0,1\})\,.$$
$$(10.3)$$

As a matter of fact, all the results reported in the present chapter could be generalized to the case of Neumann or periodic boundary conditions, possibly including cases where different boundary conditions are prescribed on different sides of the box; moreover, the methods employed here could be adapted with little effort to deal with the cases of a massive scalar field ($V = m^2$) and of a slab configuration (see Section 2.10) where

$$\Omega = \Omega_1\times\mathbf{R}^{d_2}\,, \qquad \Omega_1 = \times_{i=1}^{d_1}(0,a_i) \subset \mathbf{R}^{d_1} \quad (d_1+d_2 = d)\,. \quad (10.4)$$

None of these generalizations will be considered in the present chapter.

10.2 The heat kernel

Even in the present setting we are dealing with a product domain configuration. Indeed, working in standard Cartesian coordinates $(x^i)_{i=1,...,d}$, the Hilbert space and the fundamental operator $\mathcal{A}:=-\Delta$ can be represented as

$$L^2(\Omega) = \bigotimes_{i=1}^{d} L^2(0,a_i)\,, \qquad (10.5)$$

$$\mathcal{A} = \mathcal{A}_1\otimes\mathbf{1}\otimes...\otimes\mathbf{1} + ... + \mathbf{1}\otimes...\otimes\mathbf{1}\otimes\mathcal{A}_d\,,$$
$$\mathcal{A}_i := -\frac{d^2}{dx^2} \quad \text{on } L^2(0,a_i) \quad (i\in\{1,...,d\})\,; \qquad (10.6)$$

each one of the operators $\mathcal{A}_i$ is defined assuming Dirichlet boundary conditions in $x^i = 0$ and $x^i = a_i$.

According to the general results for product configurations presented in Section 2.9, in this situation the heat kernel associated to $\mathcal{A}$ factorizes; more precisely, there holds the identity

$$K(\mathsf{t};\mathbf{x},\mathbf{y}) = \prod_{i=1}^{d} K_i(\mathsf{t};x^i,y^i)\,, \qquad (10.7)$$

where, for $i\in\{1,...,d\}$, K_i indicates the heat kernel of $\mathcal{A}_i$.

In the forthcoming Subsections 10.2.1 and 10.2.2 we give two distinct representations for the heat kernels K_i corresponding to the reduced, one-dimensional problems. The said representations are suited to describe, respectively, the asymptotic behavior of the functions K_i for small and large t (in a sense to be made more precise in the following). Clearly, each one of these representations can be employed along with the identity (10.7) to infer alternative expressions for the total heat kernel.

10.2.1 *First representation: large t expansion*

Let us first notice that, for any $i \in \{1, ..., d\}$, a complete orthonormal set of eigenfunctions $(F_{k_i})_{k_i \in \mathcal{K}_i}$ for $\mathcal{A}_i$ in $L^2(0, a_i)$, with corresponding eigenvalues $(\omega_{k_i}^2)_{k_i \in \mathcal{K}_i}$, is given by

$$F_{k_i}(x^i) := \sqrt{\frac{2}{a_i}} \sin(k_i x^i) ,$$

$$\omega_{k_i}^2 := k_i^2 \quad \text{for } k_i \in \mathcal{K}_i \equiv \left\{ \frac{n_i \pi}{a_i} \,\middle|\, n_i = 1, 2, 3, ... \right\}.$$

(10.8)

Using the eigenfunction expansion (2.44), we obtain for the one-dimensional heat kernel K_i the following expression:

$$K_i(t; x^i, y^i) = \frac{2}{a_i} \sum_{n_i=1}^{+\infty} e^{-\frac{n_i^2 \pi^2}{a_i^2} t} \sin\left(\frac{n_i \pi}{a_i} y^i\right) \sin\left(\frac{n_i \pi}{a_i} x^i\right) . \quad (10.9)$$

The above relation, along with the identity in Eq. (10.7), yields

$$K(t; \mathbf{x}, \mathbf{y}) = \frac{2^d}{a_1 \dots a_d} \sum_{\mathbf{n} \in \mathbf{N}^d} e^{-\omega_{\mathbf{n}}^2 t} C_{\mathbf{n}}(\mathbf{x}, \mathbf{y}) , \quad (10.10)$$

where, for the sake of brevity, we have put

$$\mathbf{N} := \{1, 2, 3, ...\} , \quad \omega_{\mathbf{n}}^2 := \sum_{i=1}^{d} \frac{n_i^2 \pi^2}{a_i^2} ,$$

(10.11)

$$C_{\mathbf{n}}(\mathbf{x}, \mathbf{y}) := \prod_{i=1}^{d} \sin\left(\frac{n_i \pi}{a_i} y^i\right) \sin\left(\frac{n_i \pi}{a_i} x^i\right) \quad \text{for } \mathbf{n} := (n_i)_{i=1,...,d} .$$

Equation (10.10) is easily seen to give a large t expansion for the heat kernel of $\mathcal{A}$ meaning that, when t is large, the series over $\mathbf{n} \in \mathbf{N}^d$ in the cited equation is mainly determined by few terms corresponding to small values of n_i, for $i \in \{1, ..., d\}$.

10.2.2 Second representation: small t expansion

Consider the expression in Eq. (10.9) for the one-dimensional heat kernels K_i ($i \in \{1, ..., d\}$); writing the sines appearing therein in terms of complex exponentials, using the Poisson summation formula and evaluating explicitly some elementary Gaussian integrals,[a] this expression can be rephrased as

$$K_i(t; x^i, y^i) = \frac{1}{\sqrt{4\pi t}} \sum_{h_i=-\infty}^{+\infty} \left[e^{-\frac{(2a_i h_i - (x^i - y^i))^2}{4t}} - e^{-\frac{(2a_i h_i - (x^i + y^i))^2}{4t}} \right].$$

(10.12)

The above identity can be equivalently restated as follows:

$$K_i(t; x^i, y^i) = \frac{1}{\sqrt{4\pi t}} \sum_{h_i \in \mathbf{Z}, \, l_i \in \{1,2\}} \delta_{l_i} \, e^{-\frac{1}{t} a_i^2 (h_i - D_{l_i}(x^i, y^i))^2} \quad (10.13)$$

where

$$\delta_{l_i} := \begin{cases} 1 & \text{for } l_i = 1 \\ -1 & \text{for } l_i = 2 \end{cases}, \quad D_{l_i}(x^i, y^i) := \begin{cases} \frac{x^i - y^i}{2a_i} & \text{for } l_i = 1 \\ \frac{x^i + y^i}{2a_i} & \text{for } l_i = 2 \end{cases}.$$

(10.14)

Equation (10.13), along with Eq. (10.7), allows us to infer for the heat kernel the alternative representation

$$K(t; \mathbf{x}, \mathbf{y}) = \frac{1}{(4\pi t)^{d/2}} \sum_{\mathbf{h} \in \mathbf{Z}^d, \, \mathbf{l} \in \{1,2\}^d} \delta_\mathbf{l} \, e^{-\frac{1}{t} b_{\mathbf{h}\mathbf{l}}(\mathbf{x}, \mathbf{y})}, \quad (10.15)$$

---

[a] Let us give some more details on this computations. On the one hand, the Poisson summation formula [151] states that, for any sufficiently regular function $f : \mathbf{R} \to \mathbf{C}$, there holds

$$\sum_{n=-\infty}^{+\infty} f(n) = \sum_{h=-\infty}^{+\infty} \hat{f}(h), \quad \text{where} \quad \hat{f}(h) := \int_{-\infty}^{+\infty} dz \, e^{-2i\pi h z} f(z).$$

On the other hand, let us remark that the Gaussian integrals arising from the application of this formula can be evaluated according to the general identity

$$\int_{-\infty}^{+\infty} dz \, e^{-\frac{\pi^2}{a_i^2} t z^2} e^{i\frac{\pi}{a_i}((x^i \pm y^i) - 2a_i h_i)z}$$

$$= \frac{a_i}{\sqrt{\pi t}} e^{-\frac{(2a_i h_i - (x^i \pm y^i))^2}{4t}} \quad \text{for } h_i \in \mathbf{Z} = \{0, \pm 1, \pm 2, ...\}.$$

Let us also remark that the very same result of Eq. (10.12) could be obtained via the method of reflections, starting with the heat kernel associated to the Laplacian $-\partial_{x^1 x^1}$ on $\mathbf{R}$ (see Eq. (2.56)).

where, for simplicity of notation, we have put (recall the definitions given in Eq. (10.14))

$$\mathbf{h} := (h_i)_{i=1,\dots,d} \;, \quad \mathbf{l} := (l_i)_{i=1,\dots,d} \;, \quad \delta_{\mathbf{l}} := \prod_{i=1}^{d} \delta_{l_i} \;,$$

$$b_{\mathbf{hl}}(\mathbf{x},\mathbf{y}) := \sum_{i=1}^{d} a_i^2 (h_i - D_{l_i}(x^i, y^i))^2 \;.$$

(10.16)

Notice that, for small t, the sum of the series appearing in Eq. (10.15) is mainly determined by the terms corresponding to small values of $|h_i|$, for $i \in \{1, \dots, d\}$; in this sense, the mentioned equation yields a small t expansion for the full heat kernel of $\mathcal{A}$.

10.3 The Dirichlet kernel

In the present section we construct the analytic continuation of the Dirichlet kernel D_s in the style of Minakshisundaram (see [106]). To this purpose, let us fix arbitrarily

$$T \in (0, +\infty)$$

(10.17)

and re-express the representation (2.85) of D_s in terms of the heat kernel K of $\mathcal{A}$ as

$$D_s(\mathbf{x},\mathbf{y}) = D_s^{(>)}(\mathbf{x},\mathbf{y}) + D_s^{(<)}(\mathbf{x},\mathbf{y}) \;,$$

(10.18)

$$D_s^{(>)}(\mathbf{x},\mathbf{y}) := \frac{1}{\Gamma(s)} \int_T^{+\infty} d\mathsf{t}\, \mathsf{t}^{s-1}\, K(\mathsf{t};\mathbf{x},\mathbf{y}) \;,$$

(10.19)

$$D_s^{(<)}(\mathbf{x},\mathbf{y}) := \frac{1}{\Gamma(s)} \int_0^T d\mathsf{t}\, \mathsf{t}^{s-1}\, K(\mathsf{t};\mathbf{x},\mathbf{y}) \;.$$

(10.20)

Let us stress that, despite the fact that $D_s^{(>)}$ and $D_s^{(<)}$ do depend on the choice of T, their sum D_s does not.

In the forthcoming Subsections 10.3.1 and 10.3.2 we report the final results which can be obtained substituting the heat kernel K in Eqs. (10.19), (10.20), respectively, with the large and small t expansions (10.10) (10.15). We refer to [61] for more details on the related computations. In the final Subsection 10.3.3 we use the above mentioned results to determine the analytic continuation of the full Dirichlet kernel D_s.

10.3.1 *Series representation and analytic continuation of $D_s^{(>)}$*

Using Eqs. (10.10), (10.19) and recalling the integral representation of the upper incomplete gamma function $\Gamma(\ ,\ )$ (see [117], page 174, Eq. (8.2.2)), one infers the following result:

$$D_s^{(>)}(\mathbf{x}, \mathbf{y}) = \frac{2^d}{a_1 ... a_d\, \Gamma(s)} \sum_{\mathbf{n} \in \mathbf{N}^d} \omega_\mathbf{n}^{-2s}\, \Gamma(s, \omega_\mathbf{n}^2\, T)\, C_\mathbf{n}(\mathbf{x}, \mathbf{y})\ . \quad (10.21)$$

The above expression allows to evaluate the derivatives of any order of the function $D_s^{(>)}$; for example, for any pair of spatial variables z, w, one has

$$\partial_{zw} D_s^{(>)}(\mathbf{x}, \mathbf{y}) = \frac{2^d}{a_1 ... a_d\, \Gamma(s)} \sum_{\mathbf{n} \in \mathbf{N}^d} \omega_\mathbf{n}^{-2s}\, \Gamma(s, \omega_\mathbf{n}^2\, T)\, \partial_{zw} C_\mathbf{n}(\mathbf{x}, \mathbf{y})\ .$$

$$(10.22)$$

With some effort, the series in Eqs. (10.21), (10.22) can be proved to converge with exponential rate for all $s \in \mathbf{C}$ and any fixed $\mathbf{x}, \mathbf{y} \in \Omega$, even along the diagonal $\mathbf{y} = \mathbf{x}$. We will account for this statement in the subsequent Section 10.8; here, we limit ourselves to stress that in view of the above remark Eqs. (10.21), (10.22) yield automatically the analytic continuations of the maps $s \mapsto D_s^{(>)}(\mathbf{x}, \mathbf{y}), \partial_{zw} D_s^{(>)}(\mathbf{x}, \mathbf{y})$ to the whole complex plane for any fixed $\mathbf{x}, \mathbf{y} \in \Omega$.

10.3.2 *Series representation and analytic continuation of $D_s^{(<)}$*

After noting that $b_{\mathbf{h}\mathbf{l}}(\mathbf{x}, \mathbf{y}) \geqslant 0$ for any $\mathbf{h} \in \mathbf{Z}^d, \mathbf{l} \in \{1, 2\}^d$ (see Eq. (10.16)), one can use Eqs. (10.15), (10.20) to infer the following:

$$D_s^{(<)}(\mathbf{x}, \mathbf{y}) = \frac{T^{s - \frac{d}{2}}}{(4\pi)^{d/2} \Gamma(s)} \sum_{\mathbf{h} \in \mathbf{Z}^d,\, \mathbf{l} \in \{1,2\}^d} \delta_\mathbf{l}\, \mathcal{P}_{s - \frac{d}{2}} \left(\frac{b_{\mathbf{h}\mathbf{l}}(\mathbf{x}, \mathbf{y})}{T} \right),$$

$$(10.23)$$

where we have introduced the function

$$\mathcal{P}_s(\beta) := \begin{cases} s^{-1} & \text{for } \beta = 0,\ \Re s > 0, \\ \beta^s\, \Gamma(-s, \beta) & \text{for } \beta > 0,\ s \in \mathbf{C}. \end{cases} \quad (10.24)$$

To proceed, let us notice that from some basic properties of the upper incomplete gamma function $\Gamma(\ ,\ )$ (see [117], page 178, Eq. (8.8.16)) it follows that

$$\partial_\beta^\ell \mathcal{P}_s(\beta) = (-1)^\ell \mathcal{P}_{s-\ell}(\beta) \quad \text{for all } \ell \in \{0,1,2,...\} \,. \qquad (10.25)$$

The above identity, along with the result of Eq. (10.23), allows to derive analogous series representations for the derivatives of any order of $D_s^{(<)}$; in particular, if z,w are any two spatial variables, differentiating term by term the series in Eq. (10.23) we obtain

$$\partial_{zw} D_s^{(<)}(\mathbf{x},\mathbf{y}) = \frac{T^{s-\frac{d}{2}-2}}{(4\pi)^{d/2}\Gamma(s)} \sum_{\mathbf{h}\in\mathbf{Z}^d,\,\mathbf{l}\in\{1,2\}^d} \delta_{\mathbf{l}} \left[\mathcal{P}_{s-\frac{d}{2}-2}\left(\frac{b_{\mathbf{hl}}}{T}\right) \partial_z b_{\mathbf{hl}}\,\partial_w b_{\mathbf{hl}} \right.$$

$$\left. - T\,\mathcal{P}_{s-\frac{d}{2}-1}\left(\frac{b_{\mathbf{hl}}}{T}\right) \partial_{zw} b_{\mathbf{hl}} \right](\mathbf{x},\mathbf{y}) \,. \qquad (10.26)$$

Let us point out that, in view of the definitions in Eqs. (10.14), (10.16), we have

$$b_{\mathbf{hl}}(\mathbf{x},\mathbf{y}) = 0 \text{ only for } \mathbf{y} = \mathbf{x} \text{ and for}$$

a finite number of terms in the series of Eqs. (10.23), (10.26) .

$$(10.27)$$

The above mentioned terms of Eqs. (10.23), (10.26) deserve special attention, and must be evaluated according to the first line in (10.24); on the contrary, for the infinitely many terms with $b_{\mathbf{hl}}(\mathbf{x},\mathbf{y}) > 0$, the second line in Eq. (10.24) gives expressions in terms of upper incomplete gamma functions. In this way we obtain

$$D_s^{(<)}(\mathbf{x},\mathbf{y}) = \frac{T^{s-\frac{d}{2}}}{(4\pi)^{d/2}\Gamma(s)(s-\frac{d}{2})} \left(\sum_{\substack{\mathbf{h}\in\mathbf{Z}^d,\,\mathbf{l}\in\{1,2\}^d \\ \text{s.t. } b_{\mathbf{hl}}(\mathbf{x},\mathbf{y})=0}} \delta_{\mathbf{l}} \right)$$

$$+ \frac{1}{(4\pi)^{d/2}\Gamma(s)} \sum_{\substack{\mathbf{h}\in\mathbf{Z}^d,\,\mathbf{l}\in\{1,2\}^d \\ \text{s.t. } b_{\mathbf{hl}}(\mathbf{x},\mathbf{y})>0}} \delta_{\mathbf{l}} \left(b_{\mathbf{hl}}^{s-\frac{d}{2}} \Gamma\left(\frac{d}{2}-s\,,\,\frac{b_{\mathbf{hl}}}{T}\right) \right) (\mathbf{x},\mathbf{y}) \,.$$

$$(10.28)$$

Let us repeat that the first sum in the above expression contains finitely many terms which are related to the first equality in Eq. (10.24); therefore they would, in principle, require $\Re s > d/2$. Nevertheless, it appears that the said sum makes sense for all complex s except $s = d/2$, where a simple pole appears. On the other hand, the series in the second line of Eq. (10.28) can be proved to converge with exponential rate for any given $s \in \mathbf{C}$ and all fixed $\mathbf{x}, \mathbf{y} \in \Omega$; again, we refer to Section 10.8 for a justification of this statement.

The above considerations suffice to infer the following fact: Eq. (10.28) gives automatically the analytic continuation of $D_s^{(<)}(\mathbf{x}, \mathbf{y})$ to a meromorphic function of s on the whole complex plane, with a simple pole singularity only at $s = d/2$ for any $\mathbf{x}, \mathbf{y}$ such that the first sum in Eq. (10.29) is non-empty. As a matter of fact, due to Eq. (10.14), the last condition is only fulfilled along the diagonal $\mathbf{y} = \mathbf{x}$.

A similar analysis can be performed for the derivatives of $D_s^{(<)}$. For example, if z, w are any two spatial variables, we obtain the following expression from Eq. (10.26):

$$\partial_{zw} D_s^{(<)}(\mathbf{x}, \mathbf{y}) = -\frac{T^{s-\frac{d}{2}-1}}{(4\pi)^{d/2}\Gamma(s)(s-\frac{d}{2}-1)}$$

$$\times \left(\sum_{\substack{\mathbf{h}\in\mathbf{Z}^d, l\in\{1,2\}^d \\ \text{s.t. } b_{\mathbf{hl}}(\mathbf{x},\mathbf{y})=0}} \delta_l\, \partial_{zw} b_{\mathbf{hl}}(\mathbf{x}, \mathbf{y}) \right)$$

$$+ \frac{1}{(4\pi)^{d/2}\Gamma(s)} \sum_{\substack{\mathbf{h}\in\mathbf{Z}^d, l\in\{1,2\}^d \\ \text{s.t. } b_{\mathbf{hl}}(\mathbf{x},\mathbf{y})>0}} \delta_l$$

$$\times \left[b_{\mathbf{hl}}^{s-\frac{d}{2}-2} \left(\Gamma\left(\frac{d}{2}+2-s, \frac{b_{\mathbf{hl}}}{T}\right) \partial_z b_{\mathbf{hl}}\, \partial_w b_{\mathbf{hl}} \right.\right.$$

$$\left.\left. - \Gamma\left(\frac{d}{2}+1-s, \frac{b_{\mathbf{hl}}}{T}\right) b_{\mathbf{hl}}\, \partial_{zw} b_{\mathbf{hl}} \right) \right](\mathbf{x}, \mathbf{y}) . \quad (10.29)$$

Again, the first of the two sums appearing on the right-hand side of the above equation consists of finitely many terms; besides, contrary to what one could expect from Eq. (10.26), this sum contains no term with the first order derivatives $\partial_w b_{\mathrm{hl}}(\mathbf{x}, \mathbf{y})$, $\partial_z b_{\mathrm{hl}}(\mathbf{x}, \mathbf{y})$ because they vanish if $b_{\mathrm{hl}}(\mathbf{x}, \mathbf{y}) = 0$. In addition, the series in the second and third lines of Eq. (10.29) converges with exponential rate for all $s \in \mathbf{C}$ and any fixed $\mathbf{x}, \mathbf{y} \in \Omega$ (see Section 10.8).

In view of the facts pointed out above, Eq. (10.29) gives automatically the analytic continuation of $\partial_{zw} D_s^{(<)}(\mathbf{x}, \mathbf{y})$ to a meromorphic function of s on the whole complex plane, with a simple pole singularity only for $\mathbf{y} = \mathbf{x}$ at $s = d/2 + 1$.

Before proceeding, let us stress that Eqs. (10.28), (10.29) are just the original Eqs. (10.23), (10.26), rewritten separating the terms with $b_{\mathrm{hl}}(\mathbf{x}, \mathbf{y}) = 0$ for a better understanding of the behavior with respect to the parameter s. In the sequel, even when considering analytic continuations, we will always refer to the more concise representations (10.23) and (10.26) for brevity.

10.3.3 *Conclusions for the Dirichlet kernel*

Summing up, using Eq. (10.18) and the expressions (10.21), (10.23) for the functions $D_s^{(>)}$, $D_s^{(<)}$, respectively, we obtain the analytic continuation of the full Dirichlet kernel D_s to a meromorphic function on the whole complex plane. Similar results hold for the derivatives of D_s (see Eqs. (10.22), (10.26)).

The only singularity of $D_s(\mathbf{x}, \mathbf{y})$ is a simple pole for

$$\mathbf{y} = \mathbf{x} \quad \text{and} \quad s = d/2 , \tag{10.30}$$

while $\partial_{zw} D_s(\mathbf{x}, \mathbf{y})$ (for any pair of spatial variables z, w) has a simple pole for

$$\mathbf{y} = \mathbf{x} \quad \text{and} \quad s = d/2 + 1 . \tag{10.31}$$

In particular the analytic continuation of $D_{\frac{u-1}{2}}(\mathbf{x}, \mathbf{y})|_{\mathbf{y}=\mathbf{x}}$ and $\partial_{zw} D_{\frac{u+1}{2}}(\mathbf{x}, \mathbf{y})|_{\mathbf{y}=\mathbf{x}}$, required for the evaluation of the regularized VEVs of the stress-energy tensor and pressure, are both regular at the point $u = 0$ of interest for renormalization.

10.4   The stress-energy tensor

Consider again the representations deduced in Section 10.3 for the analytic continuations of the Dirichlet kernel and of its derivatives. Resorting to the general identities (2.28)–(2.30), the said representations can be used to obtain the following expressions for the components of the regularized stress-energy VEV:

$$\langle 0|\widehat{T}^u_{\mu\nu}(\mathbf{x})|0\rangle = T^{u,(>)}_{\mu\nu}(\mathbf{x}) + T^{u,(<)}_{\mu\nu}(\mathbf{x}) , \qquad (10.32)$$

where, for $\bullet$ equal to $>$ or $<$, $T^{u,(\bullet)}_{\mu\nu}$ has the expression corresponding to Eqs. (2.28)–(2.30), with D_s replaced by $D_s^{(\bullet)}$. Thus, for $i,j \in \{1, ..., d\}$, we have

$$T^{u,(\bullet)}_{00}(\mathbf{x}) = \kappa^u \left[\left(\frac{1}{4}+\xi\right) D^{(\bullet)}_{\frac{u-1}{2}}(\mathbf{x},\mathbf{y}) + \left(\frac{1}{4}-\xi\right) \partial^{x^\ell}\partial_{y^\ell} D^{(\bullet)}_{\frac{u+1}{2}}(\mathbf{x},\mathbf{y}) \right]_{\mathbf{y}=\mathbf{x}} ,$$
$$(10.33)$$

$$T^{u,(\bullet)}_{0j}(\mathbf{x}) = T^{u,(\bullet)}_{j0}(\mathbf{x}) = 0 , \qquad (10.34)$$

$$T^{u,(\bullet)}_{ij}(\mathbf{x}) = \kappa^u \left[\left(\frac{1}{4}-\xi\right) \delta_{ij} \left(D^{(\bullet)}_{\frac{u-1}{2}}(\mathbf{x},\mathbf{y}) - \partial^{x^\ell}\partial_{y^\ell} D^{(\bullet)}_{\frac{u+1}{2}}(\mathbf{x},\mathbf{y}) \right) \right.$$
$$\left. + \left(\left(\frac{1}{2}-\xi\right) \partial_{x^i y^j} - \xi \partial_{x^i x^j} \right) D^{(\bullet)}_{\frac{u+1}{2}}(\mathbf{x},\mathbf{y}) \right]_{\mathbf{y}=\mathbf{x}} . \qquad (10.35)$$

Next let us notice that, concerning the analyticity of the above functions, there hold considerations analogous to those discussed in Section 10.3 for the Dirichlet kernel and its derivatives; in particular, one can readily infer that $u = 0$ is a regular point for each component of the regularized stress-energy VEV. Therefore, we can proceed to determine the renormalized VEV of the stress-energy tensor according to the restricted version (1.22) of the zeta approach, by putting simply

$$\langle 0|\widehat{T}^u_{\mu\nu}(\mathbf{x})|0\rangle_{\text{ren}} := \langle 0|\widehat{T}^u_{\mu\nu}(\mathbf{x})|0\rangle \Big|_{u=0} . \qquad (10.36)$$

10.5 The pressure on the boundary

Let $\mathbf{x}$ be any point interior to one of the sides $\pi_{p,\lambda}$ of the box ($p \in \{1,...,d\}$, $\lambda \in \{0,1\}$; see Eq. (10.2)). We exclude $\mathbf{x}$ to be on an edge of the box, i.e. on the intersection of two or more sides, where the outer normal is ill-defined. As an example, let us assume $\mathbf{x}$ to be an inner point of the side $\pi_{1,0}$, so that the outer unit normal at $\mathbf{x}$ is $\mathbf{n}(\mathbf{x}) = (-1,0,...,0)$.

We first consider the regularized pressure; according to Eq. (3.21) and to the general rule (3.25) for the case of Dirichlet boundary conditions, in the present case this is given by[b]

$$p_i^u(\mathbf{x}) = -\langle 0|\widehat{T}_{i1}^u(\mathbf{x})|0\rangle = -\delta_{i1}\frac{\kappa^u}{4}\,\partial_{x^1 y^1}D_{\frac{u+1}{2}}(\mathbf{x},\mathbf{y})\Big|_{\mathbf{y}=\mathbf{x}}$$

$$= -\delta_{i1}\frac{\kappa^u}{4}\left[\partial_{x^1 y^1}D_{\frac{u+1}{2}}^{(>)}(\mathbf{x},\mathbf{y})\Big|_{\mathbf{y}=\mathbf{x}} + \partial_{x^1 y^1}D_{\frac{u+1}{2}}^{(<)}(\mathbf{x},\mathbf{y})\Big|_{\mathbf{y}=\mathbf{x}}\right].$$
(10.37)

Then, keeping in mind the considerations of Section 10.3 on the derivatives of the Dirichlet functions $D_{\frac{u+1}{2}}^{(\bullet)}$, we can employ the general prescription (1.22) to define the renormalized pressure as

$$p_i^{\mathrm{ren}}(\mathbf{x}) := p_i^u(\mathbf{x})\Big|_{u=0} = -\frac{\delta_{i1}}{4}\left[\partial_{x^1 y^1}D_{1/2}^{(>)}(\mathbf{x},\mathbf{y})\Big|_{\mathbf{y}=\mathbf{x}}\right.$$

$$\left. + \partial_{x^1 y^1}D_{1/2}^{(<)}(\mathbf{x},\mathbf{y})\Big|_{\mathbf{y}=\mathbf{x}}\right].$$
(10.38)

---

[b]In the application of Eq. (3.25) to the present case, we use the previous expression for $\mathbf{n}(\mathbf{x})$ and the fact that

$$\partial_{x^i y^j}D_s(\mathbf{x},\mathbf{y})\Big|_{\mathbf{y}=\mathbf{x}} = 0 \quad \text{for all } i,j \in \{1,...,d\} \text{ such that } i \neq 1 \text{ or } j \neq 1;$$

this follows straightforwardly from the Dirichlet conditions prescribed on the boundary of Ω and from the eigenfunction expansion (2.15), taking into account the factorized structure of the eigenfunctions in the setting under analysis. The equality in the second line of Eq. (10.37) follows readily from the decomposition (10.18) of the Dirichlet kernel D_s in terms of the functions $D_s^{(>)}$, $D_s^{(<)}$.

Let us mention that, making reference to the general discussion of Section 3.2, we could consider as well the alternative prescription

$$p_i^{\mathrm{ren}}(\mathbf{x}) := \left(\lim_{\mathbf{x}' \in \Omega,\, \mathbf{x}' \to \mathbf{x}} \langle 0|\widehat{T}_{ij}(\mathbf{x}')|0\rangle_{\mathrm{ren}} \right) n^j(\mathbf{x})$$

$$= -\left(\lim_{\mathbf{x}' \in \Omega,\, \mathbf{x}' \to \mathbf{x}} \langle 0|\widehat{T}_{i1}(\mathbf{x}')|0\rangle_{\mathrm{ren}} \right). \tag{10.39}$$

By arguments similar to those mentioned in Section 7.6 dealing with the case of a scalar field confined by orthogonal hyperplanes, the latter prescription (10.39) and the definition (10.38) can be proved to be equivalent; we refer to Subsection 3.6.1 of [61] for a detailed account on this statement.

To conclude the present section, let us spend a few words on the behavior of the renormalized pressure on the sides of the box near the edges. We remark once more that at such points the outer normal and, consequently, the pressure are both ill-defined. As an example, for any given $p \in \{2, ..., d\}$, let us evaluate the pressure p_i^{ren} acting on the side $\pi_{1,0}$ (determined, equivalently, according to Eqs. (10.38) or (10.39)) near the edge

$$\mathfrak{e}_p := \pi_{1,0} \cap ... \cap \pi_{p,0} = \left\{ \mathbf{x} \in \partial\Omega \mid x^j = 0 \ \text{ for } j = 1, ..., p \right\}. \tag{10.40}$$

By means of lengthy computations, one can prove that

$$p_i^{\mathrm{ren}}(\mathbf{x}) = \delta_{i1}\, O\left(z^{-\frac{d+1}{2}} \Big|_{z = \sqrt{(x^2)^2 + ... + (x^p)^2}} \right) \qquad \text{for } \mathbf{x} \in \pi_{1,0},\ \mathbf{x} \to \mathfrak{e}_p. \tag{10.41}$$

Analogous results can be inferred for the renormalized pressure acting on any other side of the box.

The above considerations show, in particular, that the renormalized pressure evaluated at inner points of one side diverges in a non-integrable manner when moving towards anyone of the edges. This fact is of utmost importance when attempting to evaluate the total force acting on any side of the box, a topic we shall address in the subsequent Section 10.7.

10.6   The total energy

Let us recall once more that the regularized total energy consists of the sum of a bulk and a boundary contribution (see Section 3.1 of Part 1); since the latter vanishes identically due to the Dirichlet conditions assumed on the boundary (see Eq. (3.9)), we only have to discuss the bulk term. To this purpose, consider the general representation (3.6) of the regularized bulk energy E^u and substitute therein the explicit expression (10.18) for the Dirichlet kernel. This allows to infer the following:[c]

$$E^u = E^{u,(>)} + E^{u,(<)} \qquad \text{where}$$

$$E^{u,(\bullet)} := \frac{\kappa^u}{2} \int_{(0,a_1)\times...\times(0,a_d)} dx^1 ... dx^d \ D_{\frac{u-1}{2}}^{(\bullet)}(\mathbf{x},\mathbf{x}) \quad \text{for } \bullet \in \{>,<\} .$$

$$(10.42)$$

Hereafter we give series expansions for the two summands $E^{u,(>)}$ and $E^{u,(<)}$, ultimately yielding the analytic continuations of these functions to the whole complex plane. The details on the derivation of these expansions can be found in [61] (see, in particular, Section 3.7 and the related Appendices B, C therein).

On the one hand, substituting the representation (10.21) for $D_s^{(>)}$ into Eq. (10.42), one infers

$$E^{u,(>)} = \frac{\kappa^u}{2\,\Gamma(\frac{u-1}{2})} \sum_{\mathbf{n}\in\mathbf{N}^d} \omega_{\mathbf{n}}^{1-u} \ \Gamma\left(\frac{u-1}{2},\omega_{\mathbf{n}}^2 T\right) . \qquad (10.43)$$

The series on the right-hand side of the above identity can be proved to converge with exponential rate for all $u \in \mathbf{C}$; thus, Eq. (10.43) does in fact determine the analytic continuation of $E^{u,(>)}$ to the whole complex plane, including the point $u = 0$ of interest for renormalization.

---

[c]In order to evaluate the bulk energy E^u, one could proceed in an alternative manner using the representation (3.6) of E^u in terms of the trace $\mathrm{Tr}\,\mathcal{A}^{(1-u)/2}$; the analytic continuation of the latter can be determined by arguments similar to those presented in Section 10.3 for the Dirichlet kernel, using the heat trace $K(t)$ in place of the heat kernel $K(t;\mathbf{x},\mathbf{y})$. However, we prefer to avoid this approach because it would involve quite cumbersome expressions, descending from the small t expansion of $K(t)$.

On the other hand, inserting the expansion (10.23) for $D_s^{(<)}$ into Eq. (10.42), one obtains with some effort

$$E^{u,(<)} = \frac{\kappa^u\, T^{\frac{u-1}{2}}}{2^{d+1}\, \Gamma(\frac{u-1}{2})} \sum_{p=0}^{d} \frac{(-1)^{d-p}}{(d-p)!\,p!}$$

$$\times \sum_{\sigma \in S_d} \frac{\mathfrak{a}_{\sigma,p}}{(\pi T)^{p/2}} \sum_{\mathbf{h} \in \mathbf{Z}^p} \mathcal{P}_{\frac{u-p-1}{2}}\left(\frac{B_{\sigma,p}(\mathbf{h})}{T}\right), \qquad (10.44)$$

where S_d indicates the symmetric group with d elements and we have put

$$\mathfrak{a}_{\sigma,0} := 1, \quad \mathbf{Z}^0 := \{\mathbf{0}\}, \quad B_{\sigma,0}(\mathbf{0}) := 0,$$

$$\mathfrak{a}_{\sigma,p} := \prod_{i=1}^{p} a_{\sigma(i)}, \quad B_{\sigma,p}(\mathbf{h}) := \sum_{i=1}^{p} (a_{\sigma(i)} h_i)^2 \quad \text{for } \sigma \in S_d,\ p \in \{1, ..., d\}.$$
$$(10.45)$$

Let us spend a few words on the representation (10.44) for $E^{u,(<)}$. First, notice that the term with $p = 0$ in Eq. (10.44) is just $(-1)^d\, \mathcal{P}_{\frac{u-1}{2}}(0)$. This term and the other functions $\mathcal{P}_s$ appearing in Eq. (10.44) must be evaluated according to Eq. (10.24); recall, in particular, that the first relation in this equation gives $\mathcal{P}_s(0) = 1/s$. Therefore, for all $p \in \{0, ..., d\}$, the terms in the series (10.44) with $B_{\sigma,p}(\mathbf{h}) = 0$, i.e. those with $\mathbf{h} = \mathbf{0}$ (see Eq. (10.45)), are singular at $u = p+1$ where they have a simple pole. In addition, the series obtained from the right-hand side of Eq. (10.44) by removing the finitely many terms with $\mathbf{h} = \mathbf{0}$ is proved to converge with exponential rate for all $u \in \mathbf{C}$.

In view of the above considerations, the expression (10.44) gives the analytic continuation of $E^{u,(<)}$ to a meromorphic function on the whole complex plane, with simple pole singularities at $u \in \{1, 2, ..., d+1\}$.

Summing up, $u = 0$ is a regular point for the analytic continuations of both $E^{u,(>)}$ and $E^{u,(<)}$; therefore, according to the restricted zeta approach (see Eq. (1.22)), we can define the renormalized bulk energy as

$$E^{\mathrm{ren}} = E^{u,(>)}\Big|_{u=0} + E^{u,(<)}\Big|_{u=0}, \qquad (10.46)$$

where the two summands on the right-hand side simply indicate the expressions (10.43) and (10.44) evaluated at $u = 0$.

10.7 The total force on a side of the box

Again, we refer to [61] for further details on the topic of this section. As an example, let us focus on the force acting on $\pi_{1,0}$, i.e. the side contained in the hyperplane $\{x^1 = 0\}$. To this purpose, let us recall that the outer unit normal at points $\mathbf{x}$ interior to $\pi_{1,0}$ is $\mathbf{n}(\mathbf{x}) = (-1, 0, ..., 0)$ and consider the regularized integrated force defined according to Eq. (3.31) (here employed with $\mathfrak{D} = \pi_{1,0}$). This is

$$\mathfrak{F}_{1,0}^u := \int_{\pi_{1,0}} da(\mathbf{x})\, n^i(\mathbf{x})\, p_i^u(\mathbf{x}) = -\int_{\pi_{1,0}} da(\mathbf{x})\, p_1^u(\mathbf{x}) , \qquad (10.47)$$

where $p_i^u(\mathbf{x})$ indicates the regularized pressure (10.37). Using the representation in the second line of the cited equation for $p_i^u(\mathbf{x})$, we readily infer

$$\mathfrak{F}_{1,0}^u = \mathfrak{F}_{1,0}^{u,(>)} + \mathfrak{F}_{1,0}^{u,(<)} \qquad \text{where}$$

$$\mathfrak{F}_{1,0}^{u,(\bullet)} := \frac{\kappa^u}{4} \int_{(0,a_2)\times...\times(0,a_d)} dx^2 ... dx^d\, \partial_{x^1 y^1} D_{\frac{u+1}{2}}^{(\bullet)}(\mathbf{x},\mathbf{y})\Big|_{\mathbf{y}=\mathbf{x},\, x^1=0}$$

$$\text{for } \bullet \in \{>,<\} .$$

$$(10.48)$$

On the one hand, inserting into Eq. (10.48) the series representation (10.22) for $\partial_{x^1 y^1} D_s^{(>)}$ and integrating term by term, we obtain

$$\mathfrak{F}_{1,0}^{u,(>)} = \frac{\kappa^u}{2a_1 \Gamma(\frac{u+1}{2})} \sum_{\mathbf{n}\in\mathbf{N}^d} \left(\frac{n_1 \pi}{a_1}\right)^2 \omega_{\mathbf{n}}^{-(u+1)} \Gamma\left(\frac{u+1}{2}, \omega_{\mathbf{n}}^2 T\right) .$$

$$(10.49)$$

The series in the right-hand side of the above equation can be proved to converge for all $u \in \mathbf{C}$; thus, Eq. (10.49) gives the analytic continuation of the map $u \mapsto \mathfrak{F}_{1,0}^{u,(>)}$ to the whole complex plane, in particular at $u = 0$.

On the other hand, using the series expansion (10.26) for $\partial_{x^1 y^1} D_s^{(<)}$ along with the definition (10.48), it can be shown that

$$\mathfrak{F}_{1,0}^{u,(<)} = -\frac{\kappa^u T^{\frac{u-3}{2}}}{2^{d+1}\Gamma(\frac{u+1}{2})} \sum_{p=1}^{d} \frac{(-1)^{d-p}}{(d-p)!(p-1)!} \sum_{\bar\sigma \in \bar S_d} \frac{\mathfrak{a}_{\bar\sigma,p}}{(\pi T)^{\frac{p}{2}}}$$

$$\times \sum_{h \in \mathbf{Z}^p} \left[(a_1 h_1)^2 \, \mathcal{P}_{\frac{u-p-3}{2}}\left(\frac{B_{\bar\sigma,p}(\mathbf{h})}{T}\right) - \frac{T}{2}\mathcal{P}_{\frac{u-p-1}{2}}\left(\frac{B_{\bar\sigma,p}(\mathbf{h})}{T}\right) \right] ;$$

$$(10.50)$$

in the above we have put, for brevity,

$$\bar S_d := \{\bar\sigma \in S_d \mid \bar\sigma(1) = 1\}, \qquad \mathfrak{a}_{\bar\sigma,1} := 1, \qquad B_{\bar\sigma,1}(\mathbf{h}) := (a_1 h_1)^2,$$

$$(10.51)$$

$$\mathfrak{a}_{\bar\sigma,p} := \prod_{i=2}^{p} a_{\bar\sigma(i)},$$

$$B_{\bar\sigma,p}(\mathbf{h}) := (a_1 h_1)^2 + \sum_{i=2}^{p} (a_{\bar\sigma(i)} h_i)^2 \quad \text{for } \bar\sigma \in \bar S_d, \ p \in \{2,...,d\}.$$

Concerning the representation (10.50) for $\mathfrak{F}_{1,0}^{u,(<)}$, we can make considerations analogous to those following Eq. (10.44) for $E^{u,(<)}$. More precisely: the terms in the right-hand side of Eq. (10.50) with $p \in \{1,...,d\}$ and $\mathbf{h} = \mathbf{0}$ have a simple pole at $u = p+1$; after removing these finitely many terms, the series in the right-hand side of Eq. (10.50) converges with exponential rate for all $u \in \mathbf{C}$. Therefore, Eq. (10.50) gives the analytic continuation of the map $u \mapsto \mathfrak{F}_{1,0}^{u,(<)}$ to a meromorphic function on the whole complex plane, with simple pole singularities at $u \in \{2,3,...,d,d+1\}$.

Summing up, $u = 0$ is a regular point for the analytic continuations of both $\mathfrak{F}_{1,0}^{u,(>)}$ and $\mathfrak{F}_{1,0}^{u,(<)}$ so that, using the restricted zeta approach (1.22), we can proceed to define the renormalized total force on $\pi_{1,0}$ as

$$\mathfrak{F}_{1,0}^{ren} := \mathfrak{F}_{1,0}^{u,(>)}\Big|_{u=0} + \mathfrak{F}_{1,0}^{u,(<)}\Big|_{u=0} \qquad (10.52)$$

where the two summands on the right-hand side simply indicate the expressions (10.49), (10.50) evaluated at $u = 0$.[d]

10.8 On the convergence of the previous series representations

Let us consider the expressions presented in Sections 10.4–10.7 for the renormalized VEVs of the physical observables of interest in the present chapter. We recall, in particular, that the basic elements allowing to compute the renormalized stress-energy VEV and pressure are the partial Dirichlet functions $D_s^{(>)}$, $D_s^{(<)}$ (along with their spatial derivatives), for which fully explicit series representations were given in Subsections 10.3.1 and 10.3.2; similar series expansions were reported in Sections 10.6 and 10.7 for the renormalized total energy and for the integrated force acting on any side of the box.

In the previous sections we have stated the convergence, with exponential rate, of the above cited series; the proofs are given in [61] (see, in particular, Sections 3.4, 3.7, 3.8 and the related Appendices A,C). To give an idea, here we only report a result of [61] on the partial Dirichlet function $D_s^{(>)}(\mathbf{x}, \mathbf{y})$. In this chapter we have claimed that the series in Eq. (10.21) converges with exponential rate for all $s \in \mathbf{C}$ and any fixed $\mathbf{x}, \mathbf{y} \in \Omega$; concerning this statement, let $N \in \mathbf{N}$, $T \in (0, +\infty)$ and put

$$a := \min_{i \in \{1,\dots,d\}} \{a_i\}, \quad A := \max_{i \in \{1,\dots,d\}} \{a_i\}, \quad \beta := \frac{\pi^2 T}{A^2}, \quad |\mathbf{n}| := \left(\sum_{i=1}^{d} n_i^2\right)^{1/2}.$$

$$(10.53)$$

---

[d]Making reference to the general framework for boundary forces outlined in Section 3.4, in principle we could consider an alternative approach to define the renormalized total force acting on $\pi_{1,0}$. More precisely, following the prescription (3.33), we could put

$$\mathfrak{F}_{1,0}^{\text{ren}} := \int_{\pi_{1,0}} da(\mathbf{x}) \, n^i(\mathbf{x}) \, p_i^{\text{ren}}(\mathbf{x}) = -\int_{\pi_{1,0}} da(\mathbf{x}) \, p_1^{\text{ren}}(\mathbf{x}) \,,$$

where p_i^{ren} is the renormalized pressure (10.38). Nevertheless, we know from the previous Section 10.5 that p_i^{ren} diverges in a non-integrable manner near the edges of the box, so that the alternative prescription considered in this footnote gives an infinite value for the total force on $\pi_{1,0}$. Because this result is patently physically unacceptable, in this chapter we only consider the approach (10.47), (10.52).

In [61] it is shown that

$$\sum_{\mathbf{n}\in\mathbf{N}^d,\,|\mathbf{n}|>N} \left| \omega_{\mathbf{n}}^{-2s}\,\Gamma(s,\omega_{\mathbf{n}}^2 T)\,C_{\mathbf{n}}(\mathbf{x},\mathbf{y})\right|$$

$$\leq \frac{\pi^{\frac{d}{2}-2\Re s}\,\max(a^{2\Re s},A^{2\Re s})}{2^d(1-\alpha)^{\Re s}(\alpha\,\beta)^{d/2-\Re s}\,\Gamma(\frac{d}{2})}\left(\frac{N-\sqrt{d}}{N-2\sqrt{d}}\right)^{d-1}$$

$$\times\,\Gamma(\Re s,(1-\alpha)\beta N^2)\,\Gamma\left(\frac{d}{2}-\Re s,\alpha\,\beta(N-2\sqrt{d})^2\right)$$

for either $\quad \Re s\geqslant 0,\ N>2\sqrt{d}\quad$ or $\quad \Re s<0,\ N>2\sqrt{d}+\dfrac{A}{\pi}\sqrt{\dfrac{|\Re s|}{\alpha\,T}}\,.$

$$(10.54)$$

In the above, α is a parameter that can be freely chosen in $(0,1)$; of course, the best choice is the one minimizing the right-hand side of Eq. (10.54), which depends on the other parameters (e.g. N,T) involved in these considerations.

The bound in Eq. (10.54) suffices, amongst else, to infer the fact that the series expansion (10.21) for $D_s^{(>)}(\mathbf{x},\mathbf{y})$ converges (absolutely) with exponential rate. Indeed, due to the asymptotic features of the upper incomplete gamma function (see [117], page 179, Eq. (8.11.2)), the expression in the second line of Eq. (10.54) is of order $O(e^{-\beta N^2/2})$ for $N\to+\infty$.

Explicit remainder estimates analogous to the one in Eq. (10.54) can be derived for all the other series expansions reported in the present chapter. Besides establishing the already mentioned convergence properties, these estimates turn out to be extremely useful for the numerical evaluation of the said series; indeed, one can perform this evaluation with little effort simply by summing a small number of the first terms of the said series, still committing a very small error. For a comprehensive discussion of these statements, we refer again to [61].

Before proceeding, let us anticipate that in the forthcoming Section 10.10, dealing with the case of a two-dimensional box, we will use approximate expressions for the renormalized VEVs of all observables; these are obtained replacing the full series representations with the corresponding truncations of a fixed, sufficiently large order. Explicit error estimates for all these approximants, exemplified by Eq. (10.54), can be found in [61].

10.9 Scaling considerations

From Eqs. (10.32)–(10.35) and from the expressions for the Dirichlet functions given in Section 10.3, it can be easily infered the following relation for each component of the stress-energy VEV:

$$\langle 0|\widehat{T}^u_{\mu\nu}(\mathbf{x})|0\rangle = a_1^{u-d-1}\,\mathrm{T}^u_{\mu\nu}(\mathbf{x}_\star;\boldsymbol{\rho}) \qquad (\mu,\nu\in\{0,...,d\})\ , \qquad (10.55)$$

where $\mathrm{T}^u_{\mu\nu}$ is a suitable function (independent of a_1) and $\mathbf{x}_\star$, $\boldsymbol{\rho}$ are, respectively, the d-tuple and the $(d-1)$-tuple with components

$$x^i_\star := \frac{x^i}{a_i} \in (0,1) \quad \text{for } i \in \{1,...,d\}\ , \qquad \rho_i := \frac{a_i}{a_1} \quad \text{for } i \in \{2,...,d\}\ .$$
$$(10.56)$$

For $d=1$, the variables ρ_i are not defined and $\mathrm{T}^u_{\mu\nu}$ only depends on the ratio $x^1_\star = x^1/a^1$. Similarly, for the components of the regularized pressure acting on any point $\mathbf{x}$ in the interior of the side $\pi_{1,0}$, we deduce from Eqs. (10.37), (10.55) that

$$p^u_i(\mathbf{x}) = a_1^{u-d-1}\,\mathrm{p}^u_i(\mathbf{x}_\star;\boldsymbol{\rho}) \qquad (i \in \{1,...,d\}) \qquad (10.57)$$

where p^u_i is a suitable function and $\mathbf{x}_\star$ is defined as in Eq. (10.56) at points on the boundary. Clearly, the same conclusions can be drawn for the pressure on any other side $\pi_{p,\lambda}$ ($p \in \{1,...,d\}$, $\lambda \in \{0,1\}$).

Analogous considerations hold for the total energy and for the integrated force on the boundary. On the one hand, concerning the regularized bulk energy, from the expansions derived in Section 10.6 we easily infer (indicating with E^u a suitable function)

$$E^u = a_1^{u-1}\,\mathrm{E}^u(\boldsymbol{\rho})\ . \qquad (10.58)$$

On the other hand, considering as an example the regularized total force on $\pi_{1,0}$, from Eqs. (10.49), (10.50) it follows that

$$\mathfrak{F}^u_{1,0} = a_1^{u-2}\,\mathrm{F}^u_{1,0}(\boldsymbol{\rho}) \qquad (10.59)$$

for a suitable function $\mathrm{F}^u_{1,0}$; similar results hold for the force on any other side.

By analytic continuation at $u = 0$, we obtain the renormalized counterparts of the above relations: more precisely, we have

$$\langle 0|\widehat{T}_{\mu\nu}(\mathbf{x})|0\rangle_{\text{ren}} = a_1^{-(d+1)}\, \mathrm{T}_{\mu\nu}(\mathbf{x}_\star;\boldsymbol{\rho})\,,\quad p_i^{\text{ren}}(\mathbf{x}) = a_1^{-(d+1)}\, \mathrm{p}_i(\mathbf{x}_\star;\boldsymbol{\rho})\,,$$

$$E^{\text{ren}} = a_1^{-1}\, \mathrm{E}(\boldsymbol{\rho})\,,\qquad \mathfrak{F}_{1,0}^{\text{ren}} = a_1^{-2}\, \mathrm{F}_{1,0}(\boldsymbol{\rho})\,,$$

$$(10.60)$$

where $\mathrm{T}_{\mu\nu}(\mathbf{x}_\star;\boldsymbol{\rho})$, $\mathrm{p}_i(\mathbf{x}_\star;\boldsymbol{\rho})$, $\mathrm{E}(\boldsymbol{\rho})$ and $\mathrm{F}_{1,0}(\boldsymbol{\rho})$ are obtained evaluating at $u = 0$ the functions $\mathrm{T}_{\mu\nu}^u(\mathbf{x}_\star;\boldsymbol{\rho})$, $\mathrm{p}_i^u(\mathbf{x}_\star;\boldsymbol{\rho})$, $\mathrm{E}^u(\boldsymbol{\rho})$ and $\mathrm{F}_{1,0}^u(\boldsymbol{\rho})$ of Eqs. (10.55)–(10.59).

Due to the above remarks, for any spatial dimension d the analysis of the renormalized stress-energy VEV, total energy, pressure and of the integrated force can always be reduced to the case $a_1 = 1$; to say more, it appears that in this case the quantities $\langle 0|\widehat{T}_{\mu\nu}|0\rangle_{\text{ren}}$, p_1^{ren}, E^{ren} and $\mathfrak{F}_{1,0}^{\text{ren}}$ are, respectively, equal to the rescaled functions $\mathrm{T}_{\mu\nu}^{\text{ren}}$, $\mathrm{p}_i^{\text{ren}}$, E and $\mathrm{F}_{1,0}$ appearing in Eq. (10.60). We use this fact in the forthcoming Section 10.10, dealing with the case of a two-dimensional box.

10.10   A rectangular box in spatial dimension $d = 2$

As an application of the general results described in the previous sections, let us consider the two-dimensional case where

$$d = 2\,,\qquad \Omega = (0, a_1) \times (0, a_2)\quad (a_1, a_2 > 0)\,.\qquad (10.61)$$

In our computations we restrict attention to the case with

$$a_1 = 1 \qquad\qquad (10.62)$$

and consider different values of a_2. This causes no loss of generality due to the scaling relations pointed out in Section 10.9. Moreover, we present the final results in terms of the rescaled coordinates $x_\star^1 := x^1/a_1 \equiv x^1$, $x_\star^2 := x^2/a_2 \in (0,1)$, defined according to Eq. (10.56); let us also point out that the length a_2 of the second side of the box can be identified with the ratio ρ_2 of the cited equation.

Let us first consider the stress-energy VEV and the pressure on $\pi_{1,0}$; as examples, we compute these observables for the two configurations with

$$a_2 = 1 \qquad \text{and} \qquad a_2 = 5\,.\qquad (10.63)$$

The *renormalized stress-energy VEV* $\langle 0|\widehat{T}_{\mu\nu}|0\rangle_{\text{ren}}$ is obtained setting $u = 0$ in Eqs. (10.32)–(10.35), in agreement with Eq. (10.36). In reporting our results, we distinguish between the conformal and non-conformal parts of each component; these are respectively denoted, as usual, with $\langle 0|\widehat{T}_{\mu\nu}^{(\Diamond)}|0\rangle_{\text{ren}}$ and $\langle 0|\widehat{T}_{\mu\nu}^{(\blacksquare)}|0\rangle_{\text{ren}}$ (see Eq. (1.32)). Let us also notice that (since $d = 2$) the second relation in Eq. (1.32) gives

$$\xi_2 = 1/8 . \tag{10.64}$$

In the following we present the graphs for $\langle 0|\widehat{T}_{\mu\nu}^{(\Diamond)}|0\rangle_{\text{ren}}$ and $\langle 0|\widehat{T}_{\mu\nu}^{(\blacksquare)}|0\rangle_{\text{ren}}$ obtained from the framework of Section 10.4 and from numerical evaluation of the corresponding series in Section 10.3.[e] More precisely, Figs. 10.1–10.3 and Figs. 10.4–10.7 (reprinted, as a courtesy of World Scientific, from [61]) show the results obtained for $a_2 = 1$ and $a_2 = 5$, respectively. In the cited figures we refer to the variables $x_\star^i := x^i/a_i \in (0,1)$ and, keeping into account some obvious symmetry considerations,[f] we only show the graphs for $x_\star^i \in (0, 1/2)$ ($i \in \{1, 2\}$); moreover, in the case of a square box with $a_1 = a_2 = 1$ we do not report the graphs for the conformal and non-conformal parts of $\langle 0|\widehat{T}_{22}(\mathbf{x})|0\rangle_{\text{ren}}$, since these are equal to the corresponding parts of $\langle 0|\widehat{T}_{11}(\mathbf{x})|0\rangle_{\text{ren}}$.

Now, let us consider the *renormalized pressure* $p_i^{\text{ren}}(\mathbf{x})$ at points $\mathbf{x} \equiv (0, x^2)$ in the interior of the side $\pi_{1,0}$ ($x^2 \in (0, a_2)$); to this purpose, we make reference to the general prescription (10.38). Figure 10.8 (again, reprinted from [61]) shows the graphs of p_1^{ren} as a function of the rescaled coordinate $x_\star^2 := x^2/a_2 \in (0,1)$, for the choices $a_2 = 1$ and $a_2 = 5$.

Before moving on, let us briefly comment on the behavior of the renormalized pressure p_1^{ren} near the edge $\mathbf{x} = \mathbf{0}$. Indeed, specializing to the present two-dimensional case the concluding remarks of

---

[e]The numerical evaluations of these series have all been performed by truncation at order $N = 9$ (see Section 10.8), so as to ensure that the corresponding absolute errors are smaller than 10^{-10}. All the other series expansions appearing in this section have been truncated to an appropriate order, chosen so as to ensure negligible errors.

[f]Indeed, every component of the stress-energy VEV can be shown to be symmetric under the exchange $x^i \leftrightarrow a_i - x^i$ (or $x_\star^i \leftrightarrow 1 - x_\star^i$) for $i \in \{1, 2\}$.

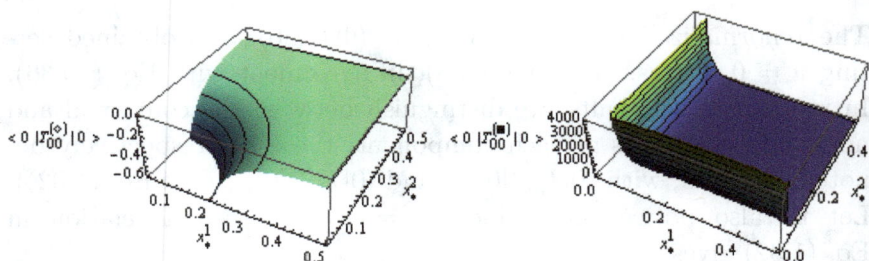

Fig. 10.1　$d = 2$: graphs of $\langle 0|\widehat{T}_{00}^{(\lozenge)}(\mathbf{x})|0\rangle_{\mathrm{ren}}$ and $\langle 0|\widehat{T}_{00}^{(\blacksquare)}(\mathbf{x})|0\rangle_{\mathrm{ren}}$ for $a_2 = 1$.

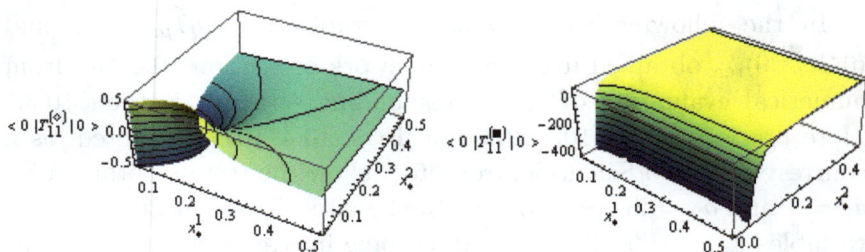

Fig. 10.2　$d = 2$: graphs of $\langle 0|\widehat{T}_{11}^{(\lozenge)}(\mathbf{x})|0\rangle_{\mathrm{ren}}$ and $\langle 0|\widehat{T}_{11}^{(\blacksquare)}(\mathbf{x})|0\rangle_{\mathrm{ren}}$ for $a_2 = 1$.

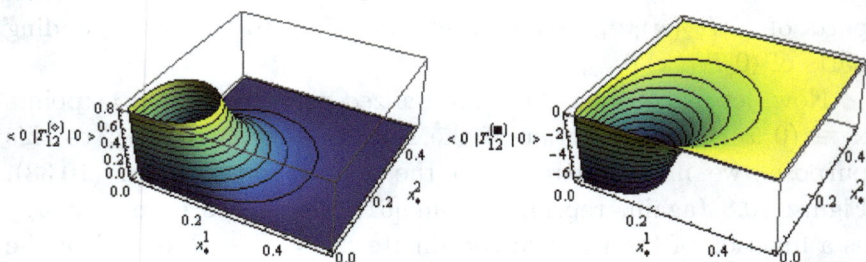

Fig. 10.3　$d = 2$: graphs of $\langle 0|\widehat{T}_{12}^{(\lozenge)}(\mathbf{x})|0\rangle_{\mathrm{ren}}$ and $\langle 0|\widehat{T}_{12}^{(\blacksquare)}(\mathbf{x})|0\rangle_{\mathrm{ren}}$ for $a_2 = 1$.

Section 10.5 one can prove that, for all $a_2 > 0$,

$$p_1^{\mathrm{ren}}(\mathbf{x}) = \frac{1}{32\pi(x^2)^3} + O((x^2)^2) \qquad \text{for } \mathbf{x} = (0, x^2) \text{ and } x^2 \to 0^+ .$$

$$(10.65)$$

Let us pass to the bulk energy and to the integrated force on the side $\pi_{1,0}$.

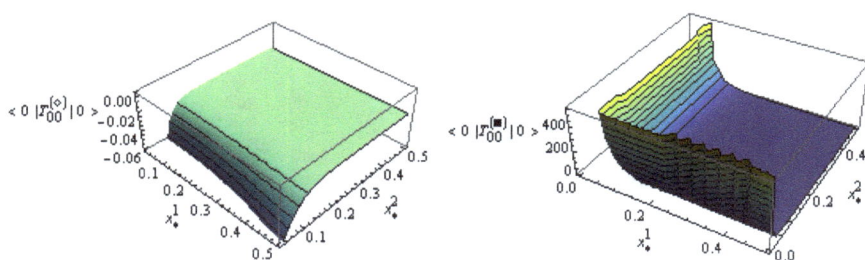

Fig. 10.4 $d = 2$: graphs of $\langle 0|\widehat{T}_{00}^{(\lozenge)}(\mathbf{x})|0\rangle_{\mathrm{ren}}$ and $\langle 0|\widehat{T}_{00}^{(\blacksquare)}(\mathbf{x})|0\rangle_{\mathrm{ren}}$ for $a_2 = 5$.

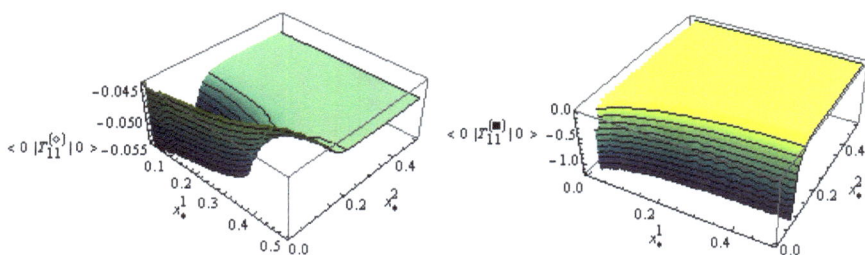

Fig. 10.5 $d = 2$: graphs of $\langle 0|\widehat{T}_{11}^{(\lozenge)}(\mathbf{x})|0\rangle_{\mathrm{ren}}$ and $\langle 0|\widehat{T}_{11}^{(\blacksquare)}(\mathbf{x})|0\rangle_{\mathrm{ren}}$ for $a_2 = 5$.

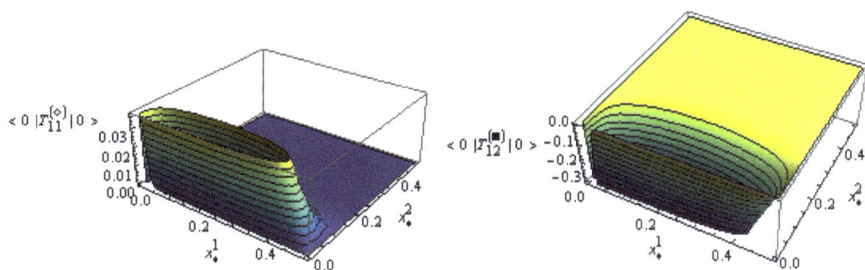

Fig. 10.6 $d = 2$: graphs of $\langle 0|\widehat{T}_{12}^{(\lozenge)}(\mathbf{x})|0\rangle_{\mathrm{ren}}$ and $\langle 0|\widehat{T}_{12}^{(\blacksquare)}(\mathbf{x})|0\rangle_{\mathrm{ren}}$ for $a_2 = 5$.

In order to obtain the *renormalized bulk energy* E^{ren}, we resort to the prescription (10.46), which allows to express E^{ren} in terms of the analytic continuations of the functions $E^{u,(>)}$ and $E^{u,(<)}$ at $u = 0$; we recall that the said continuations can be obtained simply

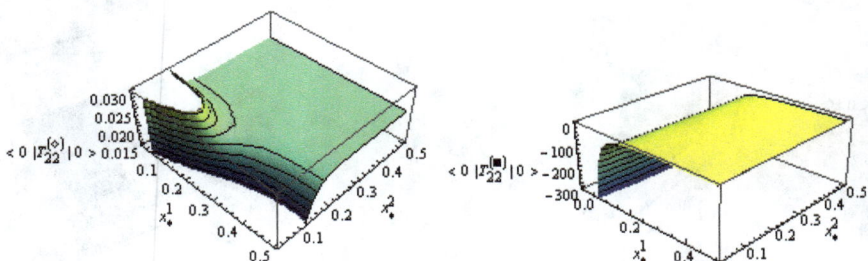

Fig. 10.7 $d = 2$: graphs of $\langle 0|\widehat{T}_{22}^{(\Diamond)}(\mathbf{x})|0\rangle_{\mathrm{ren}}$ and $\langle 0|\widehat{T}_{22}^{(\blacksquare)}(\mathbf{x})|0\rangle_{\mathrm{ren}}$ for $a_2 = 5$.

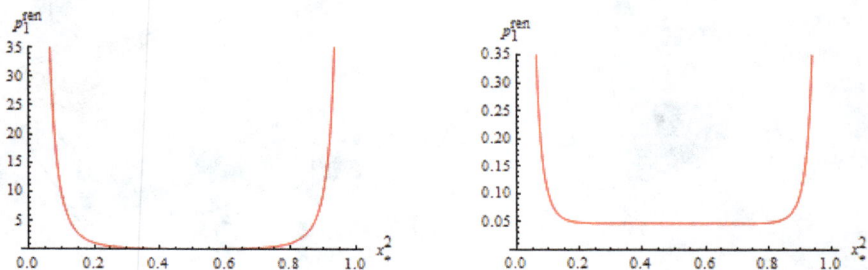

Fig. 10.8 $d = 2$: graphs of p_1^{ren} for $a_2 = 1$ (left) and $a_2 = 5$ (right).

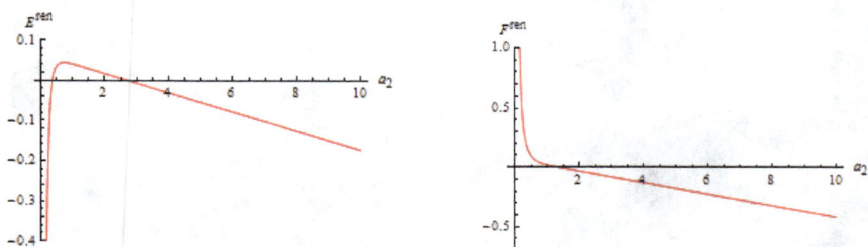

Fig. 10.9 $d = 2$: graphs of E^{ren} (left) and $\mathfrak{F}_{1,0}^{\mathrm{ren}}$ (right) as functions of a_2, for $a_1 = 1$.

by setting $u = 0$ in the series expansions (10.43), (10.44). Fixing again $a_1 = 1$, one can plot the renormalized bulk energy E^{ren} as a function of a_2; the graph in Fig. 10.9 (again, reprinted from [61]) has been obtained evaluating numerically the series (10.43), (10.44) for $a_2 \in (0.05, 10)$.

Let us discuss some facts regarding the function $a_2 \mapsto E^{\mathrm{ren}}(a_2)$, which can be extrapolated from these numerical computations.[g]

(i) There is only one point of maximum $a_2^{\max}$:

$$a_2^{\max} = 0.72719110 \pm 10^{-8} , \qquad E^{\mathrm{ren}}(a_2^{\max}) = 0.04472675 \pm 10^{-8} . \tag{10.66}$$

(ii) E^{ren} vanishes for two values $\bar{a}_2^{(1)} < \bar{a}_2^{(2)}$ of a_2; these are found to be

$$\bar{a}_2^{(1)} = 0.36538151 \pm 10^{-8} , \qquad \bar{a}_2^{(2)} = 2.73686534 \pm 10^{-8} . \tag{10.67}$$

E^{ren} is positive for $\bar{a}_2^{(1)} < a_2 < \bar{a}_2^{(2)}$ and negative elsewhere. This feature was also pointed out in [97]; therein it is stated that $\bar{a}_2^{(2)} = (\bar{a}_2^{(1)})^{-1}$, a relation (approximately) verified by the numerical values in Eq. (10.67).

(iii) For $a_2 \to 0^+$, E^{ren} has the asymptotic behavior

$$E^{\mathrm{ren}}(a_2) = \frac{e_0}{(a_2)^2} \left(1 + O(a_2)\right) \quad \text{with} \quad e_0 = -0.02391 \pm 10^{-5} . \tag{10.68}$$

(iv) There are indications that E^{ren} approaches an asymptote for $a_2 \to +\infty$; taking into account values of the abscissa up to $a_2 = 100$, we find that this asymptote is the straight line

$$y = m_E \, a_2 + q_E , \qquad \begin{cases} m_E = -0.02391416 \pm 10^{-8} \\ q_E = +0.06544985 \pm 10^{-8} \end{cases} . \tag{10.69}$$

Finally, let discuss the *renormalized total force* $\mathfrak{F}_{1,0}^{\mathrm{ren}}$ acting on the side $\pi_{1,0}$; following the analysis of Section 10.7, we consider the expression (10.52) and represent the functions $\mathfrak{F}_{1,0}^{u,(>)}$, $\mathfrak{F}_{1,0}^{u,(<)}$ appearing therein using the series expansions (10.49) (10.50). In agreement with the general prescription (10.52), $\mathfrak{F}_{1,0}^{\mathrm{ren}}$ is obtained setting $u = 0$ in these expansions.

Again for $a_1 = 1$, one can plot the renormalized force $\mathfrak{F}_{1,0}^{\mathrm{ren}}$ as a function of a_2; the graph in Fig. 10.9 (again, reprinted from [61]) has been obtained evaluating numerically the corresponding series expansions.

---

[g]The results reported in the sequel (on the maximum of E^{ren}, on its zeros and so on) have been obtained using some standard numerical methods, implemented in Mathematica.

In the following we point out a number of facts concerning the function $a_2 \mapsto \mathfrak{F}_{1,0}^{\mathrm{ren}}(a_2)$, suggested by these computations.

(i) The function under analysis is strictly decreasing for $a_2 \in (0, +\infty)$.

(ii) There is only one value of a_2 where $\mathfrak{F}_{1,0}^{\mathrm{ren}}$ vanishes; this is

$$\bar{a}_2 = 1.3751543 \pm 10^{-7} . \tag{10.70}$$

(iii) For $a_2 \to 0^+$, $\mathfrak{F}_{1,0}^{\mathrm{ren}}$ has the asymptotic behavior

$$\mathfrak{F}_{1,0}^{\mathrm{ren}}(a_2) = \frac{f_0}{(a_2)^2} \left(1 + O(a_2) \right) \quad \text{with} \quad f_0 = 0.02391 \pm 10^{-5} . \tag{10.71}$$

(iv) It appears that $\mathfrak{F}_{1,0}^{\mathrm{ren}}$ approaches an asymptote for $a_2 \to +\infty$; considering again values of the abscissa up to $a_2 = 100$, the equation of this asymptote is found to be

$$y = m\, a_2 + q , \qquad \begin{cases} m = -0.0478283 \pm 10^{-7} \\ q = +0.0654498 \pm 10^{-7} \end{cases} . \tag{10.72}$$

Let us mention that, in agreement with the general considerations of Section 3.3 in Part 1, the above results for the renormalized total force on the side $\pi_{1,0}$ of the box could be equivalently derived by differentiating the renormalized bulk energy E^{ren} with respect to the length a_2 of the edge of the box perpendicular to $\pi_{1,0}$.

To conclude, let us compare the previous results about $E^{\mathrm{ren}}, \mathfrak{F}_{1,0}^{\mathrm{ren}}$ with the calculations of Bordag *et al.* [22]. The authors of [22] derive (both by Abel–Plana formula and by zeta regularization) series expansions different from ours for the renormalized bulk energy and for the force on one side.[h] The numerical values of $E^{\mathrm{ren}}, \mathfrak{F}_{1,0}^{\mathrm{ren}}$ given by our previous analysis are in good agreement with those arising from the expansions in [22], a fact strongly indicating the equivalence between our approach and [22]. Let us also mention that the results of [22] about E^{ren} or $\mathfrak{F}_{1,0}^{\mathrm{ren}}$ are equivalent to the ones of [9, 54, 72].

---

[h] A non-trivial analysis allows to conclude that the series expansions given in [22] converge, like ours, with exponential speed.

Appendix A

The "improved" stress-energy tensor

Consider the stress-energy tensor operator introduced in Eq. (1.15) of Chapter 1. As pointed out therein, this object is obtained via second-quantization of a classical formula for the stress-energy tensor [18,30, 33,115], which we review in the present appendix for completeness.

As usual, we refer to Minkowski spacetime which, after an inertial frame has been chosen, is identified with $\mathbf{R}^{d+1} = \mathbf{R} \times \mathbf{R}^d \ni x \equiv (x^\mu) \equiv (t, \mathbf{x})$; we confine the attention to a subset of the form $\mathbf{R} \times \Omega$, where $\Omega \subset \mathbf{R}^d$ is a spatial domain. Let us consider a classical scalar field ϕ on $\mathbf{R} \times \Omega$ described by an arbitrary Lagrangian density $\mathcal{L} = \mathcal{L}(\phi, \partial\phi, x)$. The associated canonical stress-energy tensor is

$$T_{\mu\nu}^{\text{can}} := -\frac{\partial \mathcal{L}}{\partial(\partial^\mu \phi)}\, \partial_\nu \phi + \eta_{\mu\nu}\mathcal{L} \ , \qquad (A.1)$$

and fulfills

$$\partial^\mu T_{\mu\nu}^{\text{can}} = -\,\partial_\nu \mathcal{L} \qquad (A.2)$$

along the solutions of the field equations. For any spacelike hypersurface Σ with unit normal vector N^μ and volume element dv, we define the canonical momentum

$$P_\nu^{\text{can}}(\Sigma) := \int_\Sigma dv\, N^\mu\, T_{\mu\nu}^{\text{can}} \ . \qquad (A.3)$$

For simplicity we restrict the attention to the case where

$$\Sigma = \{t\} \times \Omega \ , \qquad (A.4)$$

for some fixed $t \in \mathbf{R}$, writing $P_\nu^{\mathrm{can}}(t)$ for the corresponding canonical momentum. In this case we can take $(N^\mu) = (1, 0, \dots, 0)$ and $\int_\Sigma dv$ corresponds to integration on Ω with respect to the usual volume element $d\mathbf{x}$, so

$$P_\nu^{\mathrm{can}}(t) := \int_\Omega d\mathbf{x}\, T_{0\nu}^{\mathrm{can}}(t, \mathbf{x}) \, . \tag{A.5}$$

Writing x for $(t, \mathbf{x})$, we have $dP_\nu^{\mathrm{can}}/dt(t) = \int_\Omega d\mathbf{x}\, \partial_0 T_{0\nu}^{\mathrm{can}}(x) = \int_\Omega d\mathbf{x}(-\partial^0 T_{0\nu}^{\mathrm{can}})(x) = \int_\Omega d\mathbf{x}(\partial^i T_{i\nu}^{\mathrm{can}} - \partial^\mu T_{\mu\nu}^{\mathrm{can}})(x) = \int_\Omega d\mathbf{x}\, [\partial^i T_{i\nu}^{\mathrm{can}}(x) + (\partial_\nu \mathcal{L})(\phi(x), \partial\phi(x), x)]$, i.e. by the d-dimensional divergence theorem,

$$\frac{dP_\nu^{\mathrm{can}}}{dt}(t) = \int_\Omega d\mathbf{x}\, (\partial_\nu \mathcal{L})(\phi(x), \partial\phi(x), x) + \int_{\partial\Omega} da(\mathbf{x})\, n^i(\mathbf{x})\, T_{i\nu}^{\mathrm{can}}(x) \, . \tag{A.6}$$

Here (and in the sequel) $\boldsymbol{n}(\mathbf{x}) = (n^i(\mathbf{x}))$ is the outer unit vector in $\mathbf{R}^d$ normal to the boundary $\partial\Omega$ at $\mathbf{x}$, and da is the $(d-1)$-dimensional area element.[a] For a number of reasons briefly reviewed in the sequel, it is customary (see, e.g. [33, 137]) to consider an "improved stress-energy tensor" of the form

$$T_{\mu\nu} := T_{\mu\nu}^{\mathrm{can}} + \partial^\lambda F_{\lambda\mu\nu} \, , \tag{A.7}$$

where $F_{\lambda\mu\nu}$ is a covariant tensor of rank 3 such that

$$F_{\lambda\mu\nu} = -F_{\mu\lambda\nu} \, . \tag{A.8}$$

Condition (A.8) implies $\partial^\mu(\partial^\lambda F_{\lambda\mu\nu}) = 0$, thus ensuring the identity

$$\partial^\mu T_{\mu\nu} = \partial^\mu T_{\mu\nu}^{\mathrm{can}} \, . \tag{A.9}$$

We can give definitions similar to (A.3) and (A.5) using the improved stress-energy tensor; in particular, for each $t \in \mathbf{R}$, we define the "improved momentum"

$$P_\nu(t) := \int_\Omega d\mathbf{x}\, T_{0\nu}(t, \mathbf{x}) \, . \tag{A.10}$$

---

[a] Obviously enough, if Ω is unbounded we intend $\int_{\partial\Omega} da(\mathbf{x})\, n^i(\mathbf{x}) := \lim_{\ell \to +\infty} \times \int_{\partial\Omega_\ell} da_\ell(\mathbf{x})\, n_\ell^i(\mathbf{x})$ where $\Omega_1 \subset \Omega_2 \subset \Omega_3 \subset \dots$ are bounded domains such that $\cup_{\ell=1}^{+\infty} \Omega_\ell = \Omega$.

We claim that

$$P_\nu(t) = P_\nu^{\text{can}}(t) + \int_{\partial\Omega} da(\mathbf{x}) \, n^i(\mathbf{x}) \, F_{i0\nu}(t, \mathbf{x}) \,, \qquad (A.11)$$

where da and $\mathbf{n}(\mathbf{x})$ have the same meaning as before. To prove this, note that

$$T_{0\nu} - T_{0\nu}^{\text{can}} = \partial^\lambda F_{\lambda 0\mu} = \partial^0 F_{00\nu} + \partial^i F_{i0\nu} = \partial^i F_{i0\nu} \qquad (A.12)$$

($F_{00\nu} = 0$ due to Eq. (A.8)); thus $P_\nu(t) = P_\nu^{\text{can}}(t) + \int_\Omega d\mathbf{x} \, \partial^i F_{i0\nu}(t, \mathbf{x})$, so that Eq. (A.11) follows from the d-dimensional divergence theorem.

In many cases of interest, the boundary term in Eq. (A.11) is zero. In particular, this happens if $\Omega = \mathbf{R}^d$ and $F_{i0\nu}(t, \mathbf{x})$ vanishes rapidly for $\mathbf{x} \to \infty$. In the case of a bounded domain, the boundary term can be zero if $F_{i0\nu}$ depends suitably on the field ϕ and the latter fulfills appropriate conditions on $\partial\Omega$.

Whether or not the boundary term in Eq. (A.11) vanishes, using Eq. (A.9) we prove that the improved momentum evolves according to the analogue of Eq. (A.6), i.e.

$$\frac{dP_\nu}{dt}(t) = \int_\Omega d\mathbf{x} \, (\partial_\nu \mathcal{L})(\phi(x), \partial\phi(x), x) + \int_{\partial\Omega} da(\mathbf{x}) \, n^i(\mathbf{x}) \, T_{i\nu}(x) \,.$$
$$(A.13)$$

The improved stress-energy tensor is symmetric if and only if

$$\partial^\lambda (F_{\lambda\mu\nu} - F_{\lambda\nu\mu}) = -(T_{\mu\nu}^{\text{can}} - T_{\nu\mu}^{\text{can}}) \,. \qquad (A.14)$$

When $T_{\mu\nu}^{\text{can}}$ is not symmetric and there exists a rank 3 tensor $F_{\lambda\mu\nu}$ fulfilling conditions (A.8) and (A.14), the symmetry of the improved stress-energy tensor (A.7) is itself a good reason to consider this object.

There are reasons to consider the improved stress-energy tensor even in the case when $T_{\mu\nu}^{\text{can}}$ is itself symmetric (of course, in this case Eq. (A.14) requires that $\partial^\lambda F_{\lambda\mu\nu}$ be symmetric in μ and ν). One of these reasons has been pointed out by Callan *et al.* [33]; in few words, after quantization the divergences of the improved tensor can happen to be softer than the divergences of the canonical one, especially in perturbative renormalization.

In this book we are interested in a field theory governed by the equation $0 = (-\partial_{tt} + \Delta - V)\phi = (\partial_\mu \partial^\mu - V)\phi$ which arises from the Lagrangian

$$\mathcal{L} := -\frac{1}{2}\,\partial^\mu \phi\,\partial_\mu \phi - \frac{1}{2}\,V\phi^2 \ . \tag{A.15}$$

The corresponding canonical stress-energy tensor is

$$T^{\mathrm{can}}_{\mu\nu} = \partial_\mu \phi\,\partial_\nu \phi - \frac{1}{2}\,\eta_{\mu\nu}(\partial^\lambda \phi\,\partial_\lambda \phi + V\phi^2) \ ; \tag{A.16}$$

this is symmetric and fulfills (along solutions of the field equations)

$$\partial^\mu T^{\mathrm{can}}_{\mu\nu} = -\frac{1}{2}\,(\partial_\nu V)\,\phi^2 \ . \tag{A.17}$$

Condition (A.8) is enjoyed by the tensor

$$F_{\lambda\mu\nu} := -\xi(\eta_{\lambda\nu}\partial_\mu - \eta_{\mu\nu}\partial_\lambda)\phi^2 \ , \tag{A.18}$$

where ξ is some real parameter. In this case

$$\partial^\lambda F_{\lambda\mu\nu} = -\xi(\partial_{\mu\nu} - \eta_{\mu\nu}\partial^\lambda \partial_\lambda)\phi^2 \ ; \tag{A.19}$$

this tensor is symmetric in μ and ν, so that $T_{\mu\nu}$ is symmetric as well. The derivatives of ϕ^2 in Eq. (A.19) can be re-expressed using the field equation $\partial_\mu \partial^\mu \phi = V\phi$, yielding the identity

$$\partial^\lambda F_{\lambda\mu\nu} = -2\xi\,\partial_\mu \phi\,\partial_\nu \phi + 2\xi\,\eta_{\mu\nu}(\partial^\lambda \phi\,\partial_\lambda \phi + V\phi^2) - 2\xi\,\phi\,\partial_{\mu\nu}\phi \ . \tag{A.20}$$

Thus, the improved stress-energy tensor (A.7) takes the form

$$T_{\mu\nu} = (1-2\xi)\partial_\mu \phi\,\partial_\nu \phi - \left(\frac{1}{2} - 2\xi\right)\eta_{\mu\nu}(\partial^\lambda \phi \partial_\lambda \phi + V\phi^2) - 2\xi\,\phi\,\partial_{\mu\nu}\phi \ , \tag{A.21}$$

of which (1.15) is a natural quantization.

Let us recall that the momentum P_ν corresponding to the improved tensor $T_{\mu\nu}$ is related to the canonical one via Eq. (A.11); in the present framework where $F_{\lambda\mu\nu}$ is given by Eq. (A.18), the boundary term in Eq. (A.11) vanishes under Dirichlet boundary conditions ($\phi(t, \mathbf{x}) = 0$ for all $\mathbf{x} \in \partial\Omega$); this term vanishes as well if $\Omega = \mathbf{R}^d$ and $\phi(t, \mathbf{x})$, $\partial_\lambda \phi(t, \mathbf{x})$ vanish rapidly for $\mathbf{x} \to \infty$.

In the case $V = 0$, the above improved tensor (A.21) allows another interpretation: this is the functional derivative with respect to the metric of the action describing a scalar field that interacts with a gravitational field via the scalar curvature, in the limit where ϕ is small and the metric is the Minkowski metric $\eta_{\mu\nu}$ plus a perturbation of the second order in ϕ. Concerning this, see [33, 115, 118] and [58, 59].

Finally let us remark that, assuming again $V = 0$, the action of the field coupled to gravity is conformally invariant for $\xi = (d-1)/(4d)$ [146].

Appendix B

On the regularity of some integral kernels

We refer to an operator $\mathcal{A}$ in $L^2(\Omega)$ possessing the features described by Eqs. (2.7), (2.8) or (2.9) (recall that the situation (2.7) is more general than (2.8), and (2.8) is more general than (2.9)). In this appendix we present a number of results which are useful in relation to several integral kernels associated to $\mathcal{A}$. Proving these results would require a heavy use of functional analysis, which is not among the purposes of the present book; therefore, hereafter we just sketch the basic ideas that are presented with more details (and greater mathematical rigor) in [62].

General results for the case (2.7). As in the cited equation, we assume $\mathcal{A}$ to be any strictly positive, self-adjoint operator in $L^2(\Omega)$. A basic step towards our goal consists in associating a scale of Hilbert spaces $\mathcal{H}^r$ to the powers $\mathcal{A}^r$ for $r \in \mathbf{R}$ (see [7] for a similar construction); speaking somehow loosely, we can describe $\mathcal{H}^r$ as the Hilbert space of generalized functions $f : \Omega \to \mathbf{C}$ such that $\mathcal{A}^r f \in L^2(\Omega)$, equipped with the inner product

$$\langle f|g\rangle_r := \langle \mathcal{A}^{r/2}f | \mathcal{A}^{r/2}g\rangle \tag{B.1}$$

(as usual, $\langle \ | \ \rangle$ denotes the inner product of $L^2(\Omega)$) which induces the norm

$$\|f\|_r := \sqrt{\langle f|f\rangle_r} = \|\mathcal{A}^{r/2}f\| \tag{B.2}$$

Let $(F_k)_{k \in \mathcal{K}}$ be any complete orthonormal set of (generalized) eigenfunctions[a] of $\mathcal{A}$ with eigenvalues $(\omega_k^2)_{k \in \mathcal{K}}$ ($\omega_k \geqslant \varepsilon > 0$ for all $k \in \mathcal{K}$); then $\mathcal{H}^r$ is made of the generalized function f on Ω such that

$$\int_K dk \, \omega_k^{2r} \, |\langle F_k | f \rangle|^2 < +\infty ; \tag{B.3}$$

moreover, for any pair of functions $f, g \in \mathcal{H}^r$, there holds

$$\langle f | g \rangle_r = \int_K dk \, \omega_k^{2r} \, \overline{\langle F_k | f \rangle} \, \langle F_k | g \rangle . \tag{B.4}$$

Of course, $\mathcal{H}^0$ is the usual space $L^2(\Omega)$, and it can be proved that $\mathcal{H}^{r_2} \hookrightarrow \mathcal{H}^{r_1}$ for $r_2 \geqslant r_1$.[b] Moreover, from the characterization (B.3) of $\mathcal{H}^r$ and from the identity $\mathcal{A}^s F_k = \omega_k^{2s} F_k$ it follows that

$$\begin{aligned} &\mathcal{A}^{-s} \text{ maps continuously } \mathcal{H}^r \text{ into } \mathcal{H}^{r'} \\ &\text{for } r, r' \in \mathbf{R} \text{ such that } r' - r \leqslant 2\Re s ; \end{aligned} \tag{B.5}$$

similarly, one proves that

$$e^{-t\mathcal{A}}, \, e^{-t\sqrt{\mathcal{A}}} \text{ map continuously } \mathcal{H}^r \text{ into } \mathcal{H}^{r'} \text{ for all } r, r' \in \mathbf{R} . \tag{B.6}$$

Next, consider the inner product of $L^2(\Omega)$; for all $r \in \mathbf{R}$ this can be extended to a continuous, sesquilinear map

$$\langle \, | \, \rangle : \mathcal{H}^{-r} \times \mathcal{H}^r \to \mathbf{C} , \tag{B.7}$$

---

[a]Let us sketch the precise definition of $\mathcal{H}^r$, given in [62]. For any real r we consider in $L^2(\Omega)$ the dense linear subspace D^r on which $\mathcal{A}^{r/2}$ is well defined according to the general framework for functional calculus of self-adjoint operators in Hilbert spaces (see, e.g. [123]); if $(F_k)_{k \in \mathcal{K}}$ is a (generalized) complete orthonornmal set of eigenfunctions of $\mathcal{A}$ with eigenvalues ω_k^2 ($\omega_k \geqslant \varepsilon > 0$), then D^r is formed by the functions $f \in L^2(\Omega)$ such that $\int_K dk \, \omega_k^{2r} |\langle F_k | f \rangle|^2 < +\infty$. We introduce on D^r an inner product $\langle \, | \, \rangle_r$ following Eq. (B.1), and define $\mathcal{H}^r$ to be the completion of D^r with respect to $\langle \, | \, \rangle_r$; for $r_2 \geqslant r_1$ one finds $D^{r_2} \subset D^{r_1}$, a fact implying $\mathcal{H}^{r_2} \subset \mathcal{H}^{r_1}$. Of course $\mathcal{A}^0$ coincides with the identity operator $\mathbf{1}$ of $L^2(\Omega)$; thus $D^0 = \mathcal{H}^0 = L^2(\Omega)$ and $\langle \, | \, \rangle_0$ is the usual L^2 inner product $\langle \, | \, \rangle$. For any $r \geqslant 0$, D^r is itself complete, so $\mathcal{H}^r = D^r \subset L^2(\Omega)$; for $r < 0$ the completion $\mathcal{H}^r$ is larger than D^r, and even of $L^2(\Omega) = \mathcal{H}^0$. If r is sufficiently large, $\mathcal{H}^{-r}$ contains elements which cannot even be interpreted as ordinary functions on Ω; for example, as shown later in this appendix, the Dirac delta $\delta_{\mathbf{x}}$ at any point $\mathbf{x} \in \Omega$ can be interpreted as an element of $\mathcal{H}^{-r}$ for all $r > d/2$.

[b]Given any pair of topological vector spaces $\mathcal{X}, \mathcal{Y}$, we say that $\mathcal{X}$ is continuously embedded in $\mathcal{Y}$, and we write $\mathcal{X} \hookrightarrow \mathcal{Y}$, if $\mathcal{X}$ is a linear subspace of $\mathcal{Y}$ and the identity map from $\mathcal{X}$ into $\mathcal{Y}$ is continuous.

which allows, in particular, to infer the isomorphic identification $(\mathcal{H}^r)' = \mathcal{H}^{-r}$.

Results for the case (2.8). In accordance with the cited equation, we now assume $\mathcal{A} = -\Delta + V$ on an open subset Ω of $\mathbf{R}^d$ with given boundary conditions on $\partial\Omega$ and $V : \Omega \to \mathbf{R}$ a C^∞ potential, chosen so as to ensure the self-adjointness and the strict positivity of $\mathcal{A}$.

In this case it can be proved that

$$\mathcal{H}^r \hookrightarrow C^j(\Omega) \qquad \text{for } j \in \mathbf{N}, \, r \in \mathbf{R} \text{ with } r > \frac{d}{2} + j \,, \qquad \text{(B.8)}$$

where $C^j(\Omega)$ carries the topology of uniform convergence of the derivatives up to order j on compact subsets of Ω; this is induced by the family of seminorms

$$p_{jK}(f) := \max_{|\alpha| \leqslant j, \, \mathbf{x} \in K} |\partial^\alpha f(\mathbf{x})| \qquad (j \in \mathbf{N}, \, K \subset \Omega \text{ compact}) \quad \text{(B.9)}$$

By duality, Eq. (B.8) implies[c] (under the same assumptions on j and r)

$$(C^j(\Omega))' \hookrightarrow (\mathcal{H}^r)' = \mathcal{H}^{-r} \,. \qquad \text{(B.10)}$$

Now, let $j \in \mathbf{N}$ and $\mathbf{x} \in \Omega$; the prescription

$$\langle \delta_{\mathbf{x}}, f \rangle := f(\mathbf{x}) \qquad \text{(B.11)}$$

makes sense for all $f \in C^j(\Omega)$ and allows to interpret the Dirac delta function $\delta_{\mathbf{x}}$ as an element of the dual space $(C^j(\Omega))'$. Moreover, it can be shown that the mapping $\delta : \mathbf{x} \in \Omega \mapsto \delta_{\mathbf{x}} \in (C^j(\Omega))'$ is itself

---

[c]Let us sketch the derivation of the claim stated in Eq. (B.8). Due to well-known results on elliptic operators [37], one has

$$\mathcal{H}^r \hookrightarrow H^r_{loc}(\Omega) \qquad \text{for all } r \geqslant 0 \,,$$

where $H^r_{loc}(\Omega)$ is the standard local Sobolev space of order r (that is, the space of functions $f : \Omega \to \mathbf{C}$ such that $(1 - \Delta)^{r/2}(\varphi f) \in L^2(\Omega)$ for all smooth, compactly supported functions $\varphi : \Omega \to \mathbf{C}$). On the other hand, the usual Sobolev embedding theorems [6] give

$$H^r_{loc}(\Omega) \hookrightarrow C^j(\Omega) \qquad \text{for all } r \in \mathbf{R}, \, j \in \mathbf{N} \text{ with } r > \frac{d}{2} + j \,.$$

Summing up, the embeddings discussed in this footnote yield the thesis (B.8).

of class C^j;[d] of course, this fact and Eqs. (B.10) imply that

$$\delta : \Omega \to \mathcal{H}^{-r}, \quad \mathbf{x} \mapsto \delta_{\mathbf{x}} \quad \text{is } C^j \text{ if } r \in \mathbf{R}, j \in \mathbf{N} \text{ and } r > \frac{d}{2} + j \, .$$
(B.12)

Finally, let $j \in \mathbf{N}$ and suppose that

$$\mathcal{B} : \mathcal{H}^{-(d/2+j_2+\eta)} \to \mathcal{H}^{d/2+j_1+\eta} \text{ is linear and continuous}$$
$$\text{for some } \eta > 0 \text{ and all } j_1, j_2 \in \mathbf{N} \text{ with } j_1 + j_2 \leqslant j \, ;$$
(B.13)

in this case, we claim that

$$\text{the kernel } \Omega \times \Omega \to \mathbf{C}, (\mathbf{x}, \mathbf{y}) \mapsto \mathcal{B}(\mathbf{x}, \mathbf{y}) := \langle \delta_{\mathbf{x}} | \mathcal{B} \, \delta_{\mathbf{y}} \rangle$$
$$\text{is well defined and of class } C^j(\Omega \times \Omega) \, .$$
(B.14)

In fact, each derivative $\partial \mathcal{B}(\, , \,)$ of order $\leqslant j$ involves (in an arbitrary order) α_i operations of derivation with respect to x^i and β_i operations of derivation with respect to y^i ($i = 1, ..., d$) where $j_1 := \alpha_1 + ... + \alpha_d$ and $j_2 := \beta_1 + ... + \beta_d$ are such that $j_1 + j_2 \leqslant j$; any such derivative exists and is continuous on $\Omega \times \Omega$, with the explicit expression $\partial \mathcal{B}(\mathbf{x}, \mathbf{y}) = \langle (\partial^\alpha \delta)_{\mathbf{x}} | \mathcal{B} \, (\partial^\beta \delta)_{\mathbf{y}} \rangle$.[e]

Applications to the Dirichlet, heat and cylinder kernels: regularity results. Let $s \in \mathbf{C}$, $j \in \mathbf{N}$; if $\Re s > d/2 + j/2$, using Eq. (B.5) one easily infers that the operator $\mathcal{B} := \mathcal{A}^{-s}$ fulfills the condition (B.13) (with $\eta = \Re s - d/2 - j/2$), so that (B.14) holds. In conclusion, we have the following result for the Dirichlet kernel $D_s(\mathbf{x}, \mathbf{y}) := \langle \delta_{\mathbf{x}} | \mathcal{A}^{-s} \delta_{\mathbf{y}} \rangle$:

$$D_s \in C^j(\Omega \times \Omega) \qquad \text{for } s \in \mathbf{C}, j \in \mathbf{N} \text{ and } \Re s > \frac{d}{2} + \frac{j}{2} \, .$$
(B.15)

Similarly, for any $t > 0$ and any $j \in \mathbf{N}$, due to (B.6) the operators $\mathcal{B} = e^{-t\mathcal{A}}$ and $\mathcal{B} = e^{-t\sqrt{\mathcal{A}}}$ fulfill the condition (B.13), implying

---

[d]For example, setting $j = 1$, one finds that the map $\mathbf{x} \mapsto \delta_{\mathbf{x}} \in (C^1(\Omega))'$ is C^1 with derivatives $(\partial_i \delta)_{\mathbf{x}}$ such that

$$\langle (\partial_i \delta)_{\mathbf{x}}, f \rangle = \partial_i f(\mathbf{x}) \qquad (i \in \{1, ..., d\})$$

for all $f \in C^1(\Omega)$.

[e]This claim follows from the facts stated hereafter: due to (B.12), the map $\mathbf{y} \mapsto (\partial^\beta \delta)_{\mathbf{y}}$ is continuous from Ω to $\mathcal{H}^{-(d/2+j_2+\eta)}$; due to (B.13), $\mathbf{y} \mapsto \mathcal{B}(\partial^\beta \delta)_{\mathbf{y}}$ is continuous from Ω to $\mathcal{H}^{d/2+j_1+\eta}$; due to (B.12) the map $\mathbf{x} \mapsto (\partial^\alpha \delta)_{\mathbf{x}}$ is continuous from Ω to $\mathcal{H}^{-(d/2+j_2+\eta)}$; the sesquilinear form $\langle \, | \, \rangle$ is continuous on $\mathcal{H}^{-(d/2+j_1+\eta)} \times \mathcal{H}^{d/2+j_1+\eta}$.

Eq. (B.14). Therefore, the heat and cylinder kernels $K(t; \mathbf{x}, \mathbf{y}) := \langle \delta_{\mathbf{x}} | e^{-t\mathcal{A}} \delta_{\mathbf{y}} \rangle$, $T(t; \mathbf{x}, \mathbf{y}) := \langle \delta_{\mathbf{x}} | e^{-t\sqrt{\mathcal{A}}} \delta_{\mathbf{y}} \rangle$ are of class C^j with respect to $\mathbf{x}, \mathbf{y}$, for all $j \in \mathbf{N}$; in conclusion, these kernels are C^∞ in $\mathbf{x}, \mathbf{y}$:

$$K(t;\,,\,), T(t;\,,\,) \in C^\infty(\Omega \times \Omega) \qquad \text{for all } t > 0 . \tag{B.16}$$

Results for case (2.9). We now consider the case described by the cited equation. Thus, $\mathcal{A} = -\Delta + V$ on a bounded domain Ω with C^∞ boundary $\partial\Omega$, on which Dirichlet conditions are imposed; the potential V is in $C^\infty(\overline{\Omega})$ and $V(\mathbf{x}) \geqslant 0$ for all $\mathbf{x} \in \overline{\Omega}$. Recall that, due to these assumptions, $\mathcal{A}$ is self-adjoint and strictly positive.

(a) *Smoothness up to the boundary.* In the setting described above, one can strengthen the results discussed in the previous paragraph of this appendix replacing systematically the space $C^j(\Omega)$ with $C^j(\overline{\Omega})$; this space consists of the functions which are continuous with their derivatives up to order j on the closure $\overline{\Omega} = \Omega \cup \partial\Omega$, and it is equipped with the norm

$$\|f\|_{C^j} := \max_{|\alpha| \leqslant j,\, \mathbf{x} \in \overline{\Omega}} |\partial^\alpha f(\mathbf{x})| . \tag{B.17}$$

Similarly, the space $C^j(\Omega \times \Omega)$ can be replaced with $C^j(\overline{\Omega} \times \overline{\Omega})$, endowed with the norm

$$\|f\|_{C^j} := \max_{|\alpha| + |\beta| \leqslant j,\, \mathbf{x}, \mathbf{y} \in \overline{\Omega}} |\partial_{\mathbf{x}}^\alpha \partial_{\mathbf{y}}^\beta f(\mathbf{x}, \mathbf{y})| . \tag{B.18}$$

In the present case Eq. (B.8) has the stronger version

$$\mathcal{H}^r \hookrightarrow C^j(\overline{\Omega}) \qquad \text{for } j \in \mathbf{N},\, r \in \mathbf{R} \text{ with } r > \frac{d}{2} + j ; \tag{B.19}$$

similarly, Eq. (B.14) (with the assumptions (B.13) on $\mathcal{B}$) holds with $C^j(\Omega \times \Omega)$ replaced by $C^j(\overline{\Omega} \times \overline{\Omega})$.[f] In consequence of these facts, Eqs. (B.15) and (B.16) for the Dirichlet, heat and cylinder kernels

---

[f] In order to derive Eq. (B.19) one proceeds similary to footnote c using the embeddings $\mathcal{H}^r \hookrightarrow H^r(\Omega)$ and $H^r(\Omega) \hookrightarrow C^j(\overline{\Omega})$, which involve the Sobolev space $H^r(\Omega)$; these follow from Theorem 3 on page 155 of [104] and some standard interpolation theory (see, e.g. [16, 96]).

hold in the stronger versions stated hereafter:

$$D_s \in C^j(\overline{\Omega} \times \overline{\Omega}) \qquad \text{for } s \in \mathbf{C}, \, j \in \mathbf{N} \text{ with } \Re s > \frac{d}{2} + \frac{j}{2} \, ; \quad \text{(B.20)}$$

$$K(t; \, , \,), T(t; \, , \,) \in C^\infty(\overline{\Omega} \times \overline{\Omega}) \qquad \text{for each } t > 0 \, . \qquad \text{(B.21)}$$

(b) *Estimates for the eigenfunctions and eigenvalues of* $\mathcal{A}$. Due to the boundedness of Ω, $\mathcal{A}$ has purely point spectrum; in fact, one can build for this operator a complete orthonormal set of proper eigenfunctions $(F_k)_{k \in \mathcal{K}}$ with eigenvalues $(\omega_k^2)_{k \in \mathcal{K}}$, where $\mathcal{K} = \{1, 2, 3, ...\}$ and the labels are chosen so that $0 < \omega_1 \leqslant \omega_2 \leqslant \omega_3 \leqslant ...$ (with the possibility that some of these inequalities are equalities, to deal with the case of degenerate eigenvalues).

Let us discuss the smoothness properties of the eigenfunctions and derive some norm bounds for them. Clearly, for each $k \in \{1, 2, 3, ..\}$ and $r \in \mathbf{R}$, we have $F_k \in \mathcal{H}^r$ and $\|F_k\|_r = \|\mathcal{A}^{r/2} F_k\| = \|\omega_k^r F_k\| = \omega_k^r$. If $r > \frac{d}{2} + j$ these facts and the embedding (B.19) imply $F_k \in C^j(\overline{\Omega})$; since this holds for each $j \in \mathbf{N}$, we have

$$F_k \in C^\infty(\overline{\Omega}) \, . \qquad \text{(B.22)}$$

To go on, let us note that the statement in Eq. (B.19) means the following: for each $j \in \mathbf{N}$ and $r > j + \frac{d}{2}$, there exists a constant $\Lambda_{j,r} \in (0, +\infty)$ such that

$$\|f\|_{C^j} \leqslant \Lambda_{j,r} \|f\|_r \qquad \text{for all } f \in \mathcal{H}^r \, . \qquad \text{(B.23)}$$

Using the above relation with $f = F_k$ we get $\|F_k\|_{C^j} \leqslant \Lambda_{jr} \omega_k^r$ for $j \in \mathbf{N}$, $r > j + \frac{d}{2}$ or, equivalently,

$$\|F_k\|_{C^j} \leqslant \Lambda_{j\eta} \, \omega_k^{j+d/2+\eta} \qquad \text{for all } j \in \mathbf{N}, \, \eta > 0 \qquad \text{(B.24)}$$

(where $\Lambda_{j\eta}$ stands for $\Lambda_{j,j+d/2+\eta}$). These results should be kept in mind in the sequel, together with the already mentioned Weyl asymptotics (2.10)

$$\omega_k \sim C \, k^{1/d} \qquad \text{for } k \to +\infty \, ,$$

where $C := 2\sqrt{\pi} \, \Gamma(d/2+1)^{1/d} \, \mathrm{Vol}(\Omega)^{-1/d}$.[g]

---

[g] Concerning Eq. (2.10), we have already given references [52, 104]. Eq. (B.24) is known as well in the literature, see [87, 140, 149]; here we have proposed an alternative derivation of this result just because it arises naturally from the general framework of the present appendix.

(*c*) *On the eigenfunction expansion for the Dirichlet kernel.* In the case we are considering, the eigenfunction expansion (2.15) for the Dirichlet kernel takes the form

$$D_s(\mathbf{x}, \mathbf{y}) = \sum_{k=1}^{+\infty} \frac{1}{\omega_k^{2s}} F_k(\mathbf{x}) \overline{F_k}(\mathbf{y}) \; ; \tag{B.25}$$

hereafter we discuss the absolute convergence of this expansion with respect to the norm (B.18) of $C^j(\overline{\Omega} \times \overline{\Omega})$, showing that

$$\sum_{k=1}^{+\infty} \frac{1}{|\omega_k^{2s}|} \|F_k(\cdot)\overline{F_k}(\cdots)\|_{C^j} < +\infty \qquad \text{if } \Re s > d + \tfrac{j}{2} \;. \tag{B.26}$$

To this purpose, we first notice that

$$\sum_{k=1}^{+\infty} \frac{1}{|\omega_k^{2s}|} \|F_k(\cdot)\overline{F_k}(\cdots)\|_{C^j}$$

$$= \sum_{k=1}^{+\infty} \frac{1}{\omega_k^{2\Re s}} \max_{|\alpha|+|\beta| \leqslant j} \max_{\mathbf{x} \in \overline{\Omega}, \mathbf{y} \in \overline{\Omega}} |\partial^\alpha F_k(\mathbf{x}) \partial^\beta \overline{F_k}(\mathbf{y})|$$

$$\leqslant \sum_{k=1}^{+\infty} \frac{1}{\omega_k^{2\Re s}} \max_{j_1 + j_2 \leqslant j} \|F_k\|_{C^{j_1}} \|F_k\|_{C^{j_2}} \;. \tag{B.27}$$

Let $k \in \{1, 2, 3, ...\}$, $j_1 + j_2 \leqslant j$ and $\eta > 0$. Using the estimate (B.24), we infer

$$\|F_k\|_{C^{j_1}} \|F_k\|_{C^{j_2}} \leqslant \Lambda_{j_1 \eta} \Lambda_{j_2 \eta} \, \omega_k^{d+j_1+j_2+2\eta} \;.$$

On the other hand, recalling that $\omega_k \geqslant \omega_1$, we have

$$\omega_k^{j_1+j_2} = \left(\frac{\omega_k}{\omega_1}\right)^{j_1+j_2} \omega_1^{j_1+j_2} \leqslant \left(\frac{\omega_k}{\omega_1}\right)^j \omega_1^{j_1+j_2} = \frac{\omega_k^j}{\omega_1^{j-j_1-j_2}} \;.$$

These facts suffice to ensure

$$\|F_k\|_{C^{j_1}} \|F_k\|_{C^{j_2}} \leqslant \frac{\Lambda_{j_1 \eta} \Lambda_{j_2 \eta}}{\omega_1^{j-j_1-j_2}} \, \omega_k^{d+j+2\eta} \;. \tag{B.28}$$

Inserting Eq. (B.28) into Eq. (B.27) we obtain the following, for all $\eta > 0$:

$$\sum_{k=1}^{+\infty} \frac{1}{|\omega_k^{2s}|}\, \|F_k(\cdot)\overline{F_k}(\cdot\cdot)\|_{C^j} \leqslant \left(\max_{j_1+j_2\leqslant j} \frac{\Lambda_{j_1\eta}\,\Lambda_{j_2\eta}}{\omega_1^{j-j_1-j_2}} \right) \sum_{k=1}^{+\infty} \frac{1}{\omega_k^{2\Re s-d-j-2\eta}} \,.$$
(B.29)

From here to the end of the paragraph we assume

$$\Re s > d + \frac{j}{2} \,.$$
(B.30)

    Hereafter we show that the series on the left-hand side of Eq. (B.29) converges for a suitable $\eta > 0$, a fact yielding the thesis (B.26). Indeed, due to Eq. (B.30) there exists an $\eta > 0$ such that

$$\Re s = d + \frac{j}{2} + \left(\frac{d}{2} + 1 \right) \eta \,;$$
(B.31)

expressing $\Re s$ in this way, and using the Weyl asymptotics (2.10) we get

$$\frac{1}{\omega_k^{2\Re s-d-j-2\eta}} = \frac{1}{\omega_k^{(1+\eta)d}} \sim \frac{1}{C^{(1+\eta)d}k^{1+\eta}} \qquad \text{for } k \to +\infty \,, \quad \text{(B.32)}$$

which implies convergence for the series on the left-hand side of Eq. (B.29).

    (d) *On the eigenfunction expansions for the heat and cylinder kernels.* In the type of configuration under analysis, the expansions (2.44) and (2.45) can be re-expressed, respectively, as

$$K(\mathfrak{t};\mathbf{x},\mathbf{y}) = \sum_{k=1}^{+\infty} e^{-\mathfrak{t}\omega_k^2}\, F_k(\mathbf{x})\overline{F_k}(\mathbf{y}) \,,$$
(B.33)

$$T(\mathfrak{t};\mathbf{x},\mathbf{y}) = \sum_{k=1}^{+\infty} e^{-\mathfrak{t}\omega_k}\, F_k(\mathbf{x})\overline{F_k}(\mathbf{y}) \,.$$
(B.34)

    Using considerations similar to the ones of the preceding paragraph c), one shows the absolute convergence of these expansions in the norm (B.18) of $C^j(\overline{\Omega} \times \overline{\Omega})$, for all $\mathfrak{t} > 0$ and for each $j \in \mathbf{N}$:

$$\sum_{k=1}^{+\infty} e^{-\mathfrak{t}\omega_k^2}\|F_k(\cdot)\overline{F_k}(\cdot\cdot)\|_{C^j} < +\infty \,, \qquad \sum_{k=1}^{+\infty} e^{-\mathfrak{t}\omega_k}\|F_k(\cdot)\overline{F_k}(\cdot\cdot)\|_{C^j} < +\infty \,.$$
(B.35)

(e) *Another result.* In Subsection 2.6.5 it is stated that, under the assumptions (2.9) for $\mathcal{A}$ and considering a complete orthonormal set of eigenfunctions $(F_k)_{k=1,2,3,...}$ with the usual features for this case, the function

$$\hat{T}(t;\mathbf{x},\mathbf{y}) := \sum_{k=1}^{+\infty} e^{-\omega_k t}|F_k(\mathbf{x})||F_k(\mathbf{y})| \qquad (\text{B.36})$$

admits a uniform bound

$$\hat{T}(t;\mathbf{x},\mathbf{y}) \leqslant \check{T}(t) < +\infty \qquad \text{for all } \mathbf{x},\mathbf{y} \in \overline{\Omega} \text{ and } t > 0 . \qquad (\text{B.37})$$

The proof of this statements starts from the inequality (B.24) with $j = 0$ and any $\eta > 0$; this inequality implies $|F_k(\mathbf{x})|, |F_k(\mathbf{y})| \leqslant \Lambda_{0\eta}\, \omega_k^{d/2+\eta}$, whence

$$\hat{T}(t;\mathbf{x},\mathbf{y}) \leqslant \check{T}(t) \text{ for all } \mathbf{x},\mathbf{y} \in \overline{\Omega} \text{ and } t > 0 ,$$

$$\check{T}(t) := \Lambda_{0\eta}^2 \sum_{k=1}^{+\infty} e^{-\omega_k t} \omega_k^{d+2\eta} . \qquad (\text{B.38})$$

Finally, the Weyl estimates (2.10) ensure that $\check{T}(t) < +\infty$ for each $t > 0$.

Appendix C

A contour integral representation for Mellin transforms

In the present appendix we show how to derive the identity in Eq. (2.109) (stated in Subsection 2.8.3 of Chapter 2), allowing to express the Mellin transform of a suitable function in terms of an integral along the Hankel contour in the complex plane.

Making reference to the framework of Section 2.8, let $\mathfrak{t} \mapsto h(\mathfrak{t})$ be a complex-valued function, analytic in a neighborhood of $[0, +\infty)$ and exponentially vanishing for $\Re \mathfrak{t} \to +\infty$. For any given $s \in \mathbf{C}$ with $\Re s > 0$, consider the integral

$$I(s, h) := \int_{\mathfrak{H}} d\mathfrak{t} \, \mathfrak{t}^{s-1} h(\mathfrak{t}) \, , \qquad\qquad (\text{C.1})$$

where $\mathfrak{H}$ is a Hankel contour (see below Eq. (2.108) and Fig. 2.1 for the description of this path); the complex power $\mathfrak{t}^{s-1}$ in the above equation is defined following Eqs. (1.5), (1.6). For any $\delta > 0$, $\mathfrak{H}$ is homotopic to the path $\mathfrak{H}_\delta$ described as follows:

$$\mathfrak{H}_\delta = \mathfrak{H}_\delta^+ \cup \mathfrak{H}_\delta^0 \cup \mathfrak{H}_\delta^- \, , \qquad \text{with}$$

$$\mathfrak{H}_\delta^\pm := \{\mathfrak{t} \in \mathbf{C} \mid \mathfrak{t} = v \pm i\delta, \; v \in [0, +\infty)\} \, , \qquad (\text{C.2})$$

$$\mathfrak{H}_\delta^0 := \{\mathfrak{t} \in \mathbf{C} \mid \mathfrak{t} = \delta \, e^{i\theta}, \; \theta \in (\pi/2, 3\pi/2)\} \, .$$

(Each one of these three paths must be oriented in an obvious way.) Due to this remark and to the analyticity of the function h we can replace $\mathfrak{H}$ with $\mathfrak{H}_\delta$ in Eq. (C.1); in this way, for any $\delta > 0$

we obtain

$$I(s, h) = I_\delta^+(s, h) + I_\delta^0(s, h) + I_\delta^-(s, h) \ ,$$

$$I_\delta^\pm(s, h) := \mp \int_0^{+\infty} dv \, (v \pm i\delta)^{s-1} h(v \pm i\delta) \ , \qquad \text{(C.3)}$$

$$I_\delta^0(s, h) := i \int_{\pi/2}^{3\pi/2} d\theta \, (\delta \, e^{i\theta})^s h(\delta \, e^{i\theta}) \ .$$

We are now going to consider the limit $\delta \to 0^+$. Notice that in this limit $(v+i\delta)^{s-1} \to v^{s-1}$ while $(v-i\delta)^{s-1} \to e^{2i\pi(s-1)} v^{s-1} = e^{2i\pi s} v^{s-1}$; moreover, $h(v \pm i\delta) \to h(v)$. Due to these results, we easily infer

$$\lim_{\delta \to 0^+} I_\delta^+(s, h) = -\int_0^{+\infty} dv \, v^{s-1} h(v) \ , \qquad \text{(C.4)}$$

$$\lim_{\delta \to 0^+} I_\delta^-(s, h) = e^{2i\pi s} \int_0^{+\infty} dv \, v^{s-1} h(v) \ . \qquad \text{(C.5)}$$

As for the integral $I_\delta^0(s, h)$, noting that $|h(\delta \, e^{i\theta})| \leqslant C$ for small δ and recalling that $\Re s > 0$ by hypothesis, we obtain

$$|I_\delta^0(s, h)| \leqslant C\delta^{\Re s} \int_{\pi/2}^{3\pi/2} d\theta \, e^{-(\Im s)\theta} \to 0 \qquad \text{for } \delta \to 0^+ \ . \qquad \text{(C.6)}$$

Summing up, in the limit $\delta \to 0^+$ one obtains from Eq. (C.3) that

$$I(s, h) = (e^{2i\pi s} - 1) \int_0^{+\infty} dv \, v^{s-1} h(v) \ ; \qquad \text{(C.7)}$$

noting that $e^{2i\pi s} - 1 = 2i e^{i\pi s} \sin(\pi s)$, the above relation yields

$$I(s, h) = 2i e^{i\pi s} \sin(\pi s) \int_0^{+\infty} dv \, v^{s-1} h(v) \ , \qquad \text{(C.8)}$$

yielding Eq. (2.109).

For more information concerning the Mellin transform and its contour integral representations see, e.g. [41, 63, 142].

Appendix D

Some identities for the Dirichlet kernel in a slab configuration

We refer to the framework of Section 2.10 about the slab configuration where $\Omega = \Omega_1 \times \mathbf{R}^{d_2}$; in the present appendix we retain all the assumptions and notations of the cited section and show how to derive Eqs. (2.131)–(2.133).

In particular, we indicate with $(\mathfrak{F}_{k_1})_{k_1 \in \mathcal{K}_1}$ a complete orthonormal set of eigenfunctions of $\mathcal{A}_1$ with related eigenvalues $(\varpi_{k_1}^2)_{k_1 \in \mathcal{K}_1}$; let us recall that we are assuming $\varpi_{k_1} \geqslant \varepsilon > 0$. To these objects we can associate a complete orthonormal set of eigenfunctions $(F_k)_{k \in \mathcal{K}}$ with eigenvalues $(\omega_k^2)_{k \in \mathcal{K}}$ for the operator $\mathcal{A}$ in $L^2(\Omega)$, setting

$$
F_k(\mathbf{x}) = \mathfrak{F}_{k_1}(\mathbf{x}_1) \frac{e^{i\mathbf{k}_2 \cdot \mathbf{x}_2}}{(2\pi)^{d_2/2}}, \qquad \omega_k^2 = \varpi_{k_1}^2 + |\mathbf{k}_2|^2
\tag{D.1}
$$

for $\mathbf{x} = (\mathbf{x}_1, \mathbf{x}_2)$ and $k = (k_1, \mathbf{k}_2) \in \mathcal{K}_1 \times \mathbf{R}^{d_2}$.

In the case we are considering, the eigenfunction expansion (2.15) of the Dirichlet kernel at any two points $\mathbf{x} = (\mathbf{x}_1, \mathbf{x}_2)$ and $\mathbf{y} = (\mathbf{y}_1, \mathbf{y}_2)$ can be re-expressed as follows:

$$
\begin{aligned}
D_s(\mathbf{x}_1, \mathbf{x}_2; \mathbf{y}_1, \mathbf{y}_2) &= \int_{\mathcal{K}_1 \times \mathbf{R}^{d_2}} \frac{dk_1 \, d\mathbf{k}_2}{(\varpi_{k_1}^2 + |\mathbf{k}_2|^2)^s} \, \mathfrak{F}_{k_1}(\mathbf{x}_1) \overline{\mathfrak{F}_{k_1}}(\mathbf{y}_1) \\
&\quad \times \frac{e^{i\mathbf{k}_2 \cdot (\mathbf{x}_2 - \mathbf{y}_2)}}{(2\pi)^{d_2}} \\
&= \int_{\mathcal{K}_1 \times \mathbf{R}^{d_2}} \frac{dk_1 \, d\mathbf{h}}{\varpi_{k_1}^{2s-d_2}(|\mathbf{h}|^2 + 1)^s} \, \mathfrak{F}_{k_1}(\mathbf{x}_1) \overline{\mathfrak{F}_{k_1}}(\mathbf{y}_1) \\
&\quad \times \frac{e^{i\varpi_{k_1}\mathbf{h} \cdot (\mathbf{x}_2 - \mathbf{y}_2)}}{(2\pi)^{d_2}},
\end{aligned}
\tag{D.2}
$$

where the last identity is obtained performing the change of integration variables $\mathbf{k}_2 = \varpi_{k_1}\mathbf{h}$. On the other hand, it is known that

$$\int_{\mathbf{R}^{d_2}} \frac{d\mathbf{h}}{(2\pi)^{d_2}} \frac{e^{i\mathbf{h}\cdot\mathbf{z}}}{(|\mathbf{h}|^2+1)^s} = \frac{|\mathbf{z}|^{s-\frac{d_2}{2}}}{(2\pi)^{d_2/2}\, 2^{s-1}\Gamma(s)}\, K_{s-\frac{d_2}{2}}(|\mathbf{z}|) \qquad \text{(D.3)}$$

$$\text{for any } \mathbf{z} \in \mathbf{R}^{d_2} \text{ and all } s \in \mathbf{C} \text{ with } \Re s > \tfrac{d_2}{2}$$

(K_ν is the modified Bessel function of the second kind of order $\nu \in \mathbf{C}$; see, e.g. [11, 117, 148]). Summing up, we have

$$D_s(\mathbf{x}_1, \mathbf{x}_2; \mathbf{y}_1, \mathbf{y}_2) = \int_{\mathcal{K}_1} \frac{dk_1}{\varpi_{k_1}^{2s-d_1}}\, \mathfrak{F}_{k_1}(\mathbf{x}_1)\, \overline{\mathfrak{F}_{k_1}}(\mathbf{y}_1)\, \frac{(\varpi_{k_1}|\mathbf{x}_2 - \mathbf{y}_2|)^{s-\frac{d_2}{2}}}{(2\pi)^{d_2/2}\, 2^{s-1}\Gamma(s)}$$

$$\times\, K_{s-\frac{d_2}{2}}(\varpi_{k_1}|\mathbf{x}_2 - \mathbf{y}_2|)\, . \qquad \text{(D.4)}$$

For the sake of brevity, for any $\nu \in \mathbf{C}$, we put

$$\mathfrak{G}_\nu : (0, +\infty) \to \mathbf{C}\, , \qquad z \mapsto \mathfrak{G}_\nu(z) := z^{\nu/2} K_\nu(\sqrt{z}) \qquad \text{(D.5)}$$

and notice that, due to the asymptotic behavior of the Bessel function K_ν near zero (see [117], page 252, Eq. (10.30.2)), for $\Re\nu > 0$ this function can be continuously extended up to $z = 0$ setting

$$\mathfrak{G}_\nu(0) = 2^{\nu-1}\Gamma(\nu)\, . \qquad \text{(D.6)}$$

Next, let us remark that the function $\mathfrak{G}_\nu$ allows to rephrase Eq. (D.4) as follows:

$$D_s(\mathbf{x}_1, \mathbf{x}_2; \mathbf{y}_1, \mathbf{y}_2) = \hat{D}_s(\mathbf{x}_1, \mathbf{y}_1; |\mathbf{x}_2 - \mathbf{y}_2|^2)\, ,$$

$$\hat{D}_s(\mathbf{x}_1, \mathbf{y}_1; q) = \frac{2^{1-s}}{(2\pi)^{d_2/2}\Gamma(s)} \int_{\mathcal{K}_1} \frac{dk_1}{\varpi_{k_1}^{2s-d_2}} \qquad \text{(D.7)}$$

$$\times\, \mathfrak{F}_{k_1}(\mathbf{x}_1)\, \overline{\mathfrak{F}_{k_1}}(\mathbf{y}_1)\, \mathfrak{G}_{s-\frac{d_1}{2}}(\varpi_{k_1}^2\, q)\, .$$

This proves Eq. (2.131), also giving an explicit expression for the function $\hat{D}_s$.

To proceed we note that, due to some well-known facts on the derivatives of the Bessel function K_ν for any $\nu \in \mathbf{C}$ (see [117], page 252, Eq. 10.29.4), we have

$$\frac{d\mathfrak{G}_\nu}{dz}(z) = \frac{d}{dv}\left(v^\nu K_\nu(v)\right)\Big|_{v=\sqrt{z}} \frac{1}{2\sqrt{z}} = \left(-v^\nu K_{\nu-1}(v)\right)_{v=\sqrt{z}} \frac{1}{2\sqrt{z}}$$

$$= -\frac{1}{2}\, z^{\frac{\nu-1}{2}} K_{\nu-1}(\sqrt{z}) = -\frac{1}{2}\, \mathfrak{G}_{\nu-1}(z) \ . \tag{D.8}$$

Using the above identity one can prove by induction that

$$\frac{d^n \mathfrak{G}_\nu}{dz^n}(z) = \left(-\frac{1}{2}\right)^n \mathfrak{G}_{\nu-n}(z) \qquad \text{for } n \in \{1, 2, 3, ...\} \ ; \tag{D.9}$$

this fact, along with Eq. (D.7), implies the identity

$$\frac{\partial^n \hat{D}_s}{\partial q^n}(\mathbf{x}_1, \mathbf{y}_1; q) = \frac{(-1)^n 2^{1-s-n}}{(2\pi)^{d_2/2}\Gamma(s)} \int_{\mathcal{K}_1} \frac{dk_1}{\varpi_{k_1}^{2s-d_2-2n}}$$

$$\times \mathfrak{F}_{k_1}(\mathbf{x}_1)\overline{\mathfrak{F}_{k_1}}(\mathbf{y}_1)\, \mathfrak{G}_{s-\frac{d_2}{2}-n}(\varpi_{k_1}^2 q) \ . \tag{D.10}$$

Using Eq. (D.6) we conclude that, for any $n \in \{1, 2, 3, ...\}$ and any $s \in \mathbf{C}$ with $\Re s > \frac{d}{2} + n$,

$$\frac{\partial^n \hat{D}_s}{\partial q^n}(\mathbf{x}_1, \mathbf{y}_1; 0) = \frac{(-1)^n \Gamma(s - \frac{d_2}{2} - n)}{(4\pi)^{d_2/2}\, 4^n\, \Gamma(s)} \int_{\mathcal{K}_1} \frac{dk_1}{\varpi_{k_1}^{2s-d_2-2n}}\, \mathfrak{F}_{k_1}(\mathbf{x}_1)\overline{\mathfrak{F}_{k_1}}(\mathbf{y}_1) \ . \tag{D.11}$$

Due to a representation analogous to the one in Eq. (2.15) holding for the reduced Dirichlet kernel $D_s^{(1)}$, Eq. (D.11) implies Eq. (2.133). Finally, Eq. (2.132) is just the case $n = 0$ of Eq. (2.133).

Appendix E

Derivation of some results on boundary forces

As in the final part of Subsection 3.2.1 and in Section 3.3, we work on a domain Ω and assume Dirichlet boundary conditions are prescribed on $\partial\Omega$.

E.1 The regularized pressure in the case of Dirichlet boundary conditions

Hereafter we show how to derive Eq. (3.25) of page 62 for the pressure, stating the following for all $\mathbf{x} \in \partial\Omega$ where the outer unit normal $\mathbf{n}(\mathbf{x}) = (n^i(\mathbf{x}))_{i=1,\dots,d}$ is well defined:

$$p_i^u(\mathbf{x}) = \kappa^u \left[\left(-\frac{1}{4} \delta_{ij} \, \partial^{x^\ell} \partial_{y^\ell} + \frac{1}{2} \partial_{x^i y^j} \right) D_{\frac{u+1}{2}}(\mathbf{x}, \mathbf{y}) \right]_{\mathbf{y}=\mathbf{x}} n^j(\mathbf{x}) \, .$$

To this purpose we start from Eq. (3.24), holding for general boundary conditions. In the Dirichlet case that we are considering in the present appendix, only the terms involving mixed derivatives (with respect to both $\mathbf{x}$ and $\mathbf{y}$) of the Dirichlet kernel yield non-vanishing contributions on the boundary $\partial\Omega$;[a] thus, for any $\mathbf{x} \in \partial\Omega$, Eq. (3.24) reduces to

$$p_i^u(\mathbf{x}) = \kappa^u \left[\left(-\left(\frac{1}{4} - \xi \right) \delta_{ij} \, \partial^{x^\ell} \partial_{y^\ell} + \left(\frac{1}{2} - \xi \right) \partial_{x^i y^j} \right) D_{\frac{u+1}{2}}(\mathbf{x}, \mathbf{y}) \right]_{\mathbf{y}=\mathbf{x}} n^j(\mathbf{x}) \, . \tag{E.1}$$

---

[a] One can easily infer this statement using the eigenfunction expansion (2.15).

We now claim that the terms proportional to ξ in Eq. (E.1) vanish, i.e. that

$$\left[\left(\delta_{ij}\,\partial^{x^\ell}\partial_{y^\ell} - \partial_{x^i y^j}\right) D_{\frac{u+1}{2}}(\mathbf{x},\mathbf{y})\right]_{\mathbf{y}=\mathbf{x}} n^j(\mathbf{x}) = 0 \qquad \text{for all } \mathbf{x} \in \partial\Omega\,; \tag{E.2}$$

this suffices to infer Eq. (3.25). Using the eigenfunction expansion (2.15) for the Dirichlet kernel (and its symmetry), we see that Eq. (E.2) holds if we are able to prove that

$$\left(\delta_{ij}\,\partial^\ell F_k \partial_\ell \overline{F_k} - \frac{1}{2}\,\partial_i F_k \partial_j \overline{F_k} - \frac{1}{2}\,\partial_j F_k \partial_i \overline{F_k}\right)(\mathbf{x})\, n^j(\mathbf{x})$$

$$= 0 \quad \text{for } k \in \mathcal{K}, \mathbf{x} \in \partial\Omega\,. \tag{E.3}$$

The simplest way to prove the above claim is to derive the following, equivalent statement: for all $k \in \mathcal{K}$ and all (sufficiently smooth) vector field $\mathfrak{S} \equiv (\mathfrak{S}^i) : \partial\Omega \to \mathbf{R}^d$,

$$\int_{\partial\Omega} da\, \mathfrak{S}^i \left(\delta_{ij}\,\partial^\ell F_k \partial_\ell \overline{F_k} - \frac{1}{2}\,\partial_i F_k \partial_j \overline{F_k} - \frac{1}{2}\,\partial_j F_k \partial_i \overline{F_k}\right) n^j = 0\,. \tag{E.4}$$

Let us sketch a derivation of Eq. (E.4), for any given $k \in \mathcal{K}$ and $\mathfrak{S} : \partial\Omega \to \mathbf{R}^d$. To this purpose we consider a smooth extension of $\mathfrak{S}$ to a vector field $\mathfrak{S} : \partial\Omega \cup \Omega \to \mathbf{R}^d$ and fix the attention on the integral

$$\frac{1}{2}\int_\Omega d\mathbf{x}\, (\partial^j \partial_j)(\partial_i \mathfrak{S}^i)\, |F_k|^2 = \frac{1}{2}\int_\Omega d\mathbf{x}\, \partial_i(\partial^j \partial_j \mathfrak{S}^i)\, |F_k|^2 \tag{E.5}$$

(note that $\partial^j \partial_j = \Delta$). We re-express both sides in the above identity integrating by parts with respect to all the derivatives appearing therein,[b] considering them in the two orders proposed in the two sides; some of the boundary terms arising in this way vanish since F_k is zero on $\partial\Omega$. The difference between the two expressions thus obtained, which is obviously zero, is found to coincide with the left-hand side of Eq. (E.4).

---

[b]See the footnote *a* of Chapter 3.

E.2 A variational identity

Let us stick to the framework of Section 3.3 in which the domain Ω is bounded, and Dirichlet boundary conditions are prescribed on $\partial\Omega$; the operator $\mathcal{A} = -\Delta + V$, acting in $L^2(\Omega)$, has a complete orthonormal system of eigenfunctions $(F_k)_{k\in\mathcal{K}}$ with eigenvalues $(\omega_k^2)_{k\in\mathcal{K}}$, labeled by a countable set $\mathcal{K}$. We consider, for small $\epsilon > 0$, a deformation of the domain Ω of the form (3.26)–(3.27), controlled by a vector field $\mathfrak{S}$ on $\mathbf{R}^d$. The operator $\mathcal{A}_\epsilon := -\Delta + V$ acting in $L^2(\Omega_\epsilon)$ has a complete orthonormal system of eigenfunctions $(F_{\epsilon,k})_{k\in\mathcal{K}}$ with eigenvalues $(\omega_{\epsilon,k}^2)_{k\in\mathcal{K}}$.

In Section 3.3 we have already considered the regularized bulk energy corresponding to Ω_ϵ; this is given by (see Eq. (3.28))

$$ E_\epsilon^u = \frac{\kappa^u}{2} \sum_{k\in\mathcal{K}} (\omega_{\epsilon,k}^2)^{\frac{1-u}{2}} \ . $$

We now consider the limit $\epsilon \to 0$, and expand everything to the first order in ϵ. Eq. (3.29) that we want to derive concerns the expansion in ϵ of the bulk energy E_ϵ^u; as already mentioned, Eq. (3.28) can be used to make contact with the expansion of the eigenvalues $\omega_{\epsilon,k}^2$, on which we now fix our attention.

The variation of the eigenvalues under deformations of the spatial domain for the Dirichlet Laplacian (or similar operators) has been the subject of classical investigations. Here we refer to the book of Rellich [124] (see, in particular, Chapter II at the end of §6), whose results can be expressed in the following way with our notations:

$$ \omega_{\epsilon,k}^2 = \omega_k^2 + \varepsilon\,\varpi_k^2 + O(\epsilon^2) \qquad \text{with} \qquad \varpi_k^2 := \langle F_k | \mathcal{B} F_k \rangle \ , \qquad \text{(E.6)} $$

where $\mathcal{B}$ is the self-adjoint operator in $L^2(\Omega)$ defined by

$$ \mathcal{B}f := \partial_i \Big((\partial^i \mathfrak{S}^j + \partial^j \mathfrak{S}^i)\,\partial_i f \Big) + \Big(\frac{1}{2}\,\Delta\partial_\ell \mathfrak{S}^\ell + \mathfrak{S}^\ell \partial_\ell V \Big) f \qquad \text{(E.7)} $$

(as matter of fact, [124] gives the expression of $\mathcal{B}$ for $V = 0$, but the extension to a nonzero V is straightforward). Keeping in mind these facts, we return to Eq. (3.28) for the regularized bulk energy; this implies

$$E_\epsilon^u = E^u + \epsilon \, \mathfrak{E}^u + O(\epsilon^2) \,, \qquad \mathfrak{E}^u := \left(\frac{1-u}{2}\right) \frac{\kappa^u}{2} \sum_{k \in \mathcal{K}} \omega_k^{-1-u} \, \varpi_k^2 \,.$$

(E.8)

To go on, we note that the definition of ϖ_k^2 in Eq. (E.6), with $\mathcal{B}$ as in Eq. (E.7), yields

$$\varpi_k^2 = \int_\Omega d\mathbf{x} \, \overline{F_k} \left[\partial_i \left((\partial^i \mathfrak{S}^j + \partial^j \mathfrak{S}^i) \, \partial_i F_k \right) + \left(\frac{1}{2} \Delta \partial_\ell \mathfrak{S}^\ell + \mathfrak{S}^\ell \partial_\ell V \right) F_k \right] .$$

(E.9)

The above result can be re-expressed in terms of surface integrals on $\partial\Omega$ via suitable integrations by parts;[c] to perform these computations, one must use the identity $\Delta F_k = (V - \omega_k^2) F_k$ (and its complex conjugate) and recall that F_k vanishes on $\partial\Omega$. In this way we obtain

$$\varpi_k^2 = \int_{\partial\Omega} da(\mathbf{x}) \, n^j \mathfrak{S}^i \left[\delta_{ij} \, \partial^\ell \overline{F_k} \partial_\ell F_k - (\partial_i \overline{F_k} \partial_j F_k + \partial_j \overline{F_k} \partial_i F_k) \right] .$$

(E.10)

We plug this relation into Eq. (E.8), exchange the summation with the integration and use the expansion (2.15), which in this case reads $D_s(\mathbf{x}, \mathbf{y}) = \sum_{k \in \mathcal{K}} \frac{1}{\omega_k^{2s}} F_k(\mathbf{x}) \overline{F_k}(\mathbf{y})$; this yields the identity

$$\mathfrak{E}^u = -(1-u)\kappa^u \int_{\partial\Omega} da(\mathbf{x}) \, n^j(\mathbf{x}) \, \mathfrak{S}^i(\mathbf{x})$$

$$\times \left[\left(-\frac{1}{4} \delta_{ij} \, \partial^{x^\ell} \partial_{y^\ell} + \frac{1}{2} \partial_{x^i y^j} \right) D_{\frac{u+1}{2}}(\mathbf{x}, \mathbf{y}) \right]_{\mathbf{y}=\mathbf{x}} . \qquad \text{(E.11)}$$

comparing the above result with Eq. (3.25) for the regularized pressure we see that

$$\mathfrak{E}^u = -(1-u) \int_{\partial\Omega} da(\mathbf{x}) \, \mathfrak{S}^i(\mathbf{x}) \, p_i^u(\mathbf{x}) \,; \qquad \text{(E.12)}$$

the first equality in (E.8) and Eq. (E.12) give the thesis (3.29).

---

[c]See again footnote *a* of Chapter 3.

Appendix F

An explicit expression for the renormalized Dirichlet kernel of half-integer order

Consider the framework developed in Section 4.2 for a non-negative operator $\mathcal{A} = -\Delta + V$, such that $\sigma(\mathcal{A}) \subset [0, +\infty)$ and 0 is in the continuous spectrum of $\mathcal{A}$. In the present appendix we use the deformed operator

$$\mathcal{A}_\varepsilon := (\sqrt{\mathcal{A}} + \varepsilon)^2 . \tag{F.1}$$

We already observed in Section 4.2 that the cylinder kernels T and T^ε, respectively associated to $\mathcal{A}$ and $\mathcal{A}_\varepsilon$, are related by the identity (see Eq. (4.17))

$$T^\varepsilon(\mathfrak{t}; \mathbf{x}, \mathbf{y}) = e^{-\varepsilon \mathfrak{t}} \, T(\mathfrak{t}; \mathbf{x}, \mathbf{y}) ; \tag{F.2}$$

we also showed under appropriate assumptions (see Eq. (4.18) and related comments) that

$$D_s^\varepsilon(\mathbf{x}, \mathbf{y}) = \frac{e^{-2i\pi s} \, \Gamma(1-2s)}{2\pi i} \int_{\mathfrak{H}} d\mathfrak{t} \, \mathfrak{t}^{2s-1} \, T^\varepsilon(\mathfrak{t}; \mathbf{x}, \mathbf{y}) . \tag{F.3}$$

The integral in the right-hand side of the above equation is an analytic function of s on the whole complex plane, while the Euler gamma function is meromorphic with simple poles at positive half-integer values of s. Taking into account these facts, hereafter we show how to evaluate the renormalized kernels

$$D_{s_0}^{(\kappa)}(\mathbf{x}, \mathbf{y}) := \lim_{\varepsilon \to 0^+} RP\Big|_{s=s_0} \Big(\kappa^{2(s-s_0)} D_s^\varepsilon(\mathbf{x}, \mathbf{y})\Big) \tag{F.4}$$

considering, separately, the cases $s_0 = -n/2$ ($n \in \{0, 1, 2, ...\}$) and $s_0 = n/2$ ($n \in \{1, 2, 3...\}$). Putting together the results obtained for $s_0 = -n/2$ ($n \in \{0, 1, 2, ...\}$) and for $s_0 = 1/2$ we infer the identity stated in Eq. (4.21).

Before proceeding, let us remark that, for $s_0 = \pm 1/2$, the above renormalized kernels coincide with the functions $D_{\pm 1/2}^{(\kappa)}$ introduced in Eq. (4.9).

Case 1: $s_0 = -n/2$, $n \in \{0, 1, 2, ...\}$. The right-hand side of Eq. (F.3) is clearly an analytic function of s for $\Re s < \frac{1}{2}$; thus, for $n \in \{0, 1, 2, ...\}$ $D_s^\varepsilon(\mathbf{x}, \mathbf{y})$ has an analytic continuation at $s = -n/2$, hereafter indicated with $D_{-n/2}^\varepsilon(\mathbf{x}, \mathbf{y})$, that is simply obtained substituting this value of s in the integral representation (F.3). The resulting integrand is meromorphic so that, by the residue theorem,

$$D_{-\frac{n}{2}}^\varepsilon(\mathbf{x}, \mathbf{y}) = (-1)^n \, \Gamma(n+1) \, \mathrm{Res}\left(t^{-(n+1)} e^{-\varepsilon t} T(t; \mathbf{x}, \mathbf{y}); 0\right) . \quad (F.5)$$

Of course, in the present case the prescription of taking the regular part in Eq. (F.4) is pleonastic, and the cited equation reduces to

$$D_{-\frac{n}{2}}^{(\kappa)}(\mathbf{x}, \mathbf{y}) = \lim_{\varepsilon \to 0^+} D_{-\frac{n}{2}}^\varepsilon(\mathbf{x}, \mathbf{y}) . \quad (F.6)$$

On the other hand, the explicit computation of the above limit gives[a]

$$D_{-\frac{n}{2}}^{(\kappa)}(\mathbf{x}, \mathbf{y}) = (-1)^n \, \Gamma(n+1) \, \mathrm{Res}\left(t^{-(n+1)} T(t; \mathbf{x}, \mathbf{y}); 0\right) . \quad (F.7)$$

The above result can be reformulated in terms of the modified cylinder kernel $\tilde{T}$ associated to $\mathcal{A}$. Indeed, recall that $T(t; \mathbf{x}, \mathbf{y}) =$

---

[a]To prove the identity in Eq. (F.7), one can proceed as explained hereafter. In the case under analysis the cylinder kernel is assumed to be a meromorphic function of t; so, there is an expansion $T(t; \mathbf{x}, \mathbf{y}) = \frac{1}{t^q} \sum_{k=0}^{+\infty} e_k(\mathbf{x}, \mathbf{y}) t^k$ for some $q \in \mathbf{Z}$, converging at least for small t (note that under the assumptions for the validity of (2.75) we have $q = d$). Of course, $e^{-\varepsilon t} = \sum_{k=0}^{+\infty} \frac{(-\varepsilon)^k}{k!} t^k$, so the Cauchy formula for the product of two series yields

$$\mathrm{Res}\left(t^{-(n+1)} e^{-\varepsilon t} T(t; \mathbf{x}, \mathbf{y}); 0\right) = \sum_{k=0}^{q+n} \frac{(-\varepsilon)^k}{k!} e_{q+n-k}(\mathbf{x}, \mathbf{y}) \xrightarrow{\varepsilon \to 0} e_{q+n}(\mathbf{x}, \mathbf{y})$$

$$= \mathrm{Res}\left(t^{-(n+1)} T(t; \mathbf{x}, \mathbf{y}); 0\right).$$

This proves Eq. (F.7).

$-\partial_t \tilde{T}(t;\mathbf{x},\mathbf{y})$ (see Eq. (2.54)) and note that, for any pair of functions f, g meromorphic near a point t_0, we have $\mathrm{Res}(fg';t_0) = -\mathrm{Res}(f'g;t_0)$; these facts (and the standard identity $z\,\Gamma(z) = \Gamma(1+z)$) give

$$D^{(\kappa)}_{-\frac{n}{2}}(\mathbf{x},\mathbf{y}) = (-1)^{n+1}\,\Gamma(n+2)\,\mathrm{Res}\left(t^{-(n+2)}\,\tilde{T}(t;\mathbf{x},\mathbf{y});0\right). \quad (F.8)$$

Case 2: $s_0 = +n/2,\ n \in \{1,2,3,...\}$ (with a special attention for the subcase $n = 1$). A substantial difference occurs with respect to Case 1. In fact, due to the Euler gamma function appearing in Eq. (F.3), the function $D^\varepsilon_s(\mathbf{x},\mathbf{y})$ described by this equation has a genuine singularity at $s = n/2$. In order to remove this singularity, it is essential to retain only the regular part in Eq. (F.4); to this purpose, for s in a neighborhood of $n/2$, we introduce the variable $u := 2s - n$ and note that, for $u \to 0$,

$$\kappa^{2s-n}e^{-2i\pi s}\,\Gamma(1-2s)\,t^{2s-1}$$

$$= \frac{t^{n-1}}{(n-1)!}\left[\frac{1}{u} + \left(\ln(\kappa t) + \gamma_{EM} - i\pi - H_{n-1}\right) + O(u)\right], \quad (F.9)$$

where $\gamma_{EM} \simeq 0.577216$ is the Euler–Mascheroni constant and, for $m \in \{0,1,2,...\}$, $H_m := \sum_{k=1}^{m}\frac{1}{k}$ ($H_0 := 0$) denotes the m-th harmonic number.[b]

It follows that

$$RP\Big|_{s=\frac{n}{2}}\left(\kappa^{2s-n}D^\varepsilon_s(\mathbf{x},\mathbf{y})\right)$$

$$= \frac{1}{2\pi i}\int_{\mathfrak{H}} dt\, \frac{t^{n-1}e^{-\varepsilon t}}{(n-1)!}\left(\ln(\kappa t) + \gamma - i\pi - H_{n-1}\right)T(t;\mathbf{x},\mathbf{y}). \quad (F.10)$$

---

[b]To obtain Eq. (F.9), one uses the following well known facts [117]: for $m \in \{0,1,2,...\}$,

$$\Gamma(-u-m) = \frac{(-1)^m\,\Gamma(-u)}{(u+1)...(u+m)}\ ;\qquad (u+1)...(u+m) = m!\left[1 + H_m\,u + O(u^2)\right],$$

$$\Gamma(-u) = -\frac{1}{u} - \gamma + O(u)\ ,$$

$$e^{-i\pi u}\,(\kappa t)^u = e^{u(\ln(\kappa t)-i\pi)} = 1 + (\ln(\kappa t) - i\pi)u + O(u^2)\qquad \text{for } u \to 0.$$

According to Eq. (F.4), the renormalized function $D_{\frac{n}{2}}^{(\kappa)}(\mathbf{x},\mathbf{y})$ is the limit $\varepsilon \to 0^+$ of the above expression. Under suitable hypotheses on the behavior of T for $\Re t \to +\infty$ (namely, $|T(\mathsf{t};\mathbf{x},\mathbf{y})| \leq C\,|\mathsf{t}|^{-a-n}$ for some $C, a > 0$), we can exchange the limit and the integral to obtain

$$D_{\frac{n}{2}}^{(\kappa)}(\mathbf{x},\mathbf{y}) = \frac{1}{2\pi i}\int_{\mathfrak{H}} d\mathsf{t}\, \frac{\mathsf{t}^{n-1}}{(n-1)!}\left(\ln(\kappa\mathsf{t}) + \gamma - i\pi - H_{n-1}\right)T(\mathsf{t};\mathbf{x},\mathbf{y})\;;$$

$$\text{(F.11)}$$

the term $\ln(\kappa\mathsf{t})$ prevents us from using the residue theorem, so we must find alternative ways to evaluate explicitly the above integral. In the special case $n = 1$, we can proceed as follows. First we recall that $T(\mathsf{t};\mathbf{x},\mathbf{y}) = -\partial_\mathsf{t}\tilde{T}(\mathsf{t};\mathbf{x},\mathbf{y})$ and integrate by parts Eq. (F.11) to obtain

$$D_{\frac{1}{2}}^{(\kappa)}(\mathbf{x},\mathbf{y}) = \frac{1}{2\pi i}\int_{\mathfrak{H}} d\mathsf{t}\, \mathsf{t}^{-1}\,\tilde{T}(\mathsf{t};\mathbf{x},\mathbf{y})\;; \qquad \text{(F.12)}$$

the resulting integrand is meromorphic in t so that we can resort to the residue theorem to obtain

$$D_{\frac{1}{2}}^{(\kappa)}(\mathbf{x},\mathbf{y}) = \mathrm{Res}\left(\mathsf{t}^{-1}\,\tilde{T}(\mathsf{t};\mathbf{x},\mathbf{y});0\right)\;. \qquad \text{(F.13)}$$

Conclusions. Eq. (F.8) for $D_{-\frac{n}{2}}^{(\kappa)}$ with $n \in \{0,1,2,...\}$ and Eq. (F.13) for $D_{\frac{1}{2}}^{(\kappa)}$ prove Eq. (4.21) for $D_{-\frac{n}{2}}^{(\kappa)}$ with $n \in \{-1,0,1,2,...\}$.

Bibliography

[1] A. A. Actor, *Multiple harmonic oscillator zeta functions*, J. Phys. A: Math. Gen. **20**(4), 927–936 (1987).

[2] A. A. Actor, *Local analysis of a quantum field confined within a rectangular cavity*, Ann. Phys. **230**(2), 303–320 (1994).

[3] A. A. Actor, *Scalar quantum fields confined by rectangular boundaries*, Fortsch. Phys. **43**(3), 141–205 (1995).

[4] A. A. Actor, I. Bender, *Casimir effect for soft boundaries*, Phys. Rev. D **52**(6), 3581–3590 (1995).

[5] A. A. Actor, I. Bender, *Boundaries immersed in a scalar quantum field*, Fortsch. Phys. **44**(4), 281–322 (1996).

[6] R. A. Adams, J. J. F. Fournier, "Sobolev spaces", Academic Press (2003).

[7] S. Albeverio, G. Cognola, M. Spreafico, S. Zerbini, *Singular perturbations with boundary conditions and the Casimir effect in the half space*, J. Math. Phys. **51**(06), 063502 [38 pages] (2010).

[8] S. Albeverio, C. Cacciapuoti, M. Spreafico, *Relative partition function of Coulomb plus delta interaction*, in the book J. Dittrich, H. Kovařík, A. Laptev, "Functional Analysis and Operator Theory for Quantum Physics. A Festschrift in Honor of Pavel Exner", Europ. Math. Soc. Publ. House (2016).

[9] J. Ambjørn, S. Wolfram, *Properties of the vacuum I. Mechanical and thermodynamic*, Ann.Phys. **147**(1), 1–32 (1983).

[10] J. Ambjørn, S. Wolfram, *Properties of the vacuum II. Electrodynamic*, Ann.Phys. **147**(1), 33–56 (1983).

[11] N. Aronszajn, K. T. Smith, *Theory of Bessel potentials. I*, Ann. Inst. Fourier (Grenoble) **11**, 385–475 (1961).

[12] N. Bartolo, S. Butera, M. Lattuca, R. Passante, L. Rizzuto, S. Spagnolo, *Vacuum Casimir energy densities and field divergences at boundaries*, J. Phys.: Condens. Matter **27**(21), 214015 [8 pages] (2015).

[13] M. Beauregard, G. Fucci, K. Kirsten, P. Morales, *Casimir Effect in the Presence of External Fields*, J. Phys. A: Math. Theor. **46**(11), 115401 [15 pages] (2013).

[14] M. Benini, C. Dappiaggi, T. P. Hack, *Quantum field theory on curved backgrounds – A primer*, Int. J. Mod. Phys. A **28**(17), 1330023 [49 pages] (2013).

[15] I. M. Berezanskii, "Selfadjoint operators in spaces of functions of infinitely many variables", Am. Math. Soc. (1986).

[16] J. Bergh, J. Löfström, "Interpolation spaces. An introduction", Springer–Verlag (1976).

[17] N. Berline, E. Getzler, M. Vergne, "Heat kernels and Dirac operators", Springer Science & Business Media (1992).

[18] N. D. Birrell, P. C. W. Davies, "Quantum fields in curved space", Cambridge University Press (1984).

[19] S. K. Blau, M. Visser, A. Wipf, *Zeta functions and the Casimir energy*, Nucl. Phys. B **310**(1), 163–180 (1988).

[20] M. Bordag, D. Hennig, D. Robaschik, *Vacuum energy in quantum field theory with external potentials concentrated on planes*, J. Phys. A: Math. Gen. **25**(16), 4483–4498 (1992).

[21] M. Bordag, U. Mohideen, V. M. Mostepanenko, *New developments in the Casimir effect*, Phys. Rep. -Rev. Sect. Phys. Lett. **353**(1), 1–205 (2001).

[22] M. Bordag, G. L. Klimchitskaya, U. Mohideen, V. M. Mostepanenko, "Advances in the Casimir effect", Oxford University Press (2009).

[23] T. H. Boyer, *Van der Waals forces and zero-point energy for dielectric and permeable materials*, Phys. Rev. A **9**(5), 2078–2084 (1974).

[24] A. N. Braga, J. D. L. Silva, D. T. Alves, *Casimir force between $\delta - \delta'$ mirrors transparent at high frequencies*, Phys. Rev. D **94**(12), 125007 [7 pages] (2016).

[25] I. Brevik, M. Lygren, *Casimir effect for a perfectly conducting wedge*, Ann. Phys. **251**(2), 157–179 (1996).

[26] I. Brevik, M. Lygren, V. N. Marachevsky, *Casimir-Polder effect for a perfectly conducting wedge*, Ann. Phys. **267**(1), 134–142 (1998).

[27] L. S. Brown, G. J. Maclay, *Vacuum stress between conducting plates: an image solution*, Phys. Rev. **184**(5), 1272–1279 (1969).

[28] Yu. A. Brychkov, A.P. Prudnikov, "Integral Transforms of Generalized Functions", CRC Press (1989).

[29] S. T. Bunch, P. C. W. Davies, *Quantum field theory in de Sitter space: renormalization by point-splitting*, Proc. R. Soc. Lond. A Math. Phys. Sc. **360**(1700), 117–134 (1978).

[30] A. A. Bytsenko, G. Cognola, E. Elizalde, V. Moretti, S. Zerbini, "Analytic aspects of quantum fields", World Scientific (2003).

[31] C. Cacciapuoti, D. Fermi, A. Posilicano, *Relative-Zeta and Casimir energy for a semitransparent hyperplane selecting transverse modes*, to appear in the book G.F. Dell'Antonio, A. Michelangeli, "Advances in Quantum Mechanics: contemporary trends and open problems", Springer (2017); see arXiv:1702.05296.

[32] O. Calin, D.C. Chang, K. Furutani, C. Iwasaki, "Heat kernels for elliptic and sub-elliptic operators: methods and techniques", Springer Science & Business Media (2010).

[33] C. G. Callan, S. R. Coleman, R. Jackiw, *A new improved energy-momentum tensor*, Ann. Phys. **59**(1), 42–73 (1970).

[34] H. B. G. Casimir, *On the attraction between two perfectly conducting plates*, Proc. R. Neth. Acad. Arts Sci. **51**, 793–795 (1948).

[35] I. Cavero-Peláez, K. A. Milton, K. Kirsten, *Local and global Casimir energies for a semitransparent cylindrical shell*, J. Phys. A: Math. Theor. **40**(13), 133607 [25 pages] (2007).

[36] I. Chavel, "Eigenvalues in Riemannian geometry", Academic Press (1984).

[37] J. Chazarain, A. Piriou, "Introduction to the theory of linear partial differential equations", Elsevier (2011).

[38] G. Cognola, L. Vanzo, and S. Zerbini, *A new algorithm for asymptotic heat kernel expansion for manifolds with boundary*, Phys. Lett. B **241**(3), 381–386 (1990).

[39] G. Cognola, E. Elizalde, S. Zerbini, *Fluctuations of quantum fields via zeta function regularization*, Phys. Rev. D **65**(8), 085031 [8 pages] (2002).

[40] G. Cognola, S. Zerbini, *Variances of relativistic quantum field fluctuations*, Int. J. Mod. Phys. A **18**(12), 2067–2072 (2003).

[41] E. T. Copson, "Asymptotic Expansions", Cambridge University Press (2004).

[42] M. V. Cougo-Pinto, C. Farina, F. C. Santos and A. C. Tort, *QED vacuum between a conducting and a permeable plate*, J. Phys. A: Math. Gen. **32**(24), 4463–4474 (1999).

[43] C. Dappiaggi, G. Nosari, N. Pinamonti, *The Casimir effect from the point of view of algebraic quantum field theory*, Math. Phys. Anal. Geom. **19**(2), art. 12 [44 pages] (2016).

[44] E. B. Davies, "Heat kernels and spectral theory", Cambridge University Press (1990).

[45] D. Deutsch, P. Candelas, *Boundary effects in quantum field theory*, Phys. Rev. D **20**(12), 3063–3080 (1979).

[46] J. S. Dowker, R. Critchley, *Effective Lagrangian and energy-momentum tensor in de Sitter space*, Phys. Rev. D **13**(12), 3224–3232 (1976).

[47] J. S. Dowker, G. Kennedy, *Finite temperature and boundary effects in static space-times*, J. Phys. A: Math. Gen. **11**(5), 895–920 (1978).

[48] J. S. Dowker, *Casimir effect around a cone*, Phys. Rev. D **36**(10), 3095–3101 (1987).

[49] D. G. Duffy, "Green's functions with applications", CRC Press (2001).

[50] A. Edery, *Multidimensional cut-off technique, odd-dimensional Epstein zeta functions and Casimir energy of massless scalar fields*, J. Phys. A: Math. Gen. **39**(3), 685-712 (2006).

[51] A. Edery, *Casimir piston for massless scalar fields in three dimensions*, Phys. Rev. D **75**(10), 105012 [9 pages] (2007).

[52] Y. V. Egorov, M. A. Shubin, *Linear partial differential equations* in "Elements of the modern theory of partial differential equations", Springer–Verlag (1999).

[53] E. Elizalde, S. D. Odintsov, A. Romeo, A. A. Bytsenko, S. Zerbini, "Zeta regularization techniques with applications", World Scientific (1994).

[54] E. Elizalde, "Ten physical applications of spectral zeta functions", Springer Science & Business Media (1995).

[55] G. Esposito, G. M. Napolitano, L. Rosa, *Energy-momentum tensor of a Casimir apparatus in a weak gravitational field: scalar case*, Phys. Rev. D **77**(10), 105011 [7 pages] (2008).

[56] R. Estrada, S. A. Fulling, *Distributional asymptotic expansions of spectral functions and of the associated Green kernels*, Elect. J. Diff. Eqs. **1999**, 1–37 (1999).

[57] R. Estrada, S. A. Fulling, L. Kaplan, K. Kirsten, Z. H. Liu, K. A. Milton, *Vacuum stress-energy density and its gravitational implications*, J. Phys. A: Math. Theor. **41**(16), 164055 [11 pages] (2008).

[58] D. Fermi, L. Pizzocchero, *Local zeta regularization and the Casimir effect*, Prog. Theor. Phys. **126**(3), 419–434 (2011).

[59] D. Fermi, *L'effetto Casimir e la regolarizzazione zeta*, Master Thesis, University of Milan (2012).

[60] D. Fermi, L. Pizzocchero, *Local zeta regularization and the scalar Casimir effect III. The case with a background harmonic potential*, Int. J. Mod. Phys. A **30**(35), 1550213 [42 pages] (2015).

[61] D. Fermi, L. Pizzocchero, *Local zeta regularization and the scalar Casimir effect IV. The case of a rectangular box*, Int. J. Mod. Phys. A **31**(04 & 05), 1650003 [56 pages] (2016).

[62] D. Fermi, *A functional analytic framework for local zeta regularization and the scalar Casimir effect*, PhD Thesis, Università degli Studi di Milano (2016).

[63] P. Flajolet, X. Gourdon, P. Dumas, *Mellin transforms and asymptotics: Harmonic sums*, Theor. Comp. Sc. **144**(1 & 2), 3–58 (1995).

[64] L. H. Ford, N. F. Svaiter, *Vacuum energy density near fluctuating boundaries*, Phys. Rev. D **58**(6), 065007 [8 pages] (1998).

[65] G. Fucci, K. Kirsten, *Conical Casimir pistons with hybrid boundary conditions*, J. Phys. A: Math. Theor. **44**(29), 295403 [23 pages] (2011).

[66] G. Fucci, K. Kirsten, *The spectral zeta function for laplace operators on warped product manifolds of the type $I \times f_N$*, Comm. Math. Phys **317**(3), 635–665 (2013).

[67] S. A. Fulling, "Aspects of quantum field theory in curved space time", Cambridge University Press (1989).

[68] S. A. Fulling, R. A. Gustafson, *Some properties of Riesz means and spectral expansions*, Elect. J. Diff. Eqs. **1999**(6), 1–39 (1999).

[69] S. A. Fulling, *Systematics of the relationship between vacuum energy calculations and heat kernel coefficients*, J. Phys. A: Math. Gen. **36**, 6857–6873 (2003).

[70] S. A. Fulling, *Mass dependence of vacuum energy*, Phys. Lett. B **624**(3 & 4), 281–286 (2005).

[71] S. A. Fulling, *Vacuum energy as spectral geometry*, SIGMA **3**, 094 [23 pages] (2007).

[72] S. A. Fulling, L. Kaplan, K. Kirsten, Z. H. Liu, K. A. Milton, *Vacuum Stress and Closed Paths in Rectangles, Pistons, and Pistols*, J. Phys. A **42**(15), 155402 [33 pages] (2009).

[73] S. A. Fulling, C. S. Trendafilova, P. N. Truong, J. Wagner, *Wedges,*

cones, cosmic strings and their vacuum energy, J. Phys. A: Math. Theor. **45**(37), 374018 [24 pages] (2012).

[74] I. M. Gel'fand, G. E. Shilov, "Generalized functions. Volume III: theory of differential equations", Academic Press (1967).

[75] P. B. Gilkey, "Invariance theory: the heat equation and the Atiyah-Singer index theorem", CRC Press (1994).

[76] N. Graham, R. L. Jaffe, H. Weigel, *Casimir Effects in Renormalizable Quantum Field Theories*, Int. J. Mod. Phys. A **17**(6 & 7), 846–869 (2002).

[77] N. Graham, R. L. Jaffe, V. Khemani, M. Quandt, M. Scandurra, H. Weigel, *Calculating vacuum energies in renormalizable quantum field theories: A new approach to the Casimir problem*, Nuc. Phys. B **645**(1 & 2), 49–84 (2002).

[78] N. Graham, R. L. Jaffe, V. Khemani, M. Quandt, M. Scandurra, H. Weigel, *Casimir Energies in Light of Quantum Field Theory*, Phys. Lett. B **572**(3 & 4), 196–201 (2003).

[79] N. Graham, K. D. Olum, *Negative Energy Densities in Quantum Field Theory With a Background Potential*, Phys. Rev. D **67**(8), 085014 [16 pages] (2003).

[80] A. Grigor'yan, "Heat kernel and analysis on manifolds", AMS/IP Studies in Advanced Mathematics **47** (2009).

[81] T. P. Hack, V. Moretti, *On the stress–energy tensor of quantum fields in curved spacetimes – comparison of different regularization schemes and symmetry of the Hadamard/Seeley/DeWitt coefficients*, J. Phys. A: Math. Theor. **45**(37), 374019 [19 pages] (2012).

[82] G. H. Hardy, M. Riesz, "The general theory of Dirichlet's series", Cambridge Tracts in Mathematics **18**, Cambridge University Press (1915).

[83] J. M. Harrison, K. Kirsten, C. Texier, *Spectral determinants and zeta functions of Schrödinger operators on metric graphs*, J. Phys. A: Math. Theor. **45**(12), 125206 [14 pages] (2011).

[84] S. W. Hawking, *Zeta function regularization of path integrals in curved spacetime*, Commun. Math. Phys. **55**(2), 133–148 (1977).

[85] P. Henrici, "Applied and computational complex analysis. Volume I", Wiley & Sons (1974).

[86] D. Iellici, V. Moretti, *ζ-function regularization and one-loop renormalization of field fluctuations in curved space-time*, Phys. Lett. B **425**(1 & 2), 33–40 (1998).

[87] V. A. Il'in, I. A. Shishmarev, *Uniform estimates in a closed domain*

for the eigenfunctions of an elliptic operator and their derivatives, Izv. Akad. Nauk SSSR. Set. Mat. **24**(6), 883–896 (1960).

[88] K. Kirsten, "Spectral Functions in Mathematics and Physics", CRC Press (2001).

[89] K. Kirsten, *Basic zeta functions and some applications in physics,* in the book K. Kirsten, F. L. Williams, "A Window into Zeta and Modular Physics", Mat. Sc. Res. Inst. Pub. **57**, Cambridge University Press (2010).

[90] N. V. Krylov, "Lectures on Elliptic and Parabolic Equations in Sobolev Spaces", American Mathematical Society (2008).

[91] P. K. Kythe, "Green's Functions and Linear Differential Equations: Theory, Applications, and Computation", CRC Press (2011).

[92] X. Li, H. Cheng, J. Li, X. Zhai, *Attractive or repulsive nature of the Casimir force for rectangular cavity,* Phys. Rev. D **56**(4), 2155–2162 (1997).

[93] T. Lu, T. Jeffres, K. Kirsten, *Zeta function of self-adjoint operators on surfaces of revolution,* J. Phys. A: Math. Theor. **48**(14), 145204 [22 pages] (2015).

[94] W. Lukosz, *Electromagnetic Zero-Point Energy and Radiation Pressure for a Rectangular Cavity,* Physica **56**(1), 109–120 (1971).

[95] W. Lukosz, *Electromagnetic zero-point energy shift induced by conducting surfaces II. The infinite wedge and the rectangular cavity,* Z. Physik **262**(4), 327–348 (1973).

[96] A. Lunardi, *Interpolation theory. Appunti,* Scuola Normale Superiore, Pisa (1999).

[97] S. G. Mamaev, N. N. Trunov, *Vacuum means of energy-momentum tensor of quantized fields on manifolds of different topology and geometry. I,* Sov. Phys. J. **22**(7), 766–770 (1979).

[98] S. G. Mamaev, N. N. Trunov, *Vacuum averages of the energy-momentum tensor of quantized fields on manifolds of various topology and geometry. II,* Sov. Phys. J. **22**(9), 966–969 (1979).

[99] S. G. Mamaev, N. N. Trunov, *Vacuum expectation values of the energy-momentum tensor of quantized fields on manifolds with different topologies and gometry. III,* Sov. Phys. J. **23**(7), 551–554 (1980).

[100] S. G. Mamaev, N. N. Trunov, *Vacuum expectation values of the energy-momentum tensor of quantized fields on manifolds of different topology and geometry. IV,* Sov. Phys. J. **24**(2), 171–174 (1981).

[101] F. Mandl, G. Shaw, "Quantum field theory", John Wiley & Sons (2013).

[102] F. G. Mehler, *Ueber die Entwicklung einer Function von beliebig vielen Variabeln nach Laplaceschen Functionen höherer Ordnung*, Journal für Reine und Angewandte Mathematik **66**(3), 161–176 (1866).

[103] F. D. Mera, S. A. Fulling, *Vacuum energy density and pressure of a massive scalar field*, J. Phys. A: Math. Theor. **48**(24), 245402 [18 pages] (2015).

[104] V. P. Mikhailov, "Partial differential equations", Mir Publishers (1978).

[105] K. A. Milton, "The Casimir effect - Physical manifestations of zero-point energy", World Scientific Publishing Co. (2001).

[106] S. Minakshisundaram, *A generalization of Epstein zeta functions*, Canad. J. Math. **1**, 320–327 (1949).

[107] S. Minakshisundaram, A. Pleijel, *Some properties of the eigenfunctions of the Laplace-operator on Riemannian manifolds*, Canad. J. Math. **1**, 242–256 (1949).

[108] V. Moretti, *Direct ζ-function approach and renormalization of one-loop stress tensors in curved spacetimes*, Phys. Rev. D **56**(12), 7797–7819 (1997).

[109] V. Moretti, *Local ζ-function techniques vs point-splitting procedure: a few rigorous results*, Commun. Math. Phys. **201**(2), 327–363 (1999).

[110] V. Moretti, *A review on recent results of the ζ-function regularization procedure in curved spacetime*, in the book D. Fortunato, M. Francaviglia, A. Masiello, "Recent developments in general relativity", Springer-Verlag (1999).

[111] V. Moretti, *Local ζ-functions, stress-energy tensor, field fluctuations, and all that, in curved static spacetime*, in the book S. D. Odintsov, D. Sáez-Gómez, S. Xambó-Descamps, "Cosmology, quantum vacuum and zeta functions. In honor of Emilio Elizalde", Springer Proc. Phys. **137** (2011).

[112] J. M. Muñoz-Castañeda, J. M. Guilarte, A. M. Mosquera, *Quantum vacuum energies and Casimir forces between partially transparent δ-function plates*, Phys. Rev. D **87**(10), 105020 [22 pages] (2013).

[113] J. M. Muñoz-Castañeda, J. M. Guilarte, $\delta - \delta'$ *generalized Robin boundary conditions and quantum vacuum fluctuations*, Phys. Rev. D **91**(2), 025028 [21 pages] (2015).

[114] V. V. Nesterenko, G. Lambiase, G. Scarpetta, *Casimir effect for a perfectly conducting wedge in terms of local zeta function*, Ann. Phys. **298**(2), 403–420 (2002).

[115] H. C. Ohanian, *Gravitation and the new improved energy-momentum tensor*, J. Math. Phys. **14**(12), 1892–1897 (1973).

[116] F. W. J. Olver, "Asymptotics and special functions", Academic Press (2014).

[117] F. W. J. Olver, D. W. Lozier, R. F. Boisvert, C. W. Clark, "NIST Handbook of mathematical functions", Cambridge University Press (2010).

[118] L. E. Parker, D. J. Toms, "Quantum field theory in curved spacetime: Quantized fields and gravity", Cambridge University Press (2009).

[119] A. C. A. Pinto, T. M. Britto, R. Bunchaft, F. Pascoal, F. S. S. da Rosa, *Casimir effect for a massive scalar field under mixed boundary conditions*, Braz. J. Phys. **33**(4), 860–866 (2003).

[120] G. Plunien, B. Müller, W. Greiner, *The Casimir effect*, Phys. Rep. **134**(2 & 3), 87–193 (1986).

[121] M. J. Radzikowski, *Micro-local approach to the Hadamard condition in quantum field theory on curved space-time*, Comm. Math. Phys. **179**(3), 529–553 (1996).

[122] D. B. Ray, I. M. Singer, *R-Torsion and the Laplacian on Riemannian Manifolds*, Adv. in Math. **7**(2), 145–210 (1971).

[123] M. Reed, B. Simon, "Methods of modern mathematical physics I: Functional analysis", Academic Press (1980).

[124] F. Rellich, "Perturbation theory of eigenvalue problems", CRC Press (1969).

[125] A. H. Rezaeian, A. A. Saharian, *Local Casimir energy for a wedge with a circular outer boundary*, Class. Quant. Grav. **19**(14), 3625–3634 (2002).

[126] R. B. Rodrigues, N. F. Svaiter, *Vacuum polarization of a scalar field in a rectangular waveguide*, arXiv:hep-th/0111131 [42 pages] (2001).

[127] R. B. Rodrigues, N. F. Svaiter, R. D. M. De Paola, *Vacuum stress tensor of a scalar field in a rectangular waveguide*, arXiv:hep-th/0110290 [39 pages] (2001).

[128] A. Romeo, A. A. Saharian, *Casimir effect for scalar fields under Robin boundary conditions on plates*, J. Phys. A: Math. Gen. **35**(5), 1297–1320 (2002).

[129] J. R. Ruggiero, A. Villani, A. H. Zimerman, *Application of analytic regularization to the Casimir forces*, Rev. Bras. Fis. **7**(3), 663–687 (1977).

[130] J. R. Ruggiero, A. Villani, A. H. Zimerman, *Some comments on the*

application of analytic regularisation to the Casimir forces, J. Phys. A: Math. Gen. **13**(2), 761–766 (1980).

[131] A. A. Saharian and A. S. Tarloyan, *Whightman function and scalar Casimir densities for a wedge with a cylindrical boundary*, J. Phys. A: Math. Gen. **38**(40), 8763–8780 (2005).

[132] F. C. Santos, A. Tenório, and A. C. Tort, *Zeta function method and repulsive Casimir forces for an unusual pair of plates at finite temperature*, Phys. Rev. D **60**(10), 105022 [9 pages] (1999).

[133] F. Sauvigny, "Partial differential equations 2: functional analytic methods", Springer Science & Business Media (2012).

[134] J. Schlemmer, J. Zahn, *The current density in quantum electrodynamics in external potentials*, Annals of Physics **359**, 31–45 (2015).

[135] L. Schwartz, "Théorie des noyaux" , Proc. Internat. Congress Mathematicians (Cambridge, 1950) **1**, 220–230, Amer. Math. Soc. (1952).

[136] S. S. Schweber, "An introduction to relativistic quantum field theory", Courier Corporation (2005).

[137] J. Schwinger, "Particles, sources, and fields", Perseus Books (1998).

[138] R. T. Seeley, *Complex powers of an elliptic operator*, AMS Proc. Symp. Pure Math. **X**, 288–307 (1967).

[139] N. Shimakura, "Partial differential operators of elliptic type", American Mathematical Soc. (1992).

[140] I. A. Shishmarev, *Uniform estimates for the derivatives of solutions to the Dirichlet problem and the eigenvalue problem for the operator $Lu - div(p(x)gradu) + q(x)u$ with discontinuous coefficients*, Dokl. Akad. Nauk SSSR **137**(1), 45–47 (1961); see also, Soviet Math. Dokl. **2**(2), 244–246 (1961).

[141] J. D. L. Silva, A. N. Braga, D. T. Alves, *Dynamical Casimir effect with $\delta - \delta'$ mirrors*, Phys. Rev. D **94**(10), 105009 [10 pages] (2016).

[142] I. N. Sneddon, "The use of integral transforms", McGraw-Hill (1972).

[143] A. Sommerfeld, "Partial differential equations in physics", Academic Press (1949).

[144] M. Spreafico, M. Zerbini, *Finite temperature quantum field theory on noncompact domains and application to delta interactions*, Rep. Math. Phys. **63**(1), 163–177 (2009).

[145] I. Stakgold, M. J. Holst, "Green's functions and boundary value problems", John Wiley & Sons (2011).

[146] R. M. Wald, *On the Euclidean approach to quantum field theory in curved spacetime*, Commun. Math. Phys. **70**(3), 221–242 (1979).

[147] R. M. Wald, "General relativity", University of Chicago Press (2010).

[148] G. N. Watson, "A treatise on the theory of Bessel functions", Cambridge University Press (1995).

[149] V.Y. Yakubov, *Estimates for eigenfunctions of elliptic operators with respect to the spectral parameter*, Func. Anal. App. **33**(2), 128–136 (1999).

[150] A.H. Zemanian, "Generalized integral transformations", John Wiley & Sons (1968).

[151] A. Zygmund, "Trigonometric series. Volume I", Cambridge University Press (2002).

Index

www.ingramcontent.com/pod-product-compliance
Lightning Source LLC
Chambersburg PA
CBHW050550190326
41458CB00007B/1985